Lecture Notes in Artificial Intelligence   10053

Subseries of Lecture Notes in Computer Science

More information about this series at http://www.springer.com/series/1244

Chattrakul Sombattheera · Frieder Stolzenburg
Fangzhen Lin · Abhaya Nayak (Eds.)

Multi-disciplinary Trends in Artificial Intelligence

10th International Workshop, MIWAI 2016
Chiang Mai, Thailand, December 7–9, 2016
Proceedings

 Springer

Editors
Chattrakul Sombattheera
Mahasarakham University
Maha Sarakham
Thailand

Frieder Stolzenburg
Harz University of Applied Sciences
Wernigerode
Germany

Fangzhen Lin
University of Science and Technology
Hong Kong
China

Abhaya Nayak
Macquarie University
North Ryde, NSW
Australia

ISSN 0302-9743   ISSN 1611-3349   (electronic)
Lecture Notes in Artificial Intelligence
ISBN 978-3-319-49396-1   ISBN 978-3-319-49397-8   (eBook)
DOI 10.1007/978-3-319-49397-8

Library of Congress Control Number: 2016957641

LNCS Sublibrary: SL7 – Artificial Intelligence

Printed on acid-free paper

This Springer imprint is published by Springer Nature
The registered company is Springer International Publishing AG
The registered company address is: Gewerbestrasse 11, 6330 Cham, Switzerland

Preface

This volume contains papers selected for presentation at the 10th Multi-Disciplinary International Workshop on Artificial Intelligence (MIWAI) held during December 7–9, 2016 at Chiang Mai, Thailand.

The MIWAI workshop series started in 2007 in Thailand as the Mahasarakham International Workshop on Artificial Intelligence and is held every year. It has emerged as an international workshop with participants from around the world. From 2011 to 2015, MIWAI was held in Hyderabad (India), Ho Chi Minh City (Vietnam), Krabi (Thailand), Bangalore (India), and Fuzhou (China), respectively.

The MIWAI series of workshops serves as a forum for Artificial Intelligence (AI) researchers and practitioners to discuss and deliberate cutting-edge AI research. It also aims to elevate the standards of AI research by providing researchers and students with feedback from an internationally renowned Program Committee.

The workshop solicits papers from all areas of AI including cognitive science, computational intelligence, computational philosophy, game theory, machine learning, multi-agent systems, natural language, representation and reasoning, speech, vision and the Web, as well as applications of AI in big data, bioinformatics, biometrics, decision support, e-commerce, image processing, analysis and retrieval, industrial applications, knowledge management, privacy, recommender systems, security, software engineering, spam filtering, surveillance, telecommunications, and Web services. Submissions received by MIWAI 2016 were wide ranging and covered both theory and applications.

MIWAI 2016 received 50 papers from 23 countries and regions including Australia, Austria, Brunei Darussalam, Chile, China, Colombia, Germany, India, Italy, Japan, Korea, Malaysia, Serbia, Singapore, Slovakia, Sri Lanka, Sweden, Taiwan, Thailand, Turkey, UK, USA, and Vietnam. Following the success of previous MIWAI workshops, MIWAI 2016 continued the tradition of a rigorous review process. Every submission was reviewed by at least two Program Committee members and domain experts. Additional reviews were sought when necessary. Papers with conditional acceptances were reviewed by the program and general chairs before acceptance.

A total of 22 papers were accepted as regular papers with an acceptance rate of 44 %. In addition, five short papers were accepted representing work in progress, providing an opportunity for sharing valuable ideas, eliciting useful feedback on work at an early-stage, and fostering discussions and collaborations among researchers interested in the broad area of AI. Many of the papers that were excluded from the proceedings were promising, but had to be rejected to maintain the quality of the proceedings. We would like to thank all authors for their submissions. Without their contribution, this workshop would not have been possible.

We are grateful to Rina Dechter and Michael Thielscher for accepting our invitation to deliver invited talks. Special thanks are due to Antonis Bikakis for organizing a tutorial on "Semantic Web/Linked Data." We wish to thank the members of the Steering Committee for their support, and are indebted to the Program Committee members and external reviewers for their effort in ensuring a rich scientific program. We also thank the local organizing team from Faculty of Informatics, Mahasarakham University for their arrangement of the event.

We acknowledge the use of the EasyChair Conference System for paper submission, review, and compilation of the proceedings. We are thankful to Alfred Hofmann, Anna Kramer, and the excellent LNCS team at Springer for their support and co-operation in publishing the proceedings as a volume of the *Lecture Notes in Computer Science*.

October 2016

Chattrakul Sombattheera
Frieder Stolzenburg
Fangzhen Lin
Abhaya Nayak
Sujin Butdisuwan

Acknowledgement

We are grateful to our sponsors for their financial support. Without their generous sponsorship, it would be very difficult to organize MIWAI 2016.

Software Industry Promotion Agency (Public Organization)
The Government Complex, Ratthaprasasanabhakti Building, 9th Floor
120 Moo 3, Chaengwattana Road, Thungsonghong, Laksi
Bangkok 10210, Thailand
Tel : (+66)2 141 7101, Mobile : (+66)9 5496 4294
Homepage: http://www.sipa.or.th

Mahasarakham University
Khamriang Sub-District, Kantarawichai District
Maha Sarakham 44150 Thailand
Tel/Fax : +66 4375 4241, Email : iroffice@msu.ac.th
Homepage: htpp://www.msu.ac.th

Faculty of Informatics, Mahasarakham University
Khamriang Sub-District, Kantarawichai District
Maha Sarakham 44150 Thailand
Tel/Fax: +66 4375 4359, Email: itmsu.web@gmail.com
Homepage: http://www.it.msu.ac.th

Organization

Steering Committee

Arun Agarwal	University of Hyderabad, India
Rajkumar Buyya	University of Melbourne, Australia
Patrick Doherty	University of Linkoping, Sweden
Jérôme Lang	University of Paris-Dauphine, France
James F. Peters	University of Manitoba, Canada
Srinivasan Ramani	IIIT Bangalore, India
C. Raghavendra Rao	University of Hyderabad, India
Leon van der Torre	University of Luxembourg, Luxembourg

Conveners

Richard Booth	Cardiff University, UK
Chattrakul Sombattheera	Mahasarakham University, Thailand

Honorary Chair

Sujin Butdisuwan	Mahasarakham University, Thailand

General Co-chairs

Fangzhen Lin	Hong Kong University of Science and Technology, Hong Kong, SAR China
Abhaya Nayak	Macquarie University, Australia

Program Co-chairs

Chattrakul Sombattheera	Mahasarakham University, Thailand
Frieder Stolzenburg	Harz University of Applied Sciences, Germany

Publicity Chair

Olarik Surinta	Mahasarakham University, Thailand

Web Administrators

Panich Sudkhot	Mahasarakham University, Thailand
Chinathip Printhong	Mahasarakham University, Thailand

Local Organizing Committee

Rapeeporn Chamchong	Mahasarakham University, Thailand
Nutchanat Buasri	Mahasarakham University, Thailand
Suwicha Chaimuang	Mahasarakham University, Thailand
Phatthanaphong Chomphuwiset	Mahasarakham University, Thailand
Panatda Hamontree	Mahasarakham University, Thailand
Benjawan Intara	Mahasarakham University, Thailand
Chatklaw Jareonpon	Mahasarakham University, Thailand
Jatuphum Juanchaiyaphum	Mahasarakham University, Thailand
Manasawee Kaenampornpan	Mahasarakham University, Thailand
Sasitorn Keawmun	Mahasarakham University, Thailand
Thananchai Khamgate	Mahasarakham University, Thailand
Thawatwong Lawan	Mahasarakham University, Thailand
Khachakrit Liamthaisong	Mahasarakham University, Thailand
Chayaporn Pholphunga	Mahasarakham University, Thailand
Potchara Pruksasri	Mahasarakham University, Thailand
Umaporn Sai-Saengchan	Mahasarakham University, Thailand
Suwimon Siri	Mahasarakham University, Thailand
Gamgarn Somprasertsri	Mahasarakham University, Thailand

Program Committee

Arun Agarwal	University of Hyderabad, India
Samir Aknine	Lyon 1 University, France
Rafah Almuttairi	University of Babylon, Iraq
Grigoris Antoniou	University of Huddersfield, UK
Thien Wan Au	Institut Teknologi Brunei, Brunei
Josch Bach	MIT Media Lab, USA
Costin Badica	University of Craiova, Romania
Sotiris Batsakis	University of Huddersfield, UK
Tarek Richard Besold	University of Bremen, Germany
Raj Bhatnagar	University of Cincinnati, USA
Antonis Bikakis	University College London, UK
Hima Bindu	University of Hyderabad, India
Laor Boongasame	King Mongkut's University of Technology Thonburi, Thailand
Veera Boonjing	King Mongkut's Institute of Technology Ladkrabang, Thailand
Richard Booth	Cardiff University, UK
Rajkumar Buyya	University of Melbourne, Australia

Darko Brodic	University of Belgrade, Serbia
Patrice Caire	University of Luxembourg, Luxembourg
Rapeeporn Chamchong	Mahasarakham University, Thailand
Yu-N Cheah	Universiti Sains Malaysia, Malaysia
Zhicong Chen	Fuzhou University, China
Broderick Crawford	Pontifical Catholic University of Valparaíso, Chile
Patrick Doherty	University of Linkoping, Sweden
Vlad Estivill-Castro	Griffith University, Australia
Chun Che Fung	Murdoch University, Australia
Ulrich Furbach	University of Koblenz, Germany
Guido Governatori	NICTA, Australia
Jinguang Gu	Wuhan University of Science and Technology, China
Jingzhi Guo	University of Macau, Macau
Wenzhong Guo	Fuzhou University, China
Peter Haddawy	Mahidol University, Thailand
Christos Hadjinikolis	David Game College, UK
Christopher Henry	University of Winnipeg, Canada
Zhisheng Huang	Vrije University Amsterdam, The Netherlands
Jason Jung	Chung-Ang University, South Korea
Manasawee Kaenampornpan	Mahasarakham University, Thailand
Mohan Kankanhalli	National University of Singapore, Singapore
Jérôme Lang	University of Paris-Dauphine, France
Kittichai Lavangnananda	King Mongkut's University of Technology Thonburi, Thailand
Jimmy Lee	The Chinese University of Hong Kong, Hong Kong, SAR China
Tek Yong Lim	Multimedia University, Malaysia
Tiago de Lima	University of Artois and CNRS, France
Changlu Lin	Fujian Normal University, China
Fangzhen Lin	Hong Kong University of Science and Technology, Hong Kong, SAR China
Emiliano Lorini	University of Toulouse, France
Martin Lukac	Nazarbayev University, Kazakhstan
Dickson Lukose	MIMOS Berhad, Malaysia
Michael Maher	UNSW Canberra, Australia
Jérôme Mengin	University of Toulouse, France
Sebastian Moreno	Purdue University, USA
Debajyoti Mukhopadhyay	Maharashtra Institute of Technology, India
Kavi Narayana Murthy	University of Hyderabad, India
Raja Kumar Murugesan	Taylor's University, Malaysia
Sven Naumann	University of Trier, Germany
Abhaya Nayak	Macquarie University, Australia
Naveen Nekuri	Sasi Institute of Technology and Engineering, India

Additional Reviewers

Huaming Chen	Lanzhou University, China
Lin Gong	University of Virginia, Charlottesville, USA
Matthias Oelze	Harz University of Applied Sciences, Germany
Falk Schmidsberger	Harz University of Applied Sciences, Germany

Invited Talks

Probabilistic Reasoning Meets Heuristic Search

Rina Dechter

Donald Bren School of Information and Computer Sciences,
University of California, Irvine, USA
dechter@ics.uci.edu

Abstract. Graphical models, including constraint networks, Bayesian networks, Markov random fields and influence diagrams, have become a central paradigm for knowledge representation and reasoning in Artificial Intelligence, and provide powerful tools for solving problems in a variety of application domains, including coding and information theory, signal and image processing, data mining, learning, computational biology, and computer vision. Although past decades have seen considerable progress in algorithms in graphical models, many real-world problems are of such size and complexity that they remain out of reach. Advances in exact and approximate inference methods are thus crucial to address these important problems with potential impact across many computational disciplines. Exact inference is typically NP-hard, motivating the development of approximate and anytime techniques.

Existing algorithms typically take one of two approaches: Inference, expressed as message-passing schemes, or search and conditioning methods. In the past decade, my research group has developed several very successful algorithms for reasoning in graphical models based on combining heuristic search with message passing approximations. These algorithms are state-of-the-art. For example, in the past two approximate inference competitions (The Probabilistic Inference Challenge in UAI, 2012 and 2014) our algorithms were evaluated as leading at either first or second place. My plan is to describe the main principles underlying these developments.

In this talk I will describe the unifying framework of AND/OR search spaces for graphical models and show how they can facilitate problem decomposition and allow avoiding redundant specification (and computation) by exploiting the conditional independencies of the graphical model. The AND/OR search space size is bounded exponentially by the graphical models treewidth, making them comparable to inference algorithms such as variable-elimination or join/junction-tree decompositions. Yet, as search schemes they can facilitate many of the ideas developed within Heuristic Search and Operations Research communities, for all queries (e.g., satisfiability, optimization, weighted counting and their combinations) while enjoying the treewidth bound, automatically. Most significantly, they can allow trading off memory for time and time for accuracy, leading to anytime solvers.

I will subsequently show how approximate inference can be used for generating heuristic information for guiding the AND/OR search. This can be accomplished using the two ideas of mini-bucket partitioning schemes which relaxes the input problem by node duplication only, combined with linear

programming relaxations ideas which optimize cost-shifting re-parameteriza-
tion, to yield tight bounding heuristic information within systematic, anytime
search.

AND/OR search guided by such inference-based heuristics has yielded
some of the best state-of-the-art solvers for combinatorial optimization. Current
research centers on extending these ideas beyond pure optimization, to weighted
counting and to hybrid queries (i.e., max-product or min-sum queries such as
marginal map and maximizing expected utilities in influence diagrams), aiming
for anytime behavior that generates not only an approximation, but also upper
and lower bounds which become tighter with more time.

Knowledge Representation for General Problem-Solving Robots

Michael Thielscher

University of New South Wales, Sydney, Australia
mit@cse.unsw.edu.au

Abstract. A general problem-solving robot is able to understand the representation of a new task and to successfully tackle it without human intervention. This requires architectures for cognitive robotics that integrate symbolic and sub-symbolic representations. We present a formal framework for the design of control hierarchies along with an instantiation for a real Baxter robot that combines high-level reasoning and planning with a physics simulator to provide spatial knowledge and low-level control nodes for motors and sensor processing.

As an example of a general problem representation technique, we present and discuss a formal language for describing to a general problem-solving robot so-called epistemic games. These are characterised by rules that depend on what players can and cannot deduce from the information they get during gameplay.

Acknowledgements. This research has been supported by the Australian Research Council (ARC) under the *Discovery Projects* funding scheme (project DP 150103035). The author is also affiliated with the University of Western Sydney.

Contents

Short Papers

Regular Papers

BoWT: A Hybrid Text Representation Model for Improving Text Categorization Based on AdaBoost.MH

Bassam Al-Salemi[✉], Mohd. Juzaiddin Ab Aziz,
and Shahrul Azman Mohd Noah

Knowledge Technology Research Group, Faculty of Information Science
and Technology, Universiti Kebangsaan Malaysia, Bangi, Malaysia
bassalemi@siswa.ukm.edu.my,
{juzaiddin,shahrul}@ukm.edu.my

Abstract. Text representation is the fundamental task in text categorization system. The BAG-OF-WORDS (BoW) is a typical model for representing the texts into vectors of single words. Even though it is a simple representation model, BoW has been criticized for its disregard of the relationships between the words. Alternatively, the Latent Dirichlet Allocation (LDA) topic model has been proposed to represent the texts into a BAG-OF-TOPICS (BoT). In LDA, the words in the corpus are statistically grouped into a small number of themes called "latent topics" in which the topics capture the semantic relationships between the words. Thus, representing the documents using BoT will dramatically accelerate the training time; as well improve the classification performance. However, BoT has been proven to not be effective for imbalanced datasets. Accordingly, this paper presents a hybrid text representation model as a combination of BoW and BoT, namely BoWT. In BoWT, the high weighted BoW's features are merged with the BoT's features to produce a new feature space. The proposed representation model BoWT is evaluated for multi-label text categorization based on the well-known boosting algorithm AdaBoost.MH. The experimental results on four benchmarks demonstrated that the BoWT representation model notably outperforms both BoW and BoT and dramatically improves the classification performance of AdaBoost.MH for text categorization.

Keywords: Text representation · Bowt · Text categorization · AdaBoost.MH · Topic modeling

1 Introduction

Text representation is an essential part of any text categorization system in which the text documents are converted into a compact representation in order to be recognized by the classification algorithms. The BoW model is a standard technique of representing the documents as vectors of single words that they contain and using them as elements in the feature space. The advantage of BoW is its simplicity, as it ignores the text logical structure and layout. However, BoW has been criticized for its disregard of

C. Sombattheera et al. (Eds.): MIWAI 2016, LNAI 10053, pp. 3–11, 2016.
DOI: 10.1007/978-3-319-49397-8_1

the relationships between the words and their order among the texts. Many studies had been conducted to improve on this model to capturing the word dependency and considering the words order. Instead of considering the frequencies of the features as weights in the traditional BoW model, some weighting schemes had been proposed to tackle the features correlation problem of BoW, such as Inverse-Document-Frequency (IDF) and TFIDF [8, 12]. However, for the classification algorithms that use the binary features for inducing the classification models, e.g. ADABOOST.MH [13], the feature weighting does not make any sense.

In addition to the disregard of the words' dependencies, BoW representation model generates a vast number of features (Liu et al. 2005) and using all the extracted features for inducing the weak hypotheses of ADABOOST.MH may entail a high degree of computational time complexity, especially for large-scale datasets. That is because ADABOOST.MH produces at each boosting round a set of weak hypotheses equivalent in size to the number of the training features, refer to [4] for more details.

The high dimensionality of BoW feature space can be managed by eliminating the redundant features using an appropriate feature selection technique, such as Mutual Information, Information Gain, Chi Square-statistic, Odds Ratio, GSS Coefficient [1, 5, 6, 9–11, 14, 15]. However, feature selection may eliminate some informative features and cause information loss.

Instead of using the single words for representing the texts and training ADABOOST. MH, as BoW does, an alternative text representation model using topic modeling is proposed [3] for this task. Hence, the latent Dirichlet allocation model (LDA) [7] is used to discover the latent topics among the texts. The general outputs of LDA are; topic-word index, which contains the distribution of the words over the topics, and document-topic index, which contains the distribution of the topics over the documents. Therefore, to represent the documents into BAG-OF-TOPICS (BoT), the document-topic index is used. This topics-based representation model has been extended to involve the most well-known multi-label boosting algorithms for multi-label text categorization [2].

Even though BoT representation model has proved to be efficient in improving text categorization based on ADABOOST.MH in general, its classification performance is poor comparing to BoW for imbalanced datasets [7]. That is because the number of topics assigned to the infrequent categories is much smaller than those assigned to the frequent categories.

Getting the advantage of feature selection for reducing the high dimensionality of BoW and selecting the high weighted features, and the advantage of BoT of capturing the semantic relationship between the words, this paper proposes a hybrid representation model as a combination of BoW and BoT. The hybrid model, which it called "BAG OF WORDS AND TOPICS" (BoWT) is proposed to tackle the limitations of both models, and to ensure increasing the number of features of the documents in the infrequent categories, as well the small texts, and give a chance to be classified correctly using ADABOOST.MH.

2 The Proposed Representation Model

The BoW is a simple model for the text representation in which the single words are used as elements to represent the texts in the feature space. However, BoW disregards the relationships between the words among the texts. Instead of using the single words in the feature space, the latent topics among the texts, which are estimated using LDA topic model, can be used. Thus, each document in the corpus is represented as a vector of topics. The advantage of using the topics as features is that the latent topic statistically clusters the words with similar meaning as one feature in the feature space. However, the BoT are not suitable for the imbalanced datasets [3]. That is because the number of topics assigned to the infrequent categories are very small in size, and that will negatively results in the classification performance. Accordingly, in this paper we proposed a hybrid representation model, namely BoWT, as a combination of BoW with BoT.

For a document d in a given corpus, d is represented using BoW as a set of words, $d = (w_1, w_2 \ldots w_n)$, and by using BoT, d is represented as a set of latent topics, $d = (t_1, t_2, \ldots t_m)$. Thus by combining both representations, d will be represented as $d = (w_1, w_2, \ldots, w_n, t_1, t_2, \ldots, t_m)$. Because the weights of both BoW and BoT are totally different; therefore, the binary weights are used for both models. As a result, the weighting of BoWT is also binary. While ADABOOST.MH uses binary features for inducing the classification model, the proposed representation model BoWT is an appropriate for this task.

To avoid the computational complexity of ADABOOST.MH training, not all the extracted features using BoW will be merged with the BoT features. Accordingly, the feature selection will be applied to reduce the size of BoW features. Thus, only the high weighted features of BoW will be combined with the latent topics in the new feature space.

3 Experiments and Results

3.1 Datasets and Experimental Settings

The datasets for multi-label text categorization which used for the evaluation purpose are: Reuters-21578 "ModApte", 20-Newsgroups (**20NG**) and **OHSUMED**. For more information about these datasets and their statistics, refer to [2]. For the Reuters-21578, the subset of 90 categories (**R90**) and the top 10 frequent categories (**R10**) are used. For each dataset the typical text preprocessing is performed: tokenization, stemming and feature selection. For feature selection, the label latent Dirichlet allocation (LLDA) is used [4]. The idea of using LLDA for feature selection is that, the features are selected based on the maximal conditional probabilities of the words across the labels, refer to [4] for more details. For LDA estimation and prediction, we followed the same settings used in [3]. However, in this paper the performance is evaluated for different numbers of topics, and the impact of using features selection before estimating the topics is also analysed. The evaluation measures used for evaluating the classification performance are: Macro-averaged F1 (MacroF1) and Micro-averaged F1 (MicroF1).

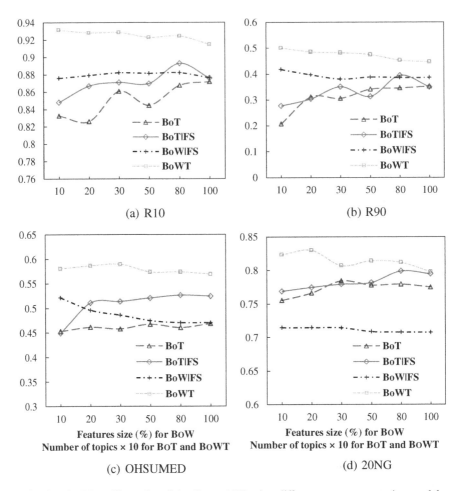

Fig. 1. The MacroF1 results of ADABOOST.MH using different text representation models

The representation models to be evaluated are:

- BoW with feature selection, dubbed (BoW|FS).
- BoT without feature selection (BoT), in which the whole extracted words from the dataset are used for LDA estimation.
- BoT after feature selection, dubbed (BoT|FS).
- BoWT, the proposed model as a combination of BoW and BoT with feature selection.

.

The BoW is evaluated on different sizes of selected features; (10, 20, 30, 50 and 80) % of the top weighted features and also 100 %, the case that all features are used without any reduction. Also BoT, BoT|FS and BoWT are evaluated with different numbers of topics: 100, 200, 300, 500, 800 and 1000 topics.

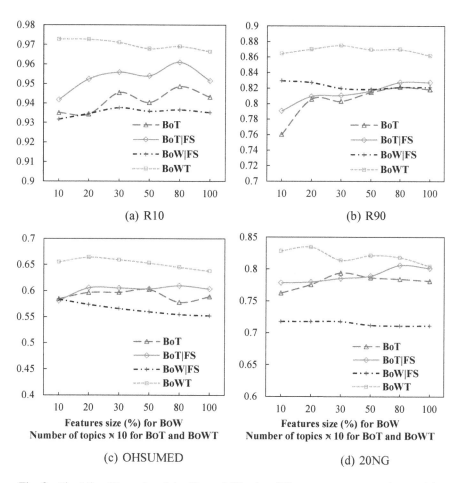

Fig. 2. The MicroF1 results of ADABOOST.MH using different text representation models

Table 1. The best MacroF1 results

Dataset	Representation	# topics	Features size	MacroF1	Rank
R10	BoT	1000	100 %	0.8717	4
	BoT|FS	800	30 %	0.8930	2
	BoWT	100	30 %	0.9315	1
	BoW|FS	–	30 %	0.8822	3
R90	BoT	1000	100 %	0.3524	4
	BoT|FS	800	10 %	0.3941	3
	BoWT	100	10 %	0.5005	1
	BoW|FS	–	10 %	0.4176	2
OHSUMED	BoT	1000	100 %	0.4686	4
	BoT|FS	800	10 %	0.5267	2

(continued)

Table 1. (*continued*)

Dataset	Representation	# topics	Features size	MacroF1	Rank
	BoWT	300	10 %	0.5896	1
	BoW|FS	–	10 %	0.5214	3
20NG	BoT	300	100 %	0.7841	3
	BoT|FS	800	10 %	0.7987	2
	BoWT	200	10 %	0.8299	1
	BoW|FS	–	10 %	0.7146	4

Table 2. The best MicroF1 results

Dataset	Representation	# topics	Features size	MacroF1	Rank
R10	BoT	800	100 %	0.9486	3
	BoT|FS	800	30 %	0.9610	2
	BoWT	100	30 %	0.9728	1
	BoW|FS	–	30 %	0.9377	4
R90	BoT	800	100 %	0.8212	4
	BoT|FS	800	10 %	0.8274	3
	BoWT	300	10 %	0.8747	1
	BoW|FS	–	10 %	0.8295	2
OHSUMED	BoT	500	100 %	0.6021	3
	BoT|FS	800	10 %	0.610	2
	BoWT	200	10 %	0.6644	1
	BoW|FS	–	10 %	0.5834	4
20NG	BoT	300	100 %	0.7937	3
	BoT|FS	800	10 %	0.8056	2
	BoWT	200	10 %	0.8345	1
	BoW|FS	–	10 %	0.7178	4

The classification algorithm used is the multi-label boosting algorithm ADABOOST. MH. The maximum number of iterations of ADABOOST.MH's weak learning is set to 2000 iterations.

The experiments are performed in two stages. In the first stage the BoW with feature selection and BoT are evaluated individually on all datasets. Then the best subset of BoW's features that yield the best performance is used for both BoT\|FS and BoWT.

3.2 Results and Discussion

The experimental results of ADABOOST.MH classification performance measured by MacroF1 using the text representation models are illustrated in Fig. 1 for all datasets. It is clear that the proposed representation model BoWT yields the best classification performance overall on all datasets. The BoW representation outperforms BoT|FS on average on both R10 and R90, while the best MacroF1 result using BoT|FS (0.8930) exceeds the finest MacroF1 of BoW on the R10 (0.8822). Except for R90 dataset, BoT| FS leads to the best MacroF1 overall compared to BoW. Using the BoT leads to the worst performance except for the 20NG where it exceeds BoW.

In terms of the MicroF1 results (Fig. 2), the combined representation model BoWT dramatically outperforms all other representation models on all datasets. The BoT exceeds the performance that achieved using BoW representation for all datasets except for the R90 where BoW representation obtained the best performance. Whereas, using feature selection to reduce the training features of LDA (BoT|FS) enhances the performance of topics-based representation.

The reason of the poor performance of BoT on the imbalanced dataset R90 is that the unsupervised topic model LDA takes all the documents under the training set without taking into account their categorical structure. Therefore, the documents under the infrequent categories will be represented into a few numbers of topics and that will result in ADABOOST.MH performance. The high impact of using BoT representation went to the balanced dataset, which the number of documents under each category is not varying in size, such as the 20NG. To tackle this matter both BoT and BoW representation are combined in the proposed representation BoWT. Therefore, merging the top most frequent features of BoW with the features of BoT will increase the number of informative features of the texts, particularly for the categories with small number of examples that gained small number of topics. While ADABOOST.MH uses the binary features, which the weights of the features among the texts are not considered; therefore, combining the latent topics with the word tokens will increase the classification performance.

Tables 1 and 2 summarize the best results of MacroF1 and MicroF1, respectively that obtained using different text representation models. The best MicroF1 results overall, on all datasets, are obtained when the BoWT representation model is used to represent the texts. The best MacroF1 results using BoW exceeds the results obtained using BoT on R10 and R90 datasets, while BoT leads to the best MacroF1 on OHSUMED and 20NG datasets. However, using feature selection before estimating the topics, the BoT yields the best results comparing with BoW, except for the R90 where BoW outperformed.

Regarding the best MicroF1 results (Table 2), ADABOOST.MH with the BoWT achieves the best results overall on all datasets. The BoT representation exceeds the performance of BoW on all datasets, except for the R90 where BoW yields a better performance. Moreover, reducing the features space dimension of LDA model by employing feature selection (BoT|FS), leads AdaBoost.MH to perform better than using LDA without reducing the training feature of LDA topic model (BoT).

4 Conclusion

The BoW is a typical representation model for most real-life classification problems. However, in text categorization, BoW does not capture the relationship between the words among the texts. In fact, this is the reason behind BoW simplicity. Nevertheless, ignoring the relevance between the words may effect negatively in the classification performance, particularly for the classification algorithms that do not consider the features' weights, such as ADABOOST.MH. An alternative method to represent the text is by using the latent topics among the texts as features for inducing the classification models. Latent topics, which estimated from the text using topic modeling, are capable of capturing the semantic similarity between the words. Thus, representing the texts as a BAG-OF-TOPICS (BoT) will improve the classification performance. However, the experimental results proved that BoT yielded a poor performance in the case of imbalanced datasets. That is because the categories with rare examples are represented into a very small number of latent topics comparing with the frequent categories. In this paper we describe a method to tackle this problem by combining the BoT's features with the high weighted features of BoW as a hybrid representation model, namely BoWT.

The experimental results demonstrate that the proposed model, BoWT, dramatically improves the classification performance of ADABOOST.MH comparing with the other models for the all datasets. The results also proved that reducing the training features of LDA topic model using feature selection increases the performance of BoT model.

References

1. Al-Salemi, B., Ab Aziz, M.J.: Statistical bayesian learning for automatic arabic text categorization. J. Comput. Sci. **7**, 39 (2010)
2. Al-Salemi, B., Ab Aziz, M.J., Noah, S.A.: Boosting algorithms with topic modeling for multi-label text categorization: a comparative empirical study. J. Inf. Sci. **41**, 732–746 (2015)
3. Al-Salemi, B., Ab Aziz, M.J., Noah, S.A.: LDA-AdaBoost.MH: Accelerated AdaBoost.MH based on latent Dirichlet allocation for text categorization. J. Inf. Sci. **41**, 27–40 (2015)
4. Al-Salemi, B., Mohd Noah, S.A., Ab Aziz, M.J.: RFBoost: an improved multi-label boosting algorithm and its application to text categorisation. Knowl.-Based Syst. **103**, 104–117 (2016)
5. Alhutaish, R., Omar, N.: Arabic text classification using k-nearest neighbour algorithm. Int. Arab J. Inf. Technol. (IAJIT) **12**, 190–195 (2015)
6. Aphinyanaphongs, Y., Fu, L.D., Li, Z., et al.: A comprehensive empirical comparison of modern supervised classification and feature selection methods for text categorization. J. Assoc. Inf. Sci. Technol. **65**, 1964–1987 (2014)
7. Blei, D.M., Ng, A.Y., Jordan, M.I.: Latent Dirichlet allocation. J. Mach. Learn. Res. **3**, 993–1022 (2003)
8. Dumais, S.T.: Improving the retrieval of information from external sources. Behav. Res. Methods Instrum. Comput. **23**, 229–236 (1991)
9. Duwairi, R., Al-Refai, M.N., Khasawneh, N.: Feature reduction techniques for arabic text categorization. J. Am. Soc. Inform. Sci. Technol. **60**, 2347–2352 (2009)
10. Galavotti, L., Sebastiani, F., Simi, M.: Experiments on the use of feature selection and negative evidence in automated text categorization. In: Borbinha, J., Baker, T. (eds.) ECDL 2000. LNCS, vol. 1923, pp. 59–68. Springer, Heidelberg (2000). doi:10.1007/3-540-45268-0_6

11. Lewis, D.D.: Feature selection and feature extraction for text categorization. In: Proceedings of the workshop on Speech and Natural Language. Association for Computational Linguistics, pp. 212–217 (1992)
12. Li, X., Liu, B.: Learning to classify texts using positive and unlabeled data. In: IJCAI, pp. 587–592 (2003)
13. Mukherjee, I., Schapire, R.E.: A theory of multiclass boosting. J. Mach. Learn. Res. **14**, 437–497 (2013)
14. Pekar, V., Krkoska, M., Staab, S.: Feature weighting for co-occurrence-based classification of words. In: Proceedings of the 20th International Conference on Computational Linguistics. Association for Computational Linguistics, p. 799 (2004)
15. Sebastiani, F.: Machine learning in automated text categorization. ACM Comput. Surv. (CSUR) **34**, 1–47 (2002)

An Improved Teaching-Learning Based Optimization for Optimization of Flatness of a Strip During a Coiling Process

Sujin Bureerat[1], Nantiwat Pholdee[1(✉)], Won-Woong Park[2], and Dong-Kyu Kim[3]

[1] Department of Mechanical Engineering, Faculty of Engineering, Sustainable and Infrastructure Research and Development Centre, Khon Kaen University, 123 Moo 16, Mittraphap Road, Tambon Muang, Khon Kaen, Thailand
nantiwat@kku.ac.th
[2] National Research Laboratory for Computer Aided Materials Processing, Department of Mechanical Engineering, KAIST, 291 Daehak-ro, Yuseong-gu, Daejeon, South Korea
[3] Neutron Science Division, Korea Atomic Energy Research Institute, 111 Daeduk-daero 989 beon-gil, Yuseong-gu, Daejeon, South Korea

Abstract. Performance enhancement of a teaching-learning basedz optimizer (TLBO) for strip flatness optimization during a coiling process is proposed. The method is termed improved teaching-learning based optimization (ITLBO). The new algorithm is achieved by modifying the teaching phase of the original TLBO. The design problem is set to find spool geometry and coiling tension in order to minimize flatness defects during the coiling process. Having implemented the new optimizer with flatness optimization for strip coiling, the results reveal that the proposed method gives a better optimum solution compared to the present state-of-the-art methods.

Keywords: Evolutionary algorithm · Flatness defect · Optimization · Strip coiling · Teaching-learning based optimization

1 Introduction

There are several processing stages during the manufacturing of a coil strip, e.g. roughing, rolling, cooling, and coiling. Based on the previous investigation by Jung and Im [1, 2], the final strip shape had non-uniform thickness profiles consisting of ∩, ∪, M, and W shapes. Generally, it is difficult to predict the final shape of the strip due to various related processing parameters in production facilities. The strip crown, while being coiled, may include imperfections that were initiated during the rolling process resulting in flatness imperfection taking place on the coil strip [3, 4].

As a result, the strip is normally welded, cut, and recoiled in the recoiling line so as to satisfy customer strip flatness requirements. However, although adding the recoiling line to the process, flatness problems sometimes cannot be avoided especially for the high-strength coil strip. In order to understand the flatness defect formation mechanism

© Springer International Publishing AG 2016
C. Sombattheera et al. (Eds.): MIWAI 2016, LNAI 10053, pp. 12–23, 2016.
DOI: 10.1007/978-3-319-49397-8_2

during the coiling process, Sims and Place [5] proposed a stress model of the coil assuming that the coil was an axial-symmetry hollow cylinder. Miller and Thornton [6] and Sarban [7] introduced a finite element method and a semi-analytical model to calculate the three-dimensional stress distribution within the coil. Nevertheless, in those models, they did not consider the physical clearance between each coiled wrap due to the strip crown as a cause of the axial inhomogeneity. Yanagi et al. [8] proposed an analytical model by wrapping the thick cylinder (the coil) with the thin-walled cylinders (the new coiling strips) to deal with inhomogeneous deformation of the cold-rolled thin-strip in the axial direction caused by the clearance and the strip crown. Moreover, Park et al. [9] studied the effect of processing parameters including a strip crown, a spool geometry, and coiling tension on the stress distribution on the strip during the coiling process where the analytical elastic model was used. In this study, it was found that enhancement of strip flatness of the cold-rolled thin-strip could be accomplished by suppressing the strip crown and lowering the coiling tension intensity compared to the measured circumferential strain distribution.

To alleviate the undesirable formation of flatness defects, manufacturing the strip coil without the strip crown is suggested as the best solution for fulfilling the strip flatness requirement. Nevertheless, suppressing the strip crown during the rolling process, as illustrated in Fig. 1, is somewhat difficult or even impossible to carry out due to many processing parameters involved. Therefore, use of optimization to find the optimum solution for a spool geometry and coiling tension was conducted [10, 11] in order to improve the strip flatness during the strip coiling process.

Optimization is a special kind of mathematical problem assigned to search for a design solution optimizing a predefined objective or merit indicator within a given feasible region. A numerical optimizer is usually employed to find such a solution. It can be categorized as an optimization method either with and without using function derivatives. The former is based on hard computing while the latter is based on a stochastic process and soft computing. The most popular non-gradient optimizer is an evolutionary algorithm (EAs) or later known as a meta-heuristic (MH). It has been implemented on a wide range of engineering applications and has shown several advantages [12–21]. For metal strip manufacturing, optimization by means of meta-heuristics has been used most commonly in the rolling process so as to control the flatness problem, whereas their use in the strip coiling process has been rarely reported [22–27].

In this study, optimization of flatness of the strips has been enhanced by an improved teaching-learning based algorithm (ITLBO). This method is compared to several well established EAs, such as simulated annealing (SA) [16], differential evolution (DE) [28], artificial bee colony optimization (ABC) [29], real code ant colony optimization (ACOR) [30], original teaching-learning based optimization (TLBO) [31], league championship algorithm (LCA) [32], charged system search (ChSS) [33], Opposition-based Differential Evolution Algorithm (OPDE) [10] and Enhanced teaching-learning based optimization with differential evolution (ETLBO-DE) [11] to determine the spool geometry and coiling tension where the objective is to minimize the axial inhomogeneity of the stress to improve the flatness of the strip. For function evaluations, the analytical elastic model proposed by Park et al. [9] similar to the one suggested by Yanagi et al. [8] was employed.

2 Formulation of the Optimization Design Problem

It is known that wavy edges occur during the strip coiling process, when the circumferential stress at the middle zone of the strip is highly compressed, while two edges are under tension or slight compression. Also, if the middle strip zone is under high tension while the two edges are compressed or slightly stretched, center buckle can happen [8, 9]. Figures 1(a) and (b) display the circumferential stress (σ_θ) distribution along the z direction within the thin strip, which respectively caused the wavy edge and center buckle.

Generally, it is impossible to obtain a flat strip after finishing a rolling process. The strip always has a crown shape. When the strips are being coiled, tension loads need to be applied, the middle zone (z = 0) of the strip at the inner coil will be considerably compressed in comparison with the two edges because of the coiling tension and the strip crown. In such a situation, the center buckle defect at the inner coil will not appear but the wavy edge defect can possibly occur. As such, the wavy edge defect at the inner coil is the major problem during the coiling process. Figure 2 depicts the circumferential stress (σ_θ) distribution in the z direction at the radius (r) of the coil (computed by the Love's elastic solution proposed by Park et al. [9]) contributing to wavy edge defect formation during the strip coiling process. It is possible to reduce the wavy edge defect by decreasing the axial inhomogeneity of the stress distribution and the maximum compressive stress at the compressive zone [10].

In this paper, optimization using the ITLBO and other well-known and newly developed EAs will be used to find the optimum solution for the processing parameters including coiling tension (σ_T) and spool geometry, as illustrated in Fig. 3.

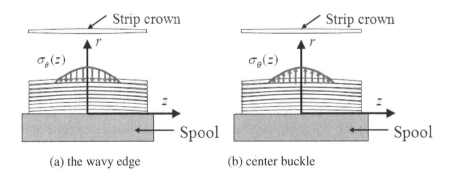

(a) the wavy edge   (b) center buckle

Fig. 1. Circumferential stress distributions for (a) the wavy edge and (b) center buckle, respectively [8, 9]

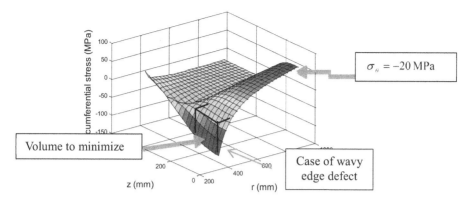

Fig. 2. Circumferential stress distribution (σ_θ) in the coil determined by Love's elastic solution [9]

To decrease the axial inhomogeneity of the stress distribution and the maximum compressive stress, minimization of the volume of the circumferential stress and maximum compressive stress (shown in Fig. 2) is defined as an objective function. In Fig. 2, the volume can only be computed for the coil, where compressive stresses were higher than 20 MPa, in order to minimize the zone that is likely to have the wavy edge defect. The objective function of the optimization problem can then be written as:

$$\text{Minimize} \qquad f(\alpha_b, \eta_b, \sigma_{T,i}) = \frac{V}{V_0} + \frac{\max(\sigma_{\theta c})}{\max(\sigma_{\theta c0})} \qquad (1)$$

minimize

$$0 \leq \alpha_b \leq 4,$$
$$0 \leq \eta_b \leq 4,$$
$$25 \leq \sigma_{T,i} \leq 50 \text{ MPa}; \qquad i = 1,\ldots n_{\max}$$
$$|\sigma_{T,i} - \sigma_{T,i-1}| \leq 2 \text{ MPa},$$

where $\sigma_{\theta c}$ and V are respectively the compressive circumferential stress higher than 20 MPa (refer to Fig. 2) and the approximate volume of the circumferential stress. $\sigma_{\theta c0}$ and V_0 are the respective values for the original design of the process. The $\sigma_{T,i}$ is the coiling tension at coil number i. The coiling tension is normally set to be constant for all coils [34]. The variable $n_{\max}$ is the maximum number of coils, which has been assigned to be 220 in this paper. η_b and α_b in Eq. (2) are spool crown exponent and the spool crown height, which were used for defining the spool geometry, as described in Fig. 3:

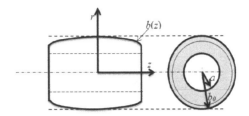

Fig. 3. Spool Geometry used in the present investigation

$$b(z) = b_0 - \alpha_b \left(\frac{|z|}{z_{\max}} \right)^{\eta_b} \tag{2}$$

where b_0 ($z = 330$ mm) and $b(z)$ are the initial value of the outer radius of the spool and the outer radius of the spool along the z direction, respectively. $z_{\max} = 525$ mm is the width of the spool. The inner radius of the spool (a) in Fig. 3 has been assigned to be 300 mm. The total number of design variables, therefore, is 222 (220 for coiling tensions and 2 for the spool geometry).

3 Improved Teaching-Learning Based Optimization

From the previous section, the optimization problem can be considered being large-scale. It has been found [10, 11], that TLBO is suitable for this type of design problem. The teaching-learning based optimization (TLBO) algorithm is an evolutionary algorithm, or an optimizer without using function derivatives, proposed by Rao et al. [31]. The concept of TLBO searching mechanism is based on mimicking a teacher on the output of learners in a classroom. Basically, the learners can improve their intellectual and knowledge by two stages i.e. learning directly from the teacher and learning among themselves. During the teacher stage, a teacher may teach the learners, however, only some learners can acquire all things presented by the teacher. Those who can accept what the teacher taught will improve their knowledge. For the second stage, which is called the learning phase, the learners can improve their knowledge during discussion with other learners. Based on the different levels of the learners' knowledge, the better learners may transfer knowledge to the inferior learners.

From the view point of optimization, the algorithm starts with a randomly created initial population, which is a group of design solutions. Learners are identical to design solutions whereas the best one is considered a teacher. The objective function is analogous to the knowledge which needs to be improved towards the optimum solution. Having identified a teacher and other learners for the current iteration, the population will be updated by two stages including "Teacher Phase" and "Learner Phase". In the "Teacher Phase", an individual (x_i) will be updated based on the best individual ($\mathbf{x}_{\text{teacher}}$) and the mean values of all populations ($\mathbf{x}_{\text{mean}}$) as follows:

$$\mathbf{x}_{\text{new},i} = \mathbf{x}_{\text{old},i} + r\{\mathbf{x}_{\text{teacher}} - (T_F \cdot \mathbf{x}_{\text{mean}})\} \tag{3}$$

Where T_F is a teaching factor, which can be either 1 or 2 and $r \in [0,1]$ is a uniform random number.

For the "Learner Phase", the members in the current population will be modified by exchanging information between themselves. Two individuals $\mathbf{x}_i$ and $\mathbf{x}_j$ will be chosen at random, where $i \neq j$. The update of the solutions can then be calculated as:

$$\mathbf{x}_{\text{new},i} = \begin{cases} \mathbf{x}_{\text{old},i} + r\left(\mathbf{x}_i - \mathbf{x}_j\right) & \text{if } f(\mathbf{x}_i) < f(\mathbf{x}_j) \\ \mathbf{x}_{\text{old},i} + r\left(\mathbf{x}_j - \mathbf{x}_i\right) & \text{if } f(\mathbf{x}_j) < f(\mathbf{x}_i) \end{cases} \tag{4}$$

At both teacher and learner phases, the new solution ($\mathbf{x}_{\text{new}}$) will replace its parent if it has better knowledge or produces better objective function value, otherwise, it will be rejected. The two phases are sequentially operated until the termination criterion is fulfilled.

For the improved teaching-learning based optimization (ITLBO), an opposition-based approach, binary crossover, and the probability of operating the learning phase are added to the original TLBO to improve the balance of search exploration and exploitation. Four random numbers including, $rand_1$, $rand_2$, $rand_3$, and $rand_4$, have been used for performing opposition-based approach, binary crossover, and the learning phase. The main search procedure starts by generating an initial population, updating the population at the teaching phase and learning phase similarly to the original TLBO. However, at the teaching phase, the updating can be done by the following equation;

$$\mathbf{x}_{\text{new},i} = \mathbf{x}_{\text{old},i} + (-1)^{rand_1} r\{\mathbf{x}_{\text{teacher}} - (T_F \cdot \mathbf{x}_{\text{mean}})\} \tag{5}$$

where $rand_1$ is a random value with either 0 or 1. Then, the binary crossover is applied if a uniform random number having an interval of 0 and 1 ($rand_2$) is lower than the crossover probability (P_r). For a new individual $\mathbf{x}_{\text{new}}^T = [x_{\text{new},1}, \ldots, x_{\text{new},D}]$ and an old individual $\mathbf{x}_{\text{old}}^T = [x_{\text{old},1}, \ldots, x_{\text{old},D}]$, the binary crossover step can be expressed as follow;

$$x_{new,j} \begin{cases} x_{old,j} & \text{if } rand_3 < CR_1 \quad j = 1, \ldots, D \\ x_{teacher,j} & \text{if } CR_1 \leq rand_3 < CR_2 \quad j = 1, \ldots, D \end{cases} \tag{6}$$

where the $rand_3$ is a uniform random number generated from 0 to 1. The CR_1 and CR_2 are the predefined crossover rates, while D is the number of design variables, respectively. Thereafter, the learning phase is conducted if a uniform random number generated from 0 to 1 ($rand_4$) is lower than the probability value (L_p), otherwise, the learning phase will be skipped. The search process will be repeated until the termination criterion is satisfied. The computational steps of the proposed algorithm are shown in Algorithm 1.

Algorithm 1 An improved TLBO

Input: Maximum iteration number (*maxiter*), population size (n_p), Crossover probability Crossover rate (CR_1 and CR_2), learning phase probability (L_p).

Output: $\mathbf{x}_{best}$, f_{best}

Initialization

1. Generate an initial population randomly.

2. Evaluate objective function values

Main algorithm

3. For $i = 1$ to *maxiter*

   3.1 Identify the best solution ($\mathbf{x}_{teacher}$)

     (Teacher Phase)

       For $j=1$ to n_p

       3.2 Update the population using equation (5)

          If $rand_2 < P_r$

             3.2.1 Applied binary crossover using equation (6)

          End

          3.2.1 Evaluate the objective function value f ($\mathbf{x}_{new,j}$)

          3.2.2 If $f(\mathbf{x}_{new,j}) < f(\mathbf{x}_{old,j})$

             Replace $\mathbf{x}_{old,j}$ by $\mathbf{x}_{new,j}$

          End

       End

       If $rand_4 < L_p$

       (Learner Phase)

       For $j=1$ to n_p

       3.3 Update the population using equation (4)

          3.3.1 Evaluate the objective function value $f(\mathbf{x}_{new,j})$

          3.2.2 If $f(\mathbf{x}_{new,j}) < f(\mathbf{x}_{old,j})$

             Replace $\mathbf{x}_{old,j}$ by $\mathbf{x}_{new,j}$

          End

       End

End

4 Numerical Experiments

In order to examine the search performance of the proposed ITLBO, several EAs have been used to solve the optimum design problem of the strip flatness as described in the previous section. The EAs used in this study are as follows:

- DE [28]: The DE/best/2/bin strategy was used. DE scaling factor was random from 0.25 to 0.7 in each calculation and crossover probability was 0.7.
- SA [16]: An annealing temperature was reduced exponentially by 10 times from the value of 10 to 0.001 in the optimization searching process. On each loop 2n children were created by means of mutation to be compared with their parent. Here, n is the number of design variables.
- ABC [29]: The number of food sources was set to be $3n_p$. A trial counter to discard a food source was 100.
- ACOR [30]: The parameters used for computing the weighting factor and the standard deviation in the algorithm were set to be $\xi = 1.0$ and $q = 0.2$, respectively.
- TLBO [31]: Parameter settings are not required.
- LCA [32]: The default parameter settings provided by the authors were used.
- ChSS [33]: The number of solutions in the charge memory was $0.2n_p$. Here, n_p is the population size. The charged moving considering rate and the parameter PAR were set to be 0.75 and 0.5, respectively.
- OPDE [10]: The DE/best/2/bin strategy was used. DE scaling factor was random from 0.25 to 0.5 in each calculation and crossover probability used was 0.7.
- ETLBO-DE [11]: Used the DE parameter setting and Latin hypercube sampling (LHS) technique to generate an initial population.
- ITLBO (Algorithm 1): The P_r, CR_1, CR_2 and L_p were set to be 0.5, 0.33, 0.66 and 0.75, respectively.

Each optimizer was employed to solve the problem for 5 optimization runs. Both the maximum number of iterations and population size were set to be 100. For the optimizers using different population sizes, such as simulated annealing, their search processes were stopped with the total number of function evaluations as 100×100. The optimal results of the various optimizers from using this limited number of function evaluations were compared. The best optimizer was used to find the optimal processing parameters of the strip coiling process.

5 Results and Discussion

After applying each optimization algorithm to solve the problem for 5 runs, the results are given in Table 1. The mean values (Mean) are used to measure the convergence rate while the standard deviation (STD) determines search consistency. The lower the mean objective function value the better, and the lower the standard deviation the more consistent. In the table, max and min stand for the maximum and minimum values of the objective function, respectively.

For the measure of convergence speed based on the mean objective value, the best method is ITLBO while the second best and the third best performers are ETLBO-DE and OPDE, respectively. The worst results came from ABC. For the measure of search consistency based on STD, the best was also ITLBO while the worst was ABC, which was similar to the measure of the search convergence. The second best and the third best for consistency were ETLBO-DE and ACOR, respectively. The minimum objective function value was obtained by the ITLBO.

Based on the results obtained, it was clearly indicated that the proposed ITLBO by adding opposition based method, binary crossover, and learning phase probability can improve the search performance of the original TLBO for solving the optimization design problem of the strip coiling process.

The optimal spool crown exponent and height obtained are 1.0822 and 2.3645, respectively. The optimal distribution of coiling tensions as a function of coil numbers is shown in Fig. 4. The results reveal that the coiling tensions start with the highest value initially and then decrease when the number of coils increases. After a few series of coiling, the tension levels become almost constant, converging to the lower bound at the end of the process. Figure 5 shows the plot of the circumferential stress distributions along the z and r directions of the original and optimum design solutions in that order. The comparison of the maximum compressive stresses and the standard deviation of stresses at the inner strip between the original and optimal designs is given in Table 2. The results show that the optimal processing parameters obtained by the proposed ITLBO algorithm can reduce the maximum compressive stress and the axial inhomogeneity of the stress distribution at the inner strip, which might cause undesirable wavy edge defects during the strip coiling process.

Table 1. Objective function values calculated

Evolutionary algorithms	Mean	STD	Max.	Min.
DE	0.9700	0.0275	1.0096	0.9354
ABC	1.7637	0.0787	1.8800	1.6751
ACOR	1.0621	0.0070	1.0705	1.0546
ChSS	1.4026	0.0289	1.4448	1.3678
LCA	1.7116	0.0408	1.7580	1.6473
SA	1.5451	0.0645	1.6323	1.4841
TLBO	0.9915	0.0132	1.0066	0.9766
OPDE	0.9539	0.0179	0.9715	0.9297
ETLBO-DE	0.8850	0.0047	0.8897	0.8784
ITLBO	0.8740	0.0025	0.8783	0.8720

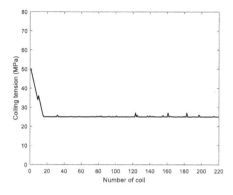

Fig. 4. Coiling tension levels as a function of number of coils

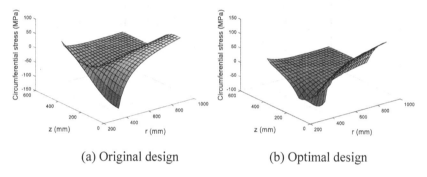

(a) Original design   (b) Optimal design

Fig. 5. Comparison of circumferential stresses along the z and r directions for the original design and optimal design, respectively

Table 2. Maximum compressive stress and the standard deviation of stresses at the inner coil

	Original design	Optimal design
Maximum compressive stress [MPa]	111.546	68.0270
Standard deviation of stresses	48.375	29.3703

6   Conclusions

The new population-based optimization algorithm obtained by improving the original TLBO for solving the flatness optimization of the strip coiling process has been proposed. The search performance of the method was compared to various established evolutionary algorithms. The numerical results show that the new optimizer ITLBO is the best performer for both convergence rate and consistency. With this, the new parameters including the spool geometry and the coiling tension distribution have been obtained and can be used in the real strip coiling process. Further studies will be made to enhance the mathematical model of the strip coiling process. A self-adaptive version of ITLBO will be investigated for search performance enhancement.

Acknowledgements. The authors gratefully acknowledge the financial support from Thailand Research Fund (TRF). The research grant from the POSCO was also appreciated.

References

1. Jung, J.Y., Im, Y.T.: Simulation of fuzzy shape control for cold-rolled strip with randomly irregular strip shape. J. Mater. Process. Tech. **63**, 248–253 (1997)
2. Jung, J.Y., Im, Y.T.: Fuzzy control algorithm for prediction of tension variations in hot rolling. J. Mater. Process. Tech. **96**, 163–172 (1999)
3. Kawanami, T., Asamura, T., Matsumoto, H.: Development of high-precision shape and crown control technology for strip rolling. J. Mater. Process. Tech. **22**, 257–275 (1990)
4. Kwon, H.C., Han, I.S., Chun, M.S.: Examination of thermal behavior of hot rolled coil based on the finite element modeling and thermal measurement. In: 10th International Conference on Technology of Plasticity, pp. 37–40 (2011)
5. Sims, R.B., Place, J.A.: The stresses in the reels of cold reduction mills. Br. J. Appl. Phys. **4**, 213–216 (1953)
6. Miller, D.B., Thornton, M.: Prediction of changes in flatness during coiling. In: 5th International Rolling Conference, pp. 73–78 (1990)
7. Sarban, A.A.: An elasto-plastic analysis for the prediction of changes in flatness during coiling. In: 2nd International Conference on Modelling of Metal Rolling Processes, pp. 92–100 (1996)
8. Yanagi, S., Hattori, S., Maeda, Y.: Analysis model for deformation of coil of thin strip under coiling process. J. JSTP. **39**, 51–55 (1998)
9. Park, W.W., Kim, D.K., Kwon, H.C., Chun, M.S., Im, Y.T.: The effect of processing parameters on elastic deformation of the coil during the thin-strip coiling process. Metals Mater. Inter. **20**, 719–726 (2014)
10. Pholdee, N., Bureerat, S., Park, W.-W., Kim, D.-K., Im, Y.-T., Kwon, H.-C., Chun, M.-S.: Optimization of flatness of strip during coiling process based on evolutionary algorithms. Int. J. Precis. Eng. Manuf. **16**(7), 1493–1499 (2015)
11. Pholdee, N., Park, W.-W., Kim, D.-K., Im, Y.-T., Bureerat, S., Kwon, H.-C., Chun, M.-S.: Efficient hybrid evolutionary algorithm for optimization of a strip coiling process. Eng. Opt. **47**(4), 521–532 (2015)
12. Pholdee, N., Bureerat, S.: Passive vibration control of an automotive component using evolutionary optimization. J. Res. Appl. Mech. Eng. **1**, 19–22 (2010)
13. Pholdee, N., Bureerat, S.: Surrogate-assisted evolutionary optimizers for multiobjective design of a torque arm structure. Appl. Mech. Mater. **101–102**, 324–328 (2012)
14. Pholdee, N., Bureerat, S.: Performance enhancement of multiobjective evolutionary optimizers for truss design using an approximate gradient. Comput. Struct. **106–107**, 115–124 (2012)
15. Pholdee, N., Bureerat, S.: Hybridisation of real-code population-based incremental learning and differential evolution for multiobjective design of trusses. Inform. Sci. **223**, 136–152 (2013)
16. Bureerat, S., Limtragool, J.: Structural topology optimisation using simulated annealing with multiresolution design variables. Finite Elem. Anal. Des. **44**, 738–747 (2008)
17. Srisoporn, S., Bureerat, S.: Geometrical design of plate-fin heat sinks using hybridization of MOEA and RSM. IEEE Trans. Compon. Packag. Technol. **31**, 351–360 (2008)
18. Yildiz, A.R.: A novel particle swarm optimization approach for product design and manufacturing. Int. J. Adv. Manuf. Tech. **40**, 617–628 (2009)

19. Goldstein, M.: DEEPSAM: a hybrid evolutionary algorithm for the prediction of biomolecules structure. In: Blesa, M.J., Blum, C., Cangelosi, A., Cutello, V., Di Nuovo, A., Pavone, M., Talbi, E.-G. (eds.) HM 2016. LNCS, vol. 9668, pp. 218–221. Springer, Heidelberg (2016). doi:10.1007/978-3-319-39636-1_16

20. Kumar, S., Dubey, A.K., Pandey, A.K.: Computer-aided genetic algorithm based multi-objective optimization of laser trepan drilling. Int. J. Precis. Eng. Manuf. **14**, 1119–1125 (2013)

21. Oladimeji, M.O., Turkey, M., Dudley, S.: A heuristic crossover enhanced evolutionary algorithm for clustering wireless sensor network. In: Squillero, G., Burelli, P. (eds.) EvoApplications 2016. LNCS, vol. 9597, pp. 251–266. Springer, Heidelberg (2016). doi:10.1007/978-3-319-31204-0_17

22. Park, H.S., Nguyen, T.T.: Optimization of roll forming process with evolutionary algorithm for green product. Int. J. Precis. Eng. Manuf. **14**, 2127–2135 (2013)

23. Kandananond, K.: The optimization of a lathing process based on neural network and factorial design method. In: Fujita, H., Ali, M., Selamat, A., Sasaki, J., Kurematsu, M. (eds.) IEA/AIE 2016. LNCS (LNAI), vol. 9799, pp. 609–619. Springer, Heidelberg (2016). doi:10.1007/978-3-319-42007-3_53

24. Hetmaniok, E., Słota, D., Zielonka, A.: Solution of the inverse continuous casting problem with the aid of modified harmony search algorithm. In: Wyrzykowski, R., Dongarra, J., Karczewski, K., Waśniewski, J. (eds.) PPAM 2013. LNCS, vol. 8384, pp. 402–411. Springer, Heidelberg (2014). doi:10.1007/978-3-642-55224-3_38

25. Tiwari, A., Oduguwa, V., Roy, R.: Rolling system design using evolutionary sequential process optimization. IEEE Trans. Evol. Comput. **12**, 196–202 (2008)

26. Chakraborti, N., Siva Kumar, B., Satish Babu, V., Moitra, S., Mukhopadhyay, A.: Optimizing surface profiles during hot rolling: a genetic algorithms based multi-objective optimization. Comp. Mater. Sci. **37**, 159–165 (2006)

27. Zhang, J., Wang, Y.: Defection recognition of cold rolling strip steel based on ACO algorithm with quantum action. In: Pan, Z., Cheok, A.D., Müller, W., Chang, M., Zhang, M. (eds.). LNCS, vol. 7145, pp. 263–271Springer, Heidelberg (2012). doi:10.1007/978-3-642-29050-3_26

28. Storn, R., Price, K.: Differential evolution – a simple and efficient heuristic for global optimisation over continuous spaces. J. Global Optim. **11**, 341–359 (1997)

29. Karaboga, D., Basturk, B.: A powerful and efficient algorithm for numerical function optimisation: artificial bee colony (ABC) algorithm. J. Global Optim. **39**, 459–471 (2007)

30. Socha, K., Dorigo, M.: Ant colony optimisation for continuous domains. Eur. J. Oper. Res. **185**, 1155–1173 (2008)

31. Rao, R.V., Savsani, V.J., Vakharia, D.P.: Teaching–learning-based optimisation: a novel method for constrained mechanical design optimisation problems. Comput. Aided Design. **43**, 303–315 (2011)

32. Kashan, A.H.: An efficient algorithm for constrained global optimisation and application to mechanical engineering design: league championship algorithm (LCA). Comput. Aided Design. **43**, 1769–1792 (2011)

33. Kaveh, A., Talatahari, S.: A novel heuristic optimisation method: charged system search. Acta Mech. **213**, 267–289 (2010)

34. Choi, Y.J., Lee, M.C.: A downcoiler simulator for high performance coiling in hot strip mill lines. Int. J. Precis. Eng. Manuf. **10**, 53–61 (2009)

An Entailment Procedure for Kleene Answer Set Programs

Patrick Doherty[1] and Andrzej Szałas[1,2(✉)]

[1] Department of Computer and Information Science,
Linköping University, SE-581 83 Linköping, Sweden
{patrick.doherty,andrzej.szalas}@liu.se
[2] Institute of Informatics, University of Warsaw, Banacha 2, 02-097 Warsaw, Poland
andrzej.szalas@mimuw.edu.pl

Abstract. Classical Answer Set Programming is a widely known knowledge representation framework based on the logic programming paradigm that has been extensively studied in the past decades. Semantic theories for classical answer sets are implicitly three-valued in nature, yet with few exceptions, computing classical answer sets is based on translations into classical logic and the use of SAT solving techniques. In this paper, we introduce a variation of Kleene three-valued logic with strong connectives, R_3, and then provide a sound and complete proof procedure for R_3 based on the use of signed tableaux. We then define a restriction on the syntax of R_3 to characterize Kleene ASPs. Strongly-supported models, which are a subset of R_3 models are then defined to characterize the semantics of Kleene ASPs. A filtering technique on tableaux for R_3 is then introduced which provides a sound and complete tableau-based proof technique for Kleene ASPs. We then show a translation and semantic correspondence between Classical ASPs and Kleene ASPs, where answer sets for normal classical ASPs are equivalent to strongly-supported models. This implies that the proof technique introduced can be used for classical normal ASPs as well as Kleene ASPs. The relation between non-normal classical and Kleene ASPs is also considered.

1 Introduction

Classical Answer Set Programs (ASP) [1,3,15–17,22] belong to the family of rule-based logic programming languages. The semantic framework for classical ASPs is based on the use of stable model semantics. There are two characteristics intrinsically associated with the construction of stable models (answer sets) for answer set programs. Any member of an answer set is supported through facts and chains of rules and those members are in the answer set only if generated minimally in such a manner. These two characteristics, supportedness and minimality, provide the essence of answer sets. Classical ASPs allows two kinds of

This work is partially supported by the Swedish Research Council (VR) Linnaeus Center CADICS, the ELLIIT network organization for Information and Communication Technology, the Swedish Foundation for Strategic Research (CUAS, SymbiCloud Project), and Vinnova NFFP6 Project 2013-01206.

C. Sombattheera et al. (Eds.): MIWAI 2016, LNAI 10053, pp. 24–37, 2016.
DOI: 10.1007/978-3-319-49397-8_3

negation, classical or strong negation and default or weak negation. Additionally, answer sets are implicitly partial and that partiality provides epistemic overtones to the interpretation of disjunctive rules and default negation.

There has been much work in providing numerous definitions [24] of what an answer set is and then relating particular definitions to other formalisms such as Circumscription [27] or Default Logic [32]. This body of research has provided much insight into the nature of answer sets. Additionally, the relationship with other formalisms has provided a basis for algorithms that generate answer sets for classical ASPs in addition to inference procedures (for implementations see, e.g., CLASP [13], ASSAT [25], CMODELS [23], SMODELS [36], DLV [21]). The majority of these approaches often use syntactic translations and encodings of classical ASPs into a classical two-valued framework using standard model theory. The benefit is that one can appeal to the broad body of techniques from classical logic.

In this paper, we are interested in Kleene Answer Set programs (Kleene ASPs) introduced in [7]. Our approach is based on an extension of the three-valued logic with strong connectives, K_3, of Kleene [19], extended with an external negation connective, $\sim$, characterizing default negation and an implication connective, $\rightarrow_s$, suitable for Kleene ASPs. This logic, which we call R_3, has been considered in [7]. Kleene ASPs are directly represented as a fragment of the R_3 language. The semantics of Kleene ASPs is based on the use of strongly supported models which turn out to be a subset of the R_3 models. It was stated previously that the intuitions behind classical ASP semantics are based on supportedness and minimality. One of the purposes in introducing Kleene ASPs and strongly supported models is to provide a weaker interpretation of non-normal ASPs where the minimality assumption on disjunction is relaxed. Separation of supportedness and minimality technically turns out to have complexity advantages when dealing with non-normal Kleene ASPs. As a derivative of this work, it can be shown that there is an equivalence between the strongly supported models of a normal Kleene ASP and the answer sets of a normal classical ASP. Consequently, these proof techniques may also be of wider interest.

The following are the main results in this paper:[1]

- We present a semantic tableaux procedure for R_3 based on the use of signed formulas that is both sound and complete for R_3. It can be used both for model generation and entailment.
- The general tableaux procedure for R_3 is only sound for Kleene ASPs. For completeness, the following method is proposed:

    A filtering procedure on tableaux branches is introduced based on the definition of strong supportedness. For both normal and non-normal Kleene ASPs, the filtering procedure is shown to be sound and complete. For normal ASP^Ks it is tractable and for non-normal ones it is in Π_1^P.

---

[1] For clarity, we restrict the results to the propositional case, but they are easily generalized to the first-order case with certain restrictions.

– Due to the equivalence between strongly supported models for a normal Kleene
  ASP and answer sets for its corresponding normal classical ASP, the proof
  techniques proposed here can be used directly for normal classical ASPs. Since
  classical answer sets are strongly supported, the proposed proof techniques
  remain sound for disjunctive classical ASPs.

Structure of the Paper. The paper is structured as follows. In Sect. 2 the
augmented Kleene three-valued logic, R$_3$, is specified. Section 3 presents the
general tableaux procedure for R$_3$. Section 4 is devoted to Kleene ASPs and its
relation to classical ASPs. It is assumed the reader is familiar with classical ASP
definitions (see, e.g., [24]). In Sect. 5, a sound and complete entailment procedure
for Kleene ASPs, using semantic tableaux and filtering, is presented. Section 6
concludes with related work and conclusions.

2   Three-Valued Logic R$_3$

The basis for our approach begins with the three-valued logic with strong connectives, K$_3$, of Kleene [19], extended with an external negation connective, $\sim$,
characterizing default negation.[2] We assume that constants T (*true*), F (*false*), U
(*unknown*) are part of the language. The implication connective, $\to_s$, is defined
using, $\sim$, as:[3]

$$A \to_s B \ = \ \sim A \vee \neg \sim B.$$

We denote this logic by R$_3$.

Truth tables for connectives of R$_3$ are shown in Table 1, where $\neg, \vee, \wedge$ are
the strong connectives of K$_3$, $\sim$ is the weak or external negation connective and
$\to_s$ is a newly defined implication for ASP rules.

Table 1. Truth tables for R$_3$ connectives.

	$\neg$	$\sim$	$\vee$	F	U	T	$\wedge$	F	U	T	$\to_s$	F	U	T
T	F	F	F	F	U	T	F	F	F	F	F	T	T	T
F	T	T	U	U	U	T	U	F	U	U	U	T	T	T
U	U	T	T	T	T	T	T	F	U	T	T	F	F	T

Let $\mathcal{P}$ be the set of propositional variables. By a *valuation of propositional
variables* we understand a mapping $\mathcal{P} \longrightarrow \{T, U, F\}$. If A is an R$_3$ formula and
v is a valuation then the *truth value* of a formula, denoted by $v(A)$, is defined
inductively extending v by using the semantics of the connectives provided in
Table 1.

By a *positive literal* (or an *atom*) we mean any propositional variable of $\mathcal{P}$.
A *negative literal* is an expression of the form $\neg r$, where $r \in \mathcal{P}$. A *(classical)*

---

[2] Strong negation, conjunction and disjunction have also been used in [26].
[3] The implication connective, $\to_s$, is in fact equivalent to that used in [35].

literal is a positive or a negative literal. For a literal $\ell = \neg p$, by $\neg \ell$ we understand p. A set of literals is *consistent* when it does not contain a propositional variable together with its negation. An *interpretation* is any consistent set of literals.

Any interpretation I defines a three-valued valuation v_I by:

$$v_I(p) \stackrel{\text{def}}{=} \begin{cases} \text{T when } p \in I; \\ \text{F when } \neg p \in I; \\ \text{U otherwise.} \end{cases} \tag{1}$$

Also, given v, one can construct a corresponding interpretation I_v by setting:

$$I_v \stackrel{\text{def}}{=} \{p \mid p \in \mathcal{P} \text{ and } v(p) = \text{T}\} \cup \{\neg p \mid p \in \mathcal{P} \text{ and } v(p) = \text{F}\}. \tag{2}$$

In the rest of the paper, we will freely switch between interpretations and valuations, using (1) and (2).

An interpretation I is a *model* of a set of formulas S iff for all $A \in S$, $v_I(A) = \text{T}$, where v_I is the valuation corresponding to I.

By convention, the empty conjunction is T and the empty disjunction is F.

3 Signed Tableaux for Three-Valued Logic R₃

We follow the signed tableaux style of [6], extended with rules for $\rightarrow_s$.[4] The generalization to multivalued logics is derived from [37].

By an *information constraint lattice* we mean the lattice $L \stackrel{\text{def}}{=} \langle \mathcal{S}; \sqcup, \sqcap \rangle$ where:

- $\mathcal{S} = \{[\text{T}], [\text{F}], [\text{U F T}], [\text{U T}], [\text{U F}], [\text{U}], [\ ]\}$ is the set of signs which also contains the element $[\ ]$, representing *contradiction*; moving upwards in the lattice should be interpreted as tightening for possible truth values for a formula;
- $\sqcup$ ($\sqcap$) are the join (meet) operation on $\mathcal{S}$ defined respectively as the least upper bound and greatest lower bound w.r.t. the ordering shown in Fig. 1.

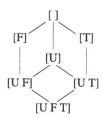

Fig. 1. Information constraint lattice.

---

[4] This signed approach to semantic tableaux for multi-valued logic was discovered independently, by [6,18] who generalized the technique for a larger class of multi-valued logics. The approach of [6] was used as a basis for [28].

A *signed formula* is any expression of the form $[s]A$, where $[s] \in \mathcal{S}$ in the lattice L. A valuation u *satisfies* a signed formula $[a \dots b]A$ if it satisfies the disjunction:

$$u(A) = a \text{ or } \dots \text{ or } u(A) = b, \tag{3}$$

where "or" in (3) is the classical disjunction.

A valuation u *satisfies* a set of signed formulas S iff u satisfies every formula in S. An interpretation is a *model* of S iff the corresponding valuation satisfies S.

3.1 Tableaux Construction Rules

Tableaux [37] are used to construct models for formulas. The construction of a tableau starts with a formula for which models (if any) are constructed. The tableau is then expanded according to rules provided in the subsequent subsections. The following types of rules are used, where ',' and '|' represent conjunction and disjunction, respectively:

$$\frac{[s]A}{[s']B} \qquad \frac{[s]A}{[s']B, [s'']C} \qquad \frac{[s]A}{[s']B \mid [s'']C}$$

The following theorem can be proved extending the corresponding proof for K_3 given in [6] by considering the new connective $\rightarrow_s$.

Theorem 1. *Tableau rules shown in Table 2 are sound and complete for R_3, i.e., a formula A is unsatisfiable iff there is a closed tableau for A.*   ◁

Table 2. Tableaux rules.

Rules for conjunction - $\wedge$.

$$\frac{[T]A \wedge B}{[T]A, [T]B} \qquad \frac{[F]A \wedge B}{[F]A \mid [F]B} \qquad \frac{[UT]A \wedge B}{[UT]A, [UT]B} \qquad \frac{[UF]A \wedge B}{[UF]A \mid [UF]B}$$

Rules for disjunction - $\vee$.

$$\frac{[T]A \vee B}{[T]A \mid [T]B} \qquad \frac{[F]A \vee B}{[F]A, [F]B} \qquad \frac{[UT]A \vee B}{[UT]A \mid [UT]B} \qquad \frac{[UF]A \vee B}{[UF]A, [UF]B}$$

Rules for strong negation - $\neg$ and weak (external) negation - $\sim$.

$$\frac{[T]\neg A}{[F]A} \quad \frac{[F]\neg A}{[T]A} \quad \frac{[UT]\neg A}{[UF]A} \quad \frac{[UF]\neg A}{[UT]A} \quad \frac{[T]\sim A}{[UF]A} \quad \frac{[F]\sim A}{[T]A} \quad \frac{[UT]\sim A}{[UF]A} \quad \frac{[UF]\sim A}{[T]A}$$

Rules for implication - $\rightarrow_s$.

$$\frac{[T]A \rightarrow_s}{[UF]A} \quad \frac{\rightarrow_s [T]B}{[T]B} \quad \frac{[T]A \rightarrow_s B}{[UF]A \mid [T]B} \quad \frac{[F]A \rightarrow_s B}{[T]A, [UF]B} \quad \frac{[UT]A \rightarrow_s B}{[UF]A \mid [T]B} \quad \frac{[UF]A \rightarrow_s B}{[T]A, [UF]B}$$

3.2   Constructing Models Using Tableaux

A path in a tableau is *completed* if no tableau rule can extend the path. A tableau is *completed* when all its paths are completed.

Let $[s]A$ and $[s']A$ be signed formulas. A path in a tableau is *closed* if both $[s]A$ and $[s']A$ are in the path and $[s] \sqcup [s'] = [\ ]$. A path is *open* when it is not closed.

A tableau $\mathcal{T}$ is *closed* if all its branches are closed. A tableau is *open* if it is not closed.

Given a tableau $\mathcal{T}$ for a formula A, to extract u satisfying A we look for an open branch. If such a branch does not exist then there is no model for A. Otherwise, for each propositional variable p appearing in the branch let:

$$\Sigma(p) \overset{\text{def}}{=} \{s \mid [s]p \text{ occurs in the branch}\} \text{ and let } \sigma(p) \overset{\text{def}}{=} \bigsqcup_{s \in \Sigma(p)} [s].$$

The satisfying valuations (models) are then defined by assigning, to each p occurring in the branch, a value from $\sigma(p)$. If for a propositional variable p, occurring in A, no formula of the form $[s]p$ occurs on a given branch then p is unconstrained and can be assigned any truth value. Note that, due to the form of rules, only signs $[F]$, $[T]$, $[U\,F]$ and $[U\,T]$ can appear in tableaux. If a proposition p does not occur in a branch, we assume that implicitly $[U\,F\,T]p$ occurs in that branch.

Example 1. Consider the following tableau:

$$\frac{[T]((p \vee \neg q) \wedge \sim q)}{[T](p \vee \neg q),\ [T](\sim q)}$$
$$\overline{[U\,F]q}$$
$$[T]p \quad \mid \quad \frac{[T]\neg q}{[F]q}$$

The first branch contains $[U\,F]q$ and $[T]p$, so $\sigma(q) = [U\,F]$ and $\sigma(p) = [T]$. The branch then encodes two models:

$$v_1(q) = U, v_1(p) = T, \text{ corresponding to } \{p\},$$
$$v_2(q) = F, v_2(p) = T, \text{ corresponding to } \{\neg q, p\}.$$

The second branch contains $[U\,F]q$ and $[F]q$, so $\sigma(q) = F$. Since p remains unconstrained, the branch encodes models:

$$u_1(q) = F, u_1(p) = T, \text{ corresponding to } \{\neg q, p\},$$
$$u_2(q) = F, u_2(p) = F, \text{ corresponding to } \{\neg q, \neg p\},$$
$$u_3(q) = F, u_3(p) = U, \text{ corresponding to } \{\neg q\}.$$

Of course, v_1, v_2, u_1, u_2, u_3 are all models for the starting formula $(p \vee \neg q) \wedge \sim q$.

◁

We have the following theorem which can be proved by extending the proof of the analogous theorem for K_3 given in [6].

Theorem 2. *Given an R_3 formula A, every interpretation satisfying A can be extracted from a completed tableau for $[T]A$.*   ◁

4   (Kleene) Answer Set Programs

The syntax for Kleene Answer Set Programs, ASP^K, is identical for that of classical ASP programs. The semantics for Kleene ASP^K programs is based on the use of the augmented three-valued Kleene logic R_3 and strongly-supported models presented in [7]. The semantics for classical ASP programs is based on stable model semantics [24]. Correspondences between classical answer sets and strongly supported models will be considered later in this section. For the sake of clarity we consider propositional programs only.

By an ASP^K rule we understand an expression ϱ of the form:

$$\ell_1 \vee \ldots \vee \ell_k \leftarrow \ell_{k+1}, \ldots, \ell_m, not\, \ell_{m+1}, \ldots, not\, \ell_n, \tag{4}$$

where $n \geq m \geq k \geq 0$ and $\ell_1, \ldots, \ell_k, \ell_{k+1}, \ldots, \ell_m, \ell_{m+1}, \ldots \ell_n$ are (positive or negative) literals. The expression at the lefthand side of '←' in (4) is called the *head* and the righthand side of '←' is called the *body* of the rule. The rule is called *disjunctive* if $k > 1$.

An ASP^K *program Π* is a finite set of rules. A program is *normal* if each of its rules has at most one literal in its head. If a program contains a disjunctive rule, we call it *disjunctive* or *non-normal*.

An interpretation I *satisfies* a rule ϱ of the form (4), denoted by $I \models \varrho$, if whenever $\ell_{k+1}, \ldots, \ell_m \in I$ and $\ell_{m+1}, \ldots, \ell_n \notin I$, we have $\ell_i \in I$ for some $1 \leq i \leq k$. An interpretation I *satisfies* an ASP^K program Π, denoted by $I \models \Pi$, if for all rules $\varrho \in \Pi$, $I \models \varrho$.

We say that an Kleene ASP program Π *entails* an R_3 formula A, and denote it by $\Pi \models A$, provided that A is true in every strongly supported model of Π.

The concept of strong supportedness [7] builds on the principle of constructing models through chains of rules grounded in facts. When evaluating the body of a rule to determine whether it is applicable, literals outside the scope of *not* must be evaluated against the set of literals for which support has already been found, represented as an interpretation I. However, literals inside the scope of *not* must be evaluated against a strongly supported model candidate J, similarly to how *not* ℓ is evaluated against an answer set candidate S when a reduct Π^S is computed classically. We therefore evaluate formulas w.r.t. two interpretations. Given interpretations I and J, the *value of a formula A w.r.t. (I, J)*, denoted by $(I, J)(A)$, is defined as follows:

$$(I, J)(A) \stackrel{\text{def}}{=} \begin{cases} \text{T when } I \models reduct^J(A); \\ \text{F when } I \models reduct^J(\neg A); \\ \text{U otherwise.} \end{cases} \tag{5}$$

where $reduct^J(A)$ (respectively, $reduct^J(\neg A)$) is a formula obtained from $A\ (\neg A)$ by substituting subformulas of the form $not\ \ell$ by their truth values evaluated in J.

Now we are ready to define strong supportedness. An interpretation N is a *strongly supported model* of an ASP^K program Π provided that N satisfies Π and there exists a sequence of interpretations $I_0 \subseteq I_1 \subseteq \ldots \subseteq I_n$ where $n \geq 0$ such that $I_0 = \emptyset$, $N = I_n$, and:

1. for every $1 \leq i \leq n$ and every rule $\ell_1 \vee \ldots \vee \ell_k \leftarrow B$ of Π,
   if $\bigl(I_{i-1}, N\bigr)(B) = \text{T}$ then a nonempty subset of $\{\ell_1, \ldots, \ell_k\}$ is included in I_i;
2. for $i = 1, \ldots, n$, I_i can only contain literals obtained by applying point 1.

A *Kleene Answer Set* for an ASP^K program is a strongly supported model. The terminology will be used interchangeably.

The following correspondences between classical answer sets with stable model semantics (for definitions see, e.g., [24]) and Kleene answer sets with strongly-supported model semantics relate the two semantics.

In [7] the following theorem is proved (see [7, point 2 of Theorem 1]).

Theorem 3. *For any normal ASP^K program Π, I is a classical answer set of Π iff I is a strongly supported model of Π.*   ◁

The following theorem clarifies the role of strong supportedness in the context of disjunctive programs.

Theorem 4. *For any ASP^K program Π, if I is a classical answer set of Π then I is a strongly supported model of Π.*   ◁

Theorem 3 allows us to use the filter-based tableau technique introduced in Sect. 5 below, not only for Kleene ASPs, but for classical normal ASPs, too. Theorem 4 shows that filtering remains sound for disjunctive classical ASPs.[5]

5 Filtering Technique for Kleene Answer Set Programs

To construct tableaux for Kleene ASP entailment we first translate Kleene ASPs into formulas of R_3 using the translation Tr defined as follows:

$$Tr(\neg A) \stackrel{\text{def}}{=} \neg A,\ Tr(not\ A) \stackrel{\text{def}}{=} {\sim} A,$$

$$Tr(A \circ B) \stackrel{\text{def}}{=} Tr(A) \circ Tr(B), \text{for } \circ \in \{\vee, \wedge\},$$

$$Tr(A \leftarrow B) \stackrel{\text{def}}{=} Tr(B) \to_s Tr(A). \tag{6}$$

Rules are then translated as follows:

$$Tr(\ell_0 \vee \ldots \vee \ell_i \leftarrow \ell_{i+1}, \ldots, \ell_j, not\ \ell_{j+1}, \ldots, not\ \ell_m) =$$
$$(\ell_{i+1} \wedge \ldots \wedge \ell_j \wedge {\sim}\ell_{j+1} \wedge \ldots \wedge {\sim}\ell_m) \to_s (\ell_0 \vee \ldots \vee \ell_i). \tag{7}$$

---

[5] In understanding the theorems, recall that the syntax for ASP^K programs and classical ASP programs is identical.

For a Kleene ASP program Π, $Tr(\Pi) \stackrel{\text{def}}{=} \bigwedge_{r \in \Pi} Tr(r)$. By a *model of a Kleene ASP program* Π we understand any model of $Tr(\Pi)$.

Note that every Kleene answer set for a program Π is an R_3 model for $Tr(\Pi)$. Therefore, we have the following theorem.

Theorem 5. *Let* Π *be a Kleene ASP program,* A *be an* R_3 *formula, and* $\mathcal{T}$ *be a tableau for:*

$$[T](Tr(\Pi) \wedge \sim A). \tag{8}$$

Then, if $\mathcal{T}$ *is closed then* $\Pi \models A$.   ◁

That is, the tableaux procedure provided in Sect. 3.1 is sound for Kleene ASP entailment.

Example 2. Let Π consists of rules:

$$q \leftarrow p.$$
$$q \leftarrow not\, p.$$

To show that $\Pi \models q$ we construct the following tableau:

$$\frac{[\mathrm{T}]\big((p \rightarrow_s q) \wedge (\sim p \rightarrow_s q) \wedge \sim q\big)}{}$$
$$\overline{[\mathrm{T}]\big((p \rightarrow_s q) \wedge (\sim p \rightarrow_s q)\big),\ [\mathrm{T}](\sim q)}$$
$$\overline{[\mathrm{T}](p \rightarrow_s q),\ [\mathrm{T}](\sim p \rightarrow_s q),\ [\mathrm{T}](\sim q)}$$
$$\overline{[\mathrm{U\,F}]q}$$
$$\frac{[\mathrm{U\,F}]p}{[\mathrm{U\,F}]\sim p \ \mid\ [\mathrm{T}]q} \quad \mid \quad [\mathrm{T}]q$$
$$\overline{[\mathrm{T}]p}$$

On the first branch $\sigma(p) = [\ ]$ and on the other two branches $\sigma(q) = [\ ]$. Thus the above tableau is closed.   ◁

For completeness of ASP^K, a filtering technique is required to filter out non-strongly supported models associated with open branches. To decide whether $\Pi \models A$ we first construct a tableau $\mathcal{T}$ for signed formula (8). Then,

1. if $\mathcal{T}$ is closed then $\Pi \models A$; otherwise
2. *filtering*: eliminate every open branch of $\mathcal{T}$ encoding only non-strongly supported models of Π. If all open branches of $\mathcal{T}$ are eliminated then $\Pi \models A$, otherwise $\Pi \not\models A$.

The filtering described in point 2. Above is sound and complete for ASP^K, as stated in the following theorem.

Theorem 6. *Let* Π *be a Kleene ASP program,* A *be an* R_3 *formula, and* $\mathcal{T}$ *be a tableau for* $[T](Tr(\Pi) \wedge \sim A)$. *Then,* $\Pi \models A$ *iff filtering eliminates every open branch of* $\mathcal{T}$.   ◁

Example 3. Let program Π consist of a single rule: $q \leftarrow not\,p$. To show that $\Pi \models q$, we construct the following tableau:

$$\frac{[\mathrm{T}]\big((\sim p \rightarrow_s q) \wedge \sim q)\big)}{[\mathrm{T}](\sim p \rightarrow_s q),\ [\mathrm{T}](\sim q)}$$

$$\frac{}{[\mathrm{U\,F}]q}$$

$$\frac{[\mathrm{U\,F}] \sim p \quad | \quad [\mathrm{T}]q}{[\mathrm{T}]p}$$

The first branch is open and the second branch is closed. Therefore we have to check whether the first branch represents a strongly supported model for Π. Since the branch contains $[\mathrm{U\,F}]q$ and $[\mathrm{T}]p$, the candidates for a strongly supported model for Π are $\{p\}$ and $\{\neg q, p\}$.[6] Of course, $\{p\}$ and $\{\neg q, p\}$ are not strongly supported, so there are no open branches representing strongly supported models for Π. Therefore, Π indeed entails q. $\triangleleft$

For non-normal programs an open branch may encode more than one strongly supported model, consequently the technique shown in Example 3 does not apply since one uses an assumption of equivalence between minimality and strong supportedness that only applies to normal programs. In this case one is required to filter all potential interpretations, including non-minimal ones, associated with an open branch.

To verify strong supportedness one can use Algorithm 1 provided in [7, p. 137]. Given a program Π and an interpretation I, this algorithm checks whether I is a strongly supported model of Π in deterministic polynomial time w.r.t. Π and I. Recall that due to the form of tableau rules shown in Table 2, only signs $[\mathrm{F}]$, $[\mathrm{T}]$, $[\mathrm{U\,F}]$ and $[\mathrm{U\,T}]$ (and implicitly $[\mathrm{U\,F\,T}]$) can appear in tableaux. Therefore, to check whether a given open branch of a tableau encodes a strongly supported model, one can extract the candidate valuation v in such a way that whenever for a given proposition p, U is in $\sigma(p)$, we set $v(p) \stackrel{\text{def}}{=} \mathrm{U}$, otherwise $v(p)$ is the (uniquely determined) truth value from $\sigma(p)$. That way, for normal programs, v represents a minimal interpretation uniquely determined by this branch. By Theorem 3, for normal programs strong supportedness is equivalent to minimality, so we have the following theorem.

Theorem 7. *For any normal ASP^K program Π checking whether a given tableau branch for $[T]\Pi$ encodes a strongly supported model is tractable.* $\triangleleft$

For non-normal programs, checking whether an open branch exists such that there is a valuation encoded by the branch defining a strongly supported model, is obviously in Σ_1^P. Therefore, checking whether a branch can be closed via filtering (does not encode a strongly supported model), is in Π_1^P (i.e., co-NP) as stated in the following theorem.

---

[6] Note that both $\{p\}$ and $\{\neg q, p\}$ are R$_3$ models for the considered formula, so the fact that q is entailed by Π cannot be proved using only rules provided in Sect. 3.1.

Theorem 8. *For any non-normal ASP^K program Π checking, in general, whether a given tableau branch for $[T]\Pi$ encodes a strongly supported model, is in Π_1^P.*   ◁

Note that Theorem 8, ensures that our entailment procedure is in Π_1^P.

Remark 1. The technique of filtering can be applied to classical answer sets, too. First, one can filter out non-strongly supported models. This can be done "locally" node by node. Normal classical ASP programs are equivalent to normal ASP^K programs so for this case our procedure remains sound and complete. For non-normal classical ASP programs, the minimality requirement results in a higher complexity of reasoning [8,9], [7, Thm. 2, p. 137] (assuming that the polynomial hierarchy does not collapse). In this case, checking for non-minimality rather than for non-supportedness is required. It calls for pairwise comparisons with all models, perhaps encoded by other nodes in the constructed tableaux. Therefore, rather than checking for minimality, we could achieve completeness for non-normal classical ASPs by suitably adding a generalization of Clark's completion and loop formulas provided in [25] to the original ASP program.   ◁

6   Related Work and Conclusions

There is a rich history of explicit use of partial interpretations and multi-valued logics as a basis for semantic theories for logic programs. Some related and additional representative examples are [5,11,12,20,30,31,34,38]. Supportedness is analyzed in many papers, starting from [10]. One of most recent generalizations, via grounded fixpoints, is investigated in [2]. However, grounded fixpoints are minimal (see [2, Proposition 3.8]) while strongly supported models do not have to be minimal.

In [33] a possible model semantics for disjunctive programs is proposed. It is formulated with the use of split programs and there can be exponentially many of them comparing to the original program. Similar semantics was independently proposed in [4] under the name of the possible world semantics. In comparison to [33], ASP^K programs allow for strong negation and a three-valued model-theoretic semantics is provided. The presence of both default and strong negation in ASP^K provides a tool to close the world locally in a contextual manner, more flexible than possible model negation proposed in [33]. Though defined independently and using different foundations, both semantics appear compatible on positive programs, so the results of the current paper apply to possible model semantics of [33], too.

Paper [14] defines a tableaux framework for classical ASP, using two (explicit) truth values T, F. They require a *cut* rule, whereas we do not. Loop formulas are explicit in some rules and supportedness is encoded as an additional set of inference rules. Our approach is very much in the spirit of Smullyan [37] and does not require special inference rules, although loop and completion formulas would have to be added to an ASP if one wanted to deal with classical non-normal ASPs.

In [29,30], the logic of here-and-there (HT) is used to define a direct declarative semantics for classical ASPs, although HT has greater generality and wider application. HT can be defined by means of a five-valued logic, N_5, defined over two worlds: h (here) and t (there), where the set of literals associated with h is included in the set of literals associated with t. N_5 uses truth values $\{-2, -1, 0, 1, 2\}$, where the values $-1, 1$ characterize literals associated with h and not associated with t. On the other hand, for classical ASP models it is assumed that these sets are equal, so $-1, 1$ become redundant. Therefore, in the context of classical ASPs one actually does not have to use full N_5 as it reduces to the three-valued logic R_3, with $-2, 0, 2$ of N_5 corresponding to F, U, T of R_3, respectively. Consequently, tableaux techniques used in [29] for N_5 could then be simplified when focus is on ASPs.

Kleene Answer set programs, ASP^K, and connectives used in R_3 have been proposed in [7]. The current paper introduces a sound and complete tableaux-based proof procedure for them. A filtering technique is introduced which, when added to the R_3 tableaux based proof procedure, provides a sound and complete proof procedure for Kleene ASPs. As a derivative result, it is shown that the proof procedure is also sound and complete for classical normal ASPs and remains sound for disjunctive classical ASPs.

References

1. Baral, C.: Knowledge Representation, Reasoning, and Declarative Problem Solving. Cambridge University Press, Cambridge (2003)
2. Bogaerts, B., Vennekens, J., Denecker, M.: Grounded fixpoints and their applications in knowledge representation. Artif. Intell. **224**, 51–71 (2015)
3. Brewka, G., Niemelä, I., Truszczynski, M.: Answer set optimization. In: Gottlob, G., Walsh, T. (eds.) Proceedings of the 18th IJCAI, pp. 867–872. Morgan Kaufmann (2003)
4. Chan, P.: A possible world semantics for disjunctive databases. IEEE Trans. Knowl. Data Eng. **5**(2), 282–292 (1993)
5. Denecker, M., Marek, V., Truszczynski, M.: Stable operators, well-founded fixpoints and applications in nonmonotonic reasoning. In: Minker, J. (ed.) Logic-Based Artificial Intelligence, pp. 127–144. Kluwer Academic Publishers, Dordrecht (2000)
6. Doherty, P.: A constraint-based approach to proof procedures for multi-valued logics. In: Proceedings of the 1st World Conference Fundamentals of AI (WOCFAI), pp. 165–178. Springer (1991)
7. Doherty, P., Szałas, A.: Stability, supportedness, minimality and Kleene Answer Set Programs. In: Eiter, T., Strass, H., Truszczyński, M., Woltran, S. (eds.) Advances in Knowledge Representation, Logic Programming, and Abstract Argumentation. LNCS (LNAI), vol. 9060, pp. 125–140. Springer, Heidelberg (2015). doi:10.1007/978-3-319-14726-0_9
8. Eiter, T., Faber, W., Fink, M., Woltran, S.: Complexity results for answer set programming with bounded predicate arities and implications. Ann. Math. Artif. Intell. **51**(2–4), 123–165 (2007)

9. Eiter, T., Gottlob, G.: Complexity results for disjunctive logic programming and application to nonmonotonic logics. In: Miller, D. (ed.) Proceedings of the Logic Programming, pp. 266–278 (1993)
10. Fages, F.: Consistency of Clark's completion and existence of stable models. Methods Logic Comput. Sci. **1**, 51–60 (1994)
11. Fitting, M.: A Kripke-Kleene semantics for logic programs. J. Logic Program. **2**(4), 295–312 (1985)
12. Fitting, M.: The family of stable models. J. Logic Program. **17**(2–4), 197–225 (1993)
13. Gebser, M., Kaufmann, B., Schaub, T.: Conflict-driven answer set solving: from theory to practice. Artif. Intell. **187**, 52–89 (2012)
14. Gebser, M., Schaub, T.: Tableau calculi for logic programs under answer set semantics. ACM Trans. Comput. Log. **14**(2), 15:1–15:40 (2013)
15. Gelfond, M., Kahl, Y.: Knowledge Representation, Reasoning, and the Design of Intelligent Agents - The Answer-Set Programming Approach. Cambridge University Press, Cambridge (2014)
16. Gelfond, M., Lifschitz, V.: The stable model semantics for logic programming. In: Kowalski, R., Bowen, K. (eds.) Proceedings of the International Logic Programming, pp. 1070–1080. MIT Press (1988)
17. Gelfond, M., Lifschitz, V.: Classical negation in logic programs and disjunctive databases. New Gener. Comput. **9**(3/4), 365–386 (1991)
18. Hähnle, R.: Uniform notation of tableau rules for multiple-valued logics. In: ISMVL, pp. 238–245 (1991)
19. Kleene, S.C.: On a notation for ordinal numbers. Symbolic Logic **3**, 150–155 (1938)
20. Kunen, K.: Negation in logic programming. J. Log. Program. **4**(4), 289–308 (1987)
21. Leone, N., Pfeifer, G., Faber, W., Eiter, T., Gottlob, G., Perri, S., Scarcello, F.: The DLV system for knowledge representation and reasoning. ACM Trans. Comput. Logic **7**(3), 499–562 (2006)
22. Leone, N., Rullo, P., Scarcello, F.: Disjunctive stable models: unfounded sets, fixpoint semantics, and computation. Inf. Comput. **135**(2), 69–112 (1997)
23. Lierler, Y.: Relating constraint answer set programming languages and algorithms. Artif. Intell. **207**, 1–22 (2014)
24. Lifschitz, V.: Thirteen definitions of a stable model. In: Blass, A., Dershowitz, N., Reisig, W. (eds.) Fields of Logic and Computation. LNCS, vol. 6300, pp. 488–503. Springer, Heidelberg (2010). doi:10.1007/978-3-642-15025-8_24
25. Lin, F., Zhao, Y.: ASSAT: computing answer sets of a logic program by SAT solvers. Artif. Intell. **157**(1–2), 115–137 (2004)
26. Łukasiewicz, J.: O logice trójwartościowej (in Polish). Ruch filozoficzny **5**, 170–171 (1920). English translation: On three-valued logic. In: Borkowski, L. (ed.) Selected works by Jan Łukasiewicz, pp. 87–88. North-Holland, Amsterdam (1970)
27. McCarthy, J.: Circumscription - a form of non-monotonic reasoning. Artif. Intell. **13**(1–2), 27–39 (1980)
28. Murray, N., Rosenthal, E.: Adapting classical inference techniques to multiple-valued logics using signed formulas. Fundam. Inform. **21**(3), 237–253 (1994)
29. Pearce, D.: Equilibrium logic. Ann. Math. AI **47**(1–2), 3–41 (2006)
30. Pearce, D., Guzmán, I.P., Valverde, A.: Computing equilibrium models using signed formulas. In: Lloyd, J., et al. (eds.) CL 2000. LNCS (LNAI), vol. 1861, pp. 688–702. Springer, Heidelberg (2000). doi:10.1007/3-540-44957-4_46
31. Przymusinski, T.: Stable semantics for disjunctive programs. New Gener. Comput. **9**(3/4), 401–424 (1991)

32. Reiter, R.: A logic for default reasoning. Artif. Intell. **13**(1–2), 81–132 (1980)
33. Sakama, C., Inoue, K.: An alternative approach to the semantics of disjunctive logic programs and deductive databases. J. Autom. Reasoning **13**(1), 145–172 (1994)
34. Seipel, D., Minker, J., Ruiz, C.: A characterization of the partial stable models for disjunctive databases. In: Małuszyński, J. (ed.) Logic Programming Symposium, pp. 245–259 (1997)
35. Shepherdson, J.C.: A sound and complete semantics for a version of negation as failure. Theor. Comput. Sci. **65**(3), 343–371 (1989)
36. Simons, P., Niemelä, I., Soininen, T.: Extending and implementing the stable model semantics. Artif. Intell. **138**(1–2), 181–234 (2002)
37. Smullyan, R.: First-Order Logic. Dover Publications, Mineola (1968)
38. Stamate, D.: Assumption based multi-valued semantics for extended logic programs. In: 36th IEEE International Symposium on ISMVL, p. 10. IEEE Computer Society (2006)

An Efficient Gaussian Kernel Based Fuzzy-Rough Set Approach for Feature Selection

Soumen Ghosh, P.S.V.S. Sai Prasad$^{(\boxtimes)}$, and C. Raghavendra Rao

School of Computer and Information Sciences,
University of Hyderabad, Hyderabad, India
hitkmca@yahoo.in, {saics,crrcs}@uohyd.ernet.in

Abstract. Fuzzy-rough set based feature selection is highly useful for reducing data dimensionality of a hybrid decision system, but the reduct computation is computationally expensive. Gaussian kernel based fuzzy rough sets merges kernel method to fuzzy-rough sets for efficient feature selection. This works aims at improving the computational performance of existing reduct computation approach in Gaussian kernel based fuzzy rough sets by incorporation of vectorized (matrix, sub-matrix) operations. The proposed approach was extensively compared by experimentation with the existing approach and also with a fuzzy rough set based reduct approaches available in Rough set R package. Results establish the relevance of proposed modifications.

Keywords: Rough set · Fuzzy-rough set · Feature selection · Reduct computation · Gaussian kernel · Hybrid decision system

1 Introduction

Rough Set Theory(RST), introduced by Prof. Pawlak [10] in 1980s, is very useful for classification and analysis of imprecise, uncertain or incomplete information and knowledge. The fundamental concept behind RST is the approximation space of a concept represented as the set. The process of knowledge discovery in a given decision system primarily consists of reduct computation (or feature selection) as the preprocessing step for dimensionality reduction. The features which are not a part of the reduct can be removed from the dataset with minimum information loss.

Feature Selection of a dataset with categorical attributes can be carried out by RST, but feature selection of a dataset with real-valued attributes is not possible through the classical RST. For handling this situation the fuzzy-rough set model is used. The Fuzzy Rough Set Theory (FRST) was introduced by Dubois and Prade [4] in 1990s and extended by several researchers [3,7,12]. Many reduct computation approaches were developed using FRST. They are categorized into fuzzy discretization based [1,6] and fuzzy similarity relation based [5,7,15,15]. In [7], it is established that fuzzy similarity relation based reduct computation are more efficient.

© Springer International Publishing AG 2016
C. Sombattheera et al. (Eds.): MIWAI 2016, LNAI 10053, pp. 38–49, 2016.
DOI: 10.1007/978-3-319-49397-8_4

Recently an extension to fuzzy rough set model, named as Gaussian Kernel-based fuzzy rough set (GK-FRS) model, by adopting a Gaussian kernel function fuzzy similarity relation was introduced by Hu et al. [5]. The GK-FRS models combine the advantages of kernel methods with fuzzy rough sets. A sequential forward selection (SFS) based reduct computation approach was also proposed [18] based on GK-FRS. In 2015, GK-FRS was adopted to systems with various type of attributes (real-valued, categorical, boolean, set, interval based) with the concept of hybrid distance by Zeng et al. [18]. FRSA-NFS-HIS algorithm was introduced in [18] based on the extended GK-FRS model.

The computation complexity of fuzzy rough set based reduct algorithm is much higher than classical rough set based reduct algorithms [14]. The reason being, in later approaches computation, is based on the granules (equivalence classes), but fuzzy rough set based reduct computation inevitably involved object based computations. The existing approaches do not attempt any possibility of imposing granular or sub-granular aspect in fuzzy rough set based reduct computations. The aim of this paper is to improve the computational performance of FRSA-NFS-HIS algorithm by bringing an aspect of granular/sub-granular computations. This study brings out the importance of modeling fuzzy rough set reduct computation using matrix, sub-matrix based vectorized operations in the efficient vector-based environment such as Matlab and R. Our proposed Modified FRSA-NFS-HIS (MFRSA-NFS-HIS) algorithm was implemented in R environment, and extensive comparative analysis with existing approaches is reported in this paper.

This paper is organized as follows. Section 2 gives the theoretical background. Section 3 discusses the Gaussian Kernel-based Fuzzy-Rough Sets and FRSA-NFS-HIS. In Sect. 4 Proposed MFRSA-NFS-HIS Algorithm is detailed. Section 5 describes Experiments, Results and Analysis of these different approaches. The paper is concluded with Sect. 6.

2 Theoretical Background

2.1 Rough Set Theory

Rough Set Theory is a useful tool to discover data dependency and to reduce the number of attributes contain in the dataset using data alone requiring no additional information. Let DS = (U, A $\cup \{d\}$) is a decision system, where d is a decision attribute and $d \notin A$, U is nonempty set of finite objects and A is a nonempty finite set of conditional attributes such that $a : U \rightarrow V_a$ for every $a \in A$. V_a is the set of domain values of attribute 'a'. For any R $\subseteq$ A, there is an associated equivalence relation, called as indiscernible relation IND(R),

$$IND(R) = \{(x, y) \in U^2 | (\forall a \in R)(a(x) = a(y))\} \tag{1}$$

The equivalence relation partitions the universe into a family of disjoint subsets, called equivalence classes. The equivalence class including x is denoted by $[x]_R$ and the set of equivalance classes induced by IND(R) is denoted by U/IND(R) or U/R in short.

Let X be a concept (X ⊆ U) and approximations to X are defined using U/R. The lower and the upper approximations of X are defined respectively as follows:

$$\underline{R}X = \{x \in U | [x]_R \subseteq X\}; \tag{2}$$
$$\overline{R}X = \{x \in U | [x]_R \cap X \neq \phi\}; \tag{3}$$

The pair $< \underline{R}X, \overline{R}X >$ is called a rough set when $\underline{R}X \neq \overline{R}X$.

Let R and Q be sets of attributes inducing equivalence relations over U, then the positive regions can be defined as:

$$POS_R(Q) = \bigcup_{X \in U/Q} \underline{R}X \tag{4}$$

The positive region $POS_R(Q)$ contains all the objects of the universe U that can be classified into different classes of U/Q using the information of attribute set R. The dependency or gamma (γ) is calculated as:

$$\gamma_R(\{d\}) = \frac{|POS_R(\{d\})|}{|U|} \tag{5}$$

2.2 Fuzzy-Rough Sets

The classical Rough Set Theory cannot deal with real-valued data and the fuzzy-rough set is a solution to that problem as it can efficiently deal with real-valued data without resorting to discretization. Let R is a fuzzy relation and decision system DS =(U, A ∪ {d}) where U is a non-empty set of objects and A is a set of conditional attributes, a ∈ A be a quantitative(real-valued) attribute. For measurement of the approximate similarity between two objects for a quantitative attributes, fuzzy similarity relations are used. Few example of the fuzzy similarity relation are [7]:

(1) $\mu_{R_a}(x, y) = 1 - \dfrac{|a(x) - a(y)|}{|a_{max} - a_{min}|}$ $\qquad$ (6)

(2) $\mu_{R_a}(x, y) = exp \left(-\dfrac{(a(x) - a(y))^2}{2\sigma_a^2} \right)$ $\qquad$ (7)

(3) $\mu_{R_a}(x, y) = max \left(min \left(\dfrac{a(y) - a(x) + \sigma_a}{\sigma_a}, \dfrac{a(x) - a(y) + \sigma_a}{\sigma_a} \right), 0 \right)$ (8)

where σ_a denotes the standard deviation of a. If attribute a ∈ A is qualitative (nominal) then R_a(x,y) = 1 for a(x) = a(y) and R_a(x,y) = 0 for a(x) ≠ a(y). The similarity relation is extended to a set of attributes A by

$$R_A(x, y) = \Im(R_{a \in A}(x, y)) \tag{9}$$

where $\Im$ represent a t-norm.

The lower and upper approximations are defined based on fuzzy similarity relations. The fuzzy R-lower and R-upper approximations are defined in Radzikowska-Kerry's fuzzy rough set model [3, 12] as:

$$\underline{R}A(y) = inf_{x \in X} \mathbb{I}(R(x, y), A(x)) \tag{10}$$

$$\overline{R}A(y) = sup_{x \in X} \Im(R(x, y), A(x)) \tag{11}$$

where for all y in U, $\mathbb{I}$ is the implicator and $\Im$ is the t-norm. The pair $< \underline{R}A, \overline{R}A >$ is called a fuzzy-rough set.

3 Gaussian Kernel Based Fuzzy-Rough Sets

The kernel methods and rough set are crucial domains for machine learning and pattern recognization. The kernel methods map the data into a higher dimensional feature space while rough set granulates the universe with the use of relations. Hu et al. [5] used the benefits of both and made a Gaussian kernel based fuzzy rough set approach for reduct computation. The content of this section is already discussed in the literature [5, 18]. For the completeness of the paper, we give a summary of the original content.

3.1 Hybrid Decision System (HDS) and Hybrid Distance (HD)

A HDS can be written as $(U, A \cup \{d\}, V, f)$, where U is the set of objects, $A = A^r \cup A^c \cup A^b$, A^r is the real-valued attribute set, A^c is the categorical attribute set and A^b is the boolean attribute set. $\{d\}$ denotes a decision attribute. $A^r \cap A^c = \phi$, $A^r \cap A^b = \phi$, and $A \cap \{d\} = \phi$.

In HDS, there may be different types of attributes. To construct the distance among the objects, different distance measurement functions are used based on attribute type in literature [18]. The Hybrid Distance(HD) for a Hybrid Decision System (HDS) based on different types of attributes is defined as:

$$HD_B(x, y) = \sqrt{\sum_{a \in B} d^2(a(x), a(y))} \tag{12}$$

where B is the set of conditional attributes of the HDS, and

$$d(a(x), a(y)) = \begin{cases} vdm(a(x), a(y)), & \text{a is a categorical attribute} \\ vdr(a(x), a(y)), & \text{a is a real-valued attribute} \\ vdb(a(x), a(y)), & \text{a is a boolean attribute} \end{cases} \tag{13}$$

3.2 Gaussian Kernel

In the literature Hu et al. [5] uses gaussian kernel function for computing the fuzzy similarity relation between the objects. The gaussian kernel function is defined as:

$$k(x_i, x_j) = exp\left(-\frac{||x_i - x_j||^2}{2\delta^2}\right) \tag{14}$$

where $||x_i - x_j||$ is the distance between the objects and δ is the kernel parameter. δ plays an important role in controlling the granularity of approximation. In [18], $||x_i - x_j||$ was taken as $HD(x_i, x_j)$ for generalised the GK-FRS to HDS.

3.3 Dependency Computation

R_G^B denotes Gaussian kernel based fuzzy similarity relation where HD is computed over $B \subseteq A$. Based on Proposition 3 in [18], $R_G^{\{a\} \cup \{b\}}$ can be computed using $R_G^{\{a\}}, R_G^{\{b\}}$ by,

$$R_G^{\{a\} \cup \{b\}}(x, y) = R_G^{\{a\}}(x, y) \times R_G^{\{b\}}(x, y) \quad \forall x, y \in U \tag{15}$$

It is essential to find the fuzzy lower and upper approximation for calculating the dependency of the attributes. This approximation is calculated from the fuzzy similarity relation R_G.

Proposition 1 [18]. *The formula for calculating the fuzzy lower and upper approximation is defined as:*

$$\underline{R_G}d_i(x) = \sqrt{1 - \left(sup_{y \notin d_i} R_G(x, y)\right)^2} \tag{16}$$

$$\overline{R_G}d_i(x) = sup_{y \in d_i} R_G(x, y) \tag{17}$$

where $\forall d_i \in U/\{d\}$.

The universe U is divided into different granules $U/\{d\} = \{d_1, d_2, ..., d_l\}$. The fuzzy positive regions of decision attribute ($\{d\}$) concerning B are defined as:

$$POS_B(\{d\}) = \bigcup_{i=1}^{l} \underline{R_G^B}d_i. \tag{18}$$

The dependency of the attribute or a set of attributes is defined as follows:

$$\gamma_B(\{d\}) = \frac{|POS_B(\{d\})|}{|U|} = \frac{|\bigcup_{i=1}^{l} \underline{R_G^B}d_i|}{|U|} \tag{19}$$

where $\bigcup_{i=1}^{l} \underline{R_G^B}d_i = \sum_i \sum_{x \in d_i} \underline{R_G^B}d_i(x)$.

3.4 The FRSA-NFS-HIS Algorithm

The FRSA-NFS-HIS algorithm uses SFS control strategy for reduct computation. This Algorithm starts with an empty reduct set R. In every iteration an attribute $a \in A - R$ is added to R based on the criteria of giving maximum gamma gain ($\gamma_{R \cup \{a\}}(\{d\}) - \gamma_R(\{d\})$). The end condition is determined by a parameter ϵ (a very small value near to zero and is a user control parameter). The algorithm completes execution and returns R when no available attributes $a \in A - R$ gives a gamma gain exceeding epsilon (ϵ).

4 Proposed MFRSA-NFS-HIS Algorithm

This section describes the proposed MFRSA-NFS-HIS algorithm by incorporating vectorized operation in the FRSA-NFS-HIS algorithm.

4.1 Vectorization in FRSA-NFS-HIS Algorithm

All fuzzy similarity relation based fuzzy rough set reduct computation involves object based computation. Starting with the computation of fuzzy similarity relation for each conditional attributes, computation of fuzzy lower approximation involves object-wise computation. Hence the implementation involves several nested looping structures over the space of objects (U). Computing environments such as R, Matlab have excellent support for matrix and sub-matrix based operation. Vectorization is the process of modeling the computation involving nested loops into matrix, sub-matrix based operations. It is established [16] that the same computation through vectorization can result in significant performance gain over implementation using loops.

The first important computation involved in FRSA-NFS-HIS algorithm is the calculation of fuzzy silmilarity relation for individual conditional attributes. This requires computation of appropriate distance function between a pair of objects requiring two loops. These are translated as matrix based operation by replication of attribute column $|U|$ times resulting in $|U| \times |U|$ matrix and finding matrix based distance computation with its transpose and applying gaussian kernel function on the resulting matrix. The fuzzy similarity relation computation between a set of attributes is computed by element-wise multiplication of individual similarity matrices by using Eq. 15.

The most frequent computation in FRSA-NFS-HIS algorithm is the computation of gamma using Algorithm 1 which computes fuzzy dependency using Eq. 19. Improving the computation efficiency of Algorithm 1 has an immense impact on the overall performance of MFRSA-NFS-HIS algorithm.

The Algorithm 1 calculates lower approximation of each concept using all the objects. For an example, let $U/\{d\} = \{d_1, d_2, d_3\}$, $d_1 = \{x_1, x_2, x_3, x_8\}$, $d_2 = \{x_4, x_5\}$, and $d_3 = \{x_6, x_7\}$.

Algorithm 1. Dependency with Gaussian Kernel Approximation (DGKA) [18]

Input: The fuzzy relation R.
Output: The fuzzy dependency $\gamma_B(\{d\})$ of $\{d\}$ to B
  1: $\gamma_B(\{d\}) = 0$
  2: **for** each $d_i \in U/\{d\}$ **do**
  3:     **for** $j = 0; j < |U|; j++$ **do**
  4:         Find the nearest sample M_j of x_j with a different class.
  5:         $\gamma_B(\{d\}) = \gamma_B(\{d\}) + \sqrt{1 - R_G^2(M_j, x_j)}$
  6:     **end for**
  7: **end for**
  8: **return** $\gamma_B(\{d\})$

$\underline{R_G}d_1(x_1) = \sqrt{1 - \left(sup_{y \notin d_1} R_G(x_1, y)\right)^2} = \sqrt{1 - \left(sup\{R_G(x_1, x_4), R_G(x_1, x_5), R_G(x_1, x_6), R_G(x_1, x_7)\}\right)^2}$

But when we calculating lower approximation for the object x_4 which is belong to the different decision class $(x_4 \in U - d_i)$

$\underline{R_G}d_1(x_4) = \sqrt{1 - \left(sup\{R_G(x_4, x_4), R_G(x_4, x_5), R_G(x_4, x_6), R_G(x_4, x_7), R_G(x_4, x_9),\}\right)^2}$

and as the value of $R_G(x_4, x_4) = 1$, the total value becomes zero. It means all the objects which do not belong to the same decision class contribute zero in gamma calculation.

Algorithm 2. Improved Dependency with Gaussian Kernel Approximation (IDGKA)

Input: The fuzzy similarity relation R^B where B$\subseteq$A of the HDS.
Output: The fuzzy dependency $\gamma_B(\{d\})$ of {d} to B
1: $\gamma_B(\{d\}) = 0$
2: $POS_G^B = zeros(|U|, 1)$
3: **for** each $d_i \in U/\{d\}$ **do**
4:     $POS_G^B(d_i) = \sqrt{1 - (rowmax(R_G^B(d_i, U - d_i)))^2}$
5: **end for**
6: $\gamma_B(\{d\}) = \frac{sum(POS_G^B)}{|U|}$
7: **return** $\gamma_B(\{d\})$

The proposed algorithm IDGKA (Algorithm 2) computes gamma using only the objects which are belonging to the same decision class. The IDGKA algorithm does not use the objects which are belonging to a different decision class. For computing gamma, the IDGKA algorithm uses all the objects only once for any number of decision classes. So, the Algorithm 2 reduces $|U - d_i|$ iterations for each decision class (d_i) and executes only $|d_i|$ times for each decision class (concept). The computation of the proposed Algorithm 2 involves sub-matrix based operation described below.

POS_G^B is a zero vector of size $|U|$ representing a fuzzy positive region membership of each object. The actual membership is incrementally assigned by considering each decision concept objects in(line no 3 to 5 of Algorithm 2) iteration. For a decision concept d_i the row-wise maximum are computed on sub-matrix $R_G^B(d_i, U - d_i)$ and the positive region membership for d_i objects is computed as a vector operation using Eq. 16. Finally, in line number 6 $\gamma_B(\{d\})$ is computed through a summation of the POS_G^B vector. Hence all the required computation of DGKA are vectorized in IDGKA using submatrix and vector operations.

4.2 Overcoming ϵ-Parameter Dependency

In SFS based reduct computation algorithm based on classical rough sets, the possibility of occurrence of a trivial ambiguous situation is identified in [14]. In such situation, no available attributes are resulting any gamma gain leading to sub reduct computation instead of reduct computation. The end condition of FRSA-NFS-HIS can result in a similar situation wherein no available attributes give a gamma gain exceeding ϵ but has a possibility for an increase in gamma if proceeded to future iterations. Such situation results in sub reduct computation in FRSA-NFS-HIS algorithm. To overcome this we have modified the end condition as $\gamma_R == \gamma_A$. This will incur computation overhead for γ_A computation but is required for having proper end condition.

The resulting MFRSA-NFS-HIS algorithm is given in Algorithm 3.

Algorithm 3. Modified FRS Approach for Naive Feature Selection in HIS (MFRSA-NFS-HIS)

Input: $HDS = (U, A \cup \{d\})$
Output: Reduct red
1: $red = \phi, \gamma_{red} = 0$
2: **for** each $a \in A$ **do**
3:     Compute $R_G^{\{a\}}$
4: **end for**
5: Compute R_G^A //*Computing using Eq. 15*
6: $\gamma_A = IDGKA(R_G^A)$ //*Computing the γ by Algorithm 2*
7: **while** $(\gamma_{red} < \gamma_A)$ **do**
8:     $\gamma_{max} = 0$
9:     $b = \phi$
10:    **for** each $a_i \in (A - red)$ **do**
11:        Compute $R_G^{red \cup \{a_i\}}$
12:        $\gamma_i = IDGKA \left(R_G^{red \cup \{a_i\}} \right)$
13:    **end for**
14:    Find the maximal dependency γ_{max} and the corrosponding attribute b
15:    $red = red \cup \{b\}$
16:    $\gamma_{red} = \gamma_{max}$
17: **end while**
18: **return** red

---

5   Experiments, Results and Analysis

The experiments are conducted on Intel (R) i5 CPU, Clock Speed 2.66 GHz, 4 GB of RAM, Ubuntu 14.04, 64 bit OS and R Studio, Version 0.99.447. In this work, we have used benchmark datasets for performing the experiments. The datasets are described in Table 1. All the datasets are from UCI Machine Learning Repository [8]. The proposed MFRSA-NFS-HIS algorithm is implemented in R environment.

5.1   Comparative Experiments with FRSA-NFS-HIS Algorithm

The experiments are performed on proposed feature selection (MFRSA-NFS-HIS) algorithm. The results obtained are compared with the result of the existing FRSA-NFS-HIS algorithm reported in [18] for all the datasets in Table 1. In [18],

Table 1. Description of datasets

No	Dataset	Objects	Attributes	Decision classes
1	Wine	178	14	3
2	Hepatitis	155	19	2
3	Horse	368	26	2
4	Ionosphere	351	34	2
5	Credit	690	15	2
6	German	1000	20	2
7	Bands	540	40	2

Table 2. Comparison with FRSA-NFS-HIS and MFRSA-NFS-HIS algorithms with respect to computational time (in seconds) and reduct size

Datasets		Computation time (s)		Reduct size		Computation gain
No	Name	FRSA-NFS-HIS	MFRSA-NFS-HIS	FRSA-NFS-HIS	MFRSA-NFS-HIS	MFRSA-NFS-HIS
1	Wine	17.5	0.144521	6	6	99.17 %
2	Hepatitis	8	0.175910	7	8	97.80 %
3	Horse	80	1.189067	4	12	98.51 %
4	Ionosphere	118	2.070024	7	18	98.24 %
5	Credit	185	2.225725	5	14	98.79 %
6	German	500	3.606263	11	11	99.27 %
7	Bands	240	1.672623	11	10	99.30 %

the results of the FRSA-NFS-HIS algorithm is reported in Fig. 1, where results are varying, and we take the lower bound of the results for comparing the result with MFRSA-NFS-HIS algorithm. The Table 2 shows the comparison of the results of FRSA-NFS-HIS and MFRSA-NFS-HIS algorithm.

Analysis of Results. The comparison of the results of existing feature selection approach and proposed feature selection approach is given in Table 2. From the Table 2, it is observed that the proposed approach takes less time than existing approach for all the datasets in Table 1 and the computation gain of the algorithm MFRSA-NFS-HIS is varying from 97 % to 99 %. The size of the reduct is almost same for both the approaches. The size of the reduct in a few datasets is higher in MFRSA-NFS-HIS due to the modification of the end condition determined by γ_A.

5.2   Comparative Experiments with L-FRFS and B-FRFS Algorithms

The feature selection algorithm Fuzzy Lower Approximation based FS (L-FRFS) and Fuzzy Boundary Approximation based FS (B-FRFS) are proposed by Jensen et al. [7] which are implemented in Rough Set package in R environment [13]. The proposed algorithm MFRSA-NFS-HIS also implemented in R environment using the vectorized operations [16]. For executing the L-FRFS and B-FRFS algorithms, we used Lukasiewicz t-norm, Lukasiewicz implicator, and fuzzy similarity measure defined in Eq. 8. The comparison of results of the algorithm L-FRFS, B-FRFS with MFRSA-NFS-HIS are given in Table 3.

Table 3. Comparison of algorithm L-FRFS and B-FRFS available in R package with MFRSA-NFS-HIS algorithms with respect to computational time (in seconds) and reduct size

Datasets		Reduct size			Time (s)			Computation_Gain
No	Name	L-FRFS	B-FRFS	MFRSA-NFS-HIS	L-FRFS	B-FRFS	MFRSA-NFS-HIS	MFRSA-NFS-HIS
1	Wine	5	5	6	4.9116	4.6624	0.144521	97.05 % & 97.04 %
2	Hepatitis	7	7	8	4.6299	4.6553	0.175910	96.20 % & 96.22 %
3	Horse	10	10	12	43.5804	43.8784	1.189067	97.27 % & 97.29 %
4	Iono	7	7	18	42.9626	43.6401	2.070024	95.18 % & 95.25 %
5	Credit	12	12	14	82.3428	81.7314	2.225725	97.29 % & 97.27 %
6	German	10	10	11	248.6718	230.3088	3.606263	98.54 % & 98.43 %
7	Bands	8	8	10	56.4242	57.0249	1.672623	97.03 % & 97.06 %

The computation time of the algorithms L-FRFS and B-FRFS reported in the literature [7], and the results of Rough Set R package executed on the above mentioned system is significantly different. This may be primarily due to the hardware configuration used in [7]. But the details of the hardware configurations are not specified in [7]. So, for the completeness of comparative analysis, we have executed MFRSA-NFS-HIS on the datasets used in [7] and the results are summarized in Table 4.

Table 4. Comparison of algorithm L-FRFS and B-FRFS reported in the literature [7] with MFRSA-NFS-HIS algorithms with respect to computational time (in seconds) and reduct size

Datasets			Reduct Size			Time(s)			Computation_Gain
No	Name	Objects	Features	L-FRFS	B-FRFS	**MFRSA-NFS-HIS**	L-FRFS	B-FRFS	**MFRSA-NFS-HIS**	**MFRSA-NFS-HIS**
1	Cleveland	297	14	9	9	**10**	3.32	8.78	**0.41058**	87.63 % & 95.32 %
2	Glass	214	10	9	9	**9**	1.53	3.30	**0.15331**	89.97 % & 95.35 %
3	Heart	270	14	8	8	**11**	2.17	3.61	**0.32041**	85.23 % & 91.12 %
4	Ionosphere	230	35	9	9	**18**	3.77	8.53	**2.02270**	46.41 % & 76.28 %
5	Olitos	120	26	6	6	**7**	0.72	1.29	**0.19042**	73.61 % & 85.23 %
6	Web	149	2557	21	20	**23**	541.85	949.69	**84.0546**	84.48 % & 91.09 %
7	Wine	178	14	6	6	**6**	0.97	1.69	**0.14312**	85.26 % & 91.53 %

Analysis of Results. From the analysis of the results, it is observed that the reduct computation algorithm MFRSA-NFS-HIS takes comparatively lesser computation time than the algorithms L-FRFS and B-FRFS which is available in Rough Set package in R platform. The last column of the Table 3 depicts the computational gain percentage obtained by MFRSA-NFS-HIS over L-FRFS and B-FRFS algorithm respectively. The computation gain with respect to L-FRFS and B-FRFS is more than 95 %.

From the analysis of the results reported in the literature for the algorithm L-FRFS and B-FRFS, it is observed that the computation gain of MFRSA-NFS-HIS algorithm with respect to algorithm L-FRFS and B-FRFS is more than 84 % for all datasets from Table 4 except the Ionosphere dataset in which a gain percentage 46 % with L-FRFS is obtained. The reduct size of MFRSA-NFS-HIS is also much higher than FRFS algorithms. This is primarily due to the difference in Ionosphere dataset size used in [7] and in our experiments. From the comparison of the results of R package and results reported in the literature with MFRSA-NFS-HIS it is observed that the proposed MFRSA-NFS-HIS algorithm achieves a significant computational gain over the existing methods.

6   Conclusion

In this paper, improvements for FRSA-NFS-HIS algorithm are proposed by incorporation of vectorized operation as MFRSA-NFS-HIS algorithm. The proposed MFRSA-NFS-HIS algorithm has a significant improvement on computation time over the existing method. The size of the reduct computed by MFRSA-NFS-HIS algorithm are almost same as FRSA-NFS-HIS. We have also compared the MFRSA-NFS-HIS with L-FRFS and B-FRFS algorithms available in R package and obtained significant computational gains. The obtained results establish the relevance and role of vectorization in fuzzy rough reduct computation. The proposed approach facilitates model construction in HDS by giving relevant reduced feature set in an effective manner. In future distributed/parallel algorithm for MFRSA-NFS-HIS will be investigated for feasible fuzzy rough reduct computation in Big data scenario.

References

1. Bhatt, R.B., Gopal, M.: On the compact computational domain of fuzzy-rough sets. Pattern Recognit. Lett. **26**(11), 1632–1640 (2005)
2. Chouchoulas, A., Shen, Q.: Rough set-aided keyword reduction for text categorization. Appl. Artif. Intell. **15**(9), 843–873 (2001)
3. Cornelis, C., Jensen, R., Hurtado, G., Śle, D., et al.: Attribute selection with fuzzy decision reducts. Inf. Sci. **180**(2), 209–224 (2010)
4. Dubois, D., Prade, H.: Rough fuzzy sets and fuzzy rough sets*. Int. J. Gen. Syst. **17**(2–3), 191–209 (1990)
5. Hu, Q., Zhang, L., Chen, D., Pedrycz, W., Yu, D.: Gaussian kernel based fuzzy rough sets: model, uncertainty measures and applications. Int. J. Approx. Reason. **51**(4), 453–471 (2010)
6. Jensen, R., Shen, Q.: Fuzzy-rough attribute reduction with application to web categorization. Fuzzy Sets Syst. **141**(3), 469–485 (2004)
7. Jensen, R., Shen, Q.: New approaches to fuzzy-rough feature selection. IEEE Trans. Fuzzy Syst. **17**(4), 824–838 (2009)
8. Lichman, M.: UCI machine learning repository (2013)
9. Moser, B.: On representing and generating kernels by fuzzy equivalence relations. J. Mach. Learn. Res. **7**, 2603–2620 (2006)

10. Pawlak, Z.: Rough sets. Int. J. Comput. Inf. Sci. **11**(5), 341–356 (1982)
11. Pawlak, Z., Skowron, A.: Rough sets: some extensions. Inf. Sci. **177**(1), 28–40 (2007)
12. Radzikowska, A.M., Kerre, E.E.: A comparative study of fuzzy rough sets. Fuzzy Sets Syst. **126**(2), 137–155 (2002)
13. Riza, L.S., Janusz, A., Slezak, D., Cornelis, C., Herrera, F., Benitez, J.M., Bergmeir, C., Stawicki, S.: Package roughsets
14. Prasad, P.S.V.S.S., Rao, C.R.: IQuickReduct: an improvement to quick reduct algorithm. In: Sakai, H., Chakraborty, M.K., Hassanien, A.E., Ślęzak, D., Zhu, W. (eds.) RSFDGrC 2009. LNCS (LNAI), vol. 5908, pp. 152–159. Springer, Heidelberg (2009). doi:10.1007/978-3-642-10646-0_18
15. Sai Prasad, P.S.V.S., Raghavendra Rao, C.: An efficient approach for fuzzy decision reduct computation. In: Peters, J.F., Skowron, A. (eds.) Transactions on Rough Sets XVII. LNCS, vol. 8375, pp. 82–108. Springer, Heidelberg (2014). doi:10.1007/978-3-642-54756-0_5
16. Wang, H., Padua, D., Wu, P.: Vectorization of apply to reduce interpretation overhead of R. In: ACM SIGPLAN Notices, vol. 50, pp. 400–415. ACM (2015)
17. Zadeh, L.A.: Fuzzy sets. Inf. Control **8**(3), 338–353 (1965)
18. Zeng, A., Li, T., Liu, D., Zhang, J., Chen, H.: A fuzzy rough set approach for incremental feature selection on hybrid information systems. Fuzzy Sets Syst. **258**, 39–60 (2015)

WSCOVER: A Tool for Automatic Composition and Verification of Web Services Using Heuristic-Guided Model Checking and Logic-Based Clustering

Khai T. Huynh$^{(\boxtimes)}$, Thang H. Bui, and Than Tho Quan

Faculty of Computer Science and Engineering,
Ho Chi Minh City University of Technology, Ho Chi Minh City, Vietnam
{htkhai,thang,qttho}@cse.hcmut.edu.vn
http://www.cse.hcmut.edu.vn/site/en/

Abstract. The paper presents the WSCOVER tool, which aims at automatic composition and verification of web services by means of model techniques. WSCOVER can support checking both hard constraints, soft constraints and also the temporal relations over those constraints on the resulted web service compositions. Especially, WSCOVER employs additional intelligent techniques including heuristic-guided searching and logic-based clustering to make the model checking process more efficient. The experiments show that those techniques really improve the verification performance of WSCOVER, allowing the tool to handle a repository of up to 1000 web services.

Keywords: WSCOVER · Web service composition · Web service verification · Logic-based clustering · Heuristic-guided model checking

1 Introduction

Web services (WS) have increasingly played an important role for software development today. They can be used for fulfilling the user requirements based on their functional and also non-functional (or Quality of Service – QoS) properties [1]. Table 1 gives example of a repository of 10 WS in the domain of Tourism. In the table, the functional property for each WS is represented as Input(s) and Output(s), and the non-functional property (QoS) is its response time (resp-Time).

To enhance usability, web services are usually very basic, just to satisfy the simple requests. However, the user requirements are usually complex, probably combining several available services together to accomplish the desired task. For example, a user who is looking for a place to travel (sightseeing) and may also looks for the information about nearby hotels. In order to do so, the user indicates the place to travel to (*Sightseeing*) and the desired date (*Date*). He also states the expected hotel price (*Price*) that he can afford. If all of those

© Springer International Publishing AG 2016
C. Sombattheera et al. (Eds.): MIWAI 2016, LNAI 10053, pp. 50–62, 2016.
DOI: 10.1007/978-3-319-49397-8_5

Table 1. The travel booking web service repository

#	Service Name	Input	Output	respTime (s)
1	HotelReserveService (HR)	Dates, Hotel	HotelReservation	5
2	CityHotelService (CH)	City	Hotel	3
3	HotelCityService (HC)	Hotel	City	3
4	HotelPriceService (HP)	Hotel	Price	10
5	SightseeingCityService (SC)	Sightseeing	City	2
6	SightseeingCityHotelService (SCH)	Sightseeing	City, Hotel	16
7	CitySightseeingService (CS)	City	Sightseeing	4
8	ActivityBeachService (ABS)	Activity	Beach	5
9	AreaWeatherService (AWS)	Area	Weather	5
10	CityWeatherService (CWS)	City	Weather	5
10	CityWeatherService (CWS)	City	Weather	5

requirements are met, the user also expects that the reservation will also be done (*HotelReservation*). When no single WS fully fits the user requirement, a composition of them should be conducted. Besides these functional requirements (*hard constraint*), the user also has certain requests for the QoS (*soft constraint*). He wants the total response time (*respTime*) of the whole composition not to exceed 30 s. In addition, the user also wants to obtain information of *Price* before the *HotelReservation* is made since the price is an important criterion to choose the hotel. This requirement can be represented as a *temporal relationship* among web services. All of user requirements, or *goal*, are summarized in Table 2.

Table 2. Requirements for Travel Booking web service

Kinds of constraint	Value		
Hard constraint:	(S1)	Input:	$Dates, Sightseeing$
		Output:	$Price, HotelReservation$
	(S2)	Input:	$Dates, Sightseeing$
		Output:	$Price, HotelReservation, Weather$
Soft constraint:	$respTime \leq 30(s)$		
Temporal relationship:	$\Box(\neg HotelReservation \cup Price)$		

Currently, there are many tools which can create the composite web services based on the user requirements. PORSCE II [5] and OWLS-XPlan [10] use the planning-based approaches to compose web services based on just hard constraints. The research in [1] composes web services based on both hard and soft constraints. However, in order to verify the results of composition, there is only tool VeriWS [3] claimed to be able to verify both functional and non-functional properties at the same time. VeriWS requires an existing *full composition schema* of all possible compositions based on hard constraints and then verifies if those compositions satisfy soft constraints. Nevertheless, building such full composition schema suffers from high computational cost as this is an NP-hard task. In addition, whenever there is a new goal, the full composition schema needs to be rebuilt, although repository of WS may be unchanged.

In this paper, we present WSCOVER (Web Service COmposition and VERification), a tool supports composition and verification of WS using heuristics-guided model checking and clustering technique. WSCOVER offers the following features.

- Using a unique LTS model to represent a WS repository. With this LTS model, a model checking technique will find a composition solution for each user-supplied goal.
- Applying a heuristic search mechanism to increase the composition performance.
- Proposing a logic-based clustering technique to divide the WSs in the repository into groups. Thereby, we can eliminate the unnecessary web services in the process of finding a suitable solution. More specifically, the logic-based clustering technique developed in WSCOVER generates more reasonable clustering result than the feature-based techniques involved in another similar works of WS clustering, e.g. [2,11,16].

The rest of the paper is organized as follows. In Sect. 2, we present the functionality of WSCOVER. We propose the architecture of our tool in Sect. 3. We compare WSCOVER to existing tools in Sect. 4. The demonstration and experimentation are presented in Sect. 5. We conclude the paper and raise the future work in Sect. 6.

2 Functionality

Aiming at automatic composition and verification of WS, WSCOVER offers the following unique features, as compared to other similar works.

2.1 On-the-Fly Composition and Verification

WSCOVER performs simultaneously the web service composition and verification by representing a WS repository as a special LTS model which consists of a single state, whereas each web service in the repository is represented as a guarded transition, as illustrated in Fig. 1a. The composition process will be carried out as a searching problem on the state space. Each state in state space corresponds to a composition step where the conditions on hard, soft constraints and the temporal relation will be verified accordingly. The path leading to the state that violates these constraints will be pruned, as illustrated in Fig. 2. Details of LTS representation and the verification process can be found on [6].

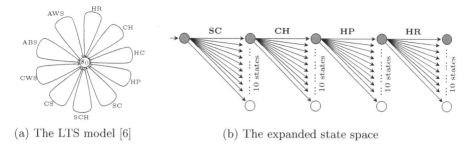

(a) The LTS model [6]   (b) The expanded state space

Fig. 1. The model and expanded state space of web service composition and verification

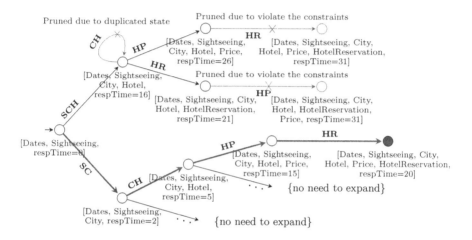

Fig. 2. The state space exploration

2.2   Heuristic-Guided Composition of WS

WSCOVER also employees a heuristic technique for finding the composition solution. At a state which is being examined, the next chosen state n will be prioritized based on the heuristics function of $F(n) = \alpha G(n) + \beta H(n)$ where $G(n)$ is the real cost from state $init$ to state n and $H(n)$ the estimated cost (heuristic) from n to $goal$. The value of $H(n)$ will be determined based on the similarity of state n to state $goal$. The factors α and β are empirically determined. Readers can refer to the details of heuristic function in [6].

As illustrated in Fig. 2, guided by the heuristic function, WSCOVER can choose a faster path to reach the goal (indicated by the thick blue line, $SC \rightarrow CH \rightarrow HP \rightarrow HR$) and skip some other unnecessary paths.

2.3   Logic-Based Clustering

In order to reduce the number of states which we have to visit in each composition step, WSCOVER adopts clustering technique, which groups the initial WSs in the repository into *clusters*. At each composition step, only WSs in the

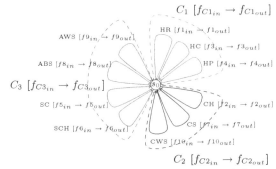

(a) The LTS4WS model and its logic-based clusters

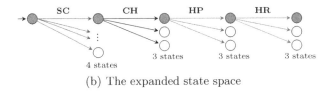

(b) The expanded state space

Fig. 3. The model and expanded state space of logic-based clustered WS repository

suitable clusters are considered to be examined. For example, in Fig. 3a, all 10 initial WSs are divided into 3 clusters, called C_1, C_2, and C_3. For the goal presented in Table 2, WSCOVER infers that only the WSs in the cluster C_1 are appropriate in the first step. Therefore, the state space is only expanded to states corresponding to 4 WSs in this cluster. In the 2nd step (after supposedly selecting SC), WSCOVER continues inferring that the cluster C_2 is suitable with the current state and it only examines 3 WSs in C_2. This process continues until the entire composition process is completed.

To implement this idea, WSCOVER uses a special clustering technique, known as *logic-based clustering*. The details are in [7]. The main characteristics of this technique are summarized as follows.

– Each WS will be represented by a First-Order Logic (FOL) formula. This formula is built from the specification of the WS hard constraint. For example, the representative logic formula of web service SC in Table 1 is $Sightseeing \rightarrow City$.
– When clustering, the similarity of WSs will be calculated based on *the semantic similarity of the corresponding represented logic formulas*. For example, let us consider three first web services in Table 1: CH: $City \rightarrow Hotel$, HC: $Hotel \rightarrow City$, and HR: $Dates \wedge Hotel \rightarrow HotelReservation$. In this similarity approach, HC and HR are similar, while CH and HC are not. In other words, if HC is a candidate at a composition step, that means the input of $City$ is expected at this step, and therefore HR is also a candidate to be considered. Meanwhile, CH is not a suitable candidate. As a result, HC and HR are grouped together in cluster C_1, which not consists of CH.

– Each cluster is represented by a FOL formula, which is synthesized from the formulas of all WSs in this cluster. Based on the representative formula of clusters, WSCOVER will determine the appropriate cluster for each step. In [7], we have proven that our composition approach using logic-based clustering is soundness and completeness, i.e. the composition result always satisfies the given goal, and this approach does not suffer from missing solution situation.

3   Architecture

WSCOVER is implemented in C#. It extends the model checker PAT [12] and is covered by a friendly interface for users. Figure 4 shows the architecture of WSCOVER, which consists of 3 layers, known as *Pre-processor*, *Composition and Verification Processor* and *Editor Interface*.

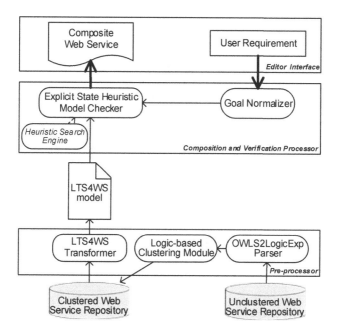

Fig. 4. WSCOVER framework architecture

3.1   Pre-processor Layer

This layer works mainly to pre-process descriptions of WSs stored in a repository before performing composition. This layer consists of the following modules.

– *OWLS2LogicExp Parser* is responsible for parsing the OWL-S [13] web service profiles and transforming them into logic expressions.[1]

---

[1] Currently WSCOVER only supports OWL-S, however it can be easily extended and adapted for other WS description language like WSDL.

- *Logic-based Clustering Module* performs the clustering on web services based on their representative logic formulas using the k-means algorithm. The clusters of web service are stored as an XML file.
- *LTS4WS Transformer* transforms the web services stored in clusters into a corresponding LTS model, as illustrated in Fig. 3a.

3.2 Composition and Verification Processor

This is the main function of WSCOVER, which undertakes the functions of composing and verifying web services, given the user's requirement as goal. This layer includes the following modules.

- *Goal Normalizer* transforms the goal into an assertion which is then put into the model checker. The assertion can be a reachability assertion or a safety LTL assertion, depending on user's purpose. Table 4 illustrates some assertions created by this module.
- *Heuristic Engine* is responsible for implementation of the heuristic function as previously discussed.
- *Heuristic-guided Model Checker* extends PAT [12] by integrating the heuristic engine when visiting new states. The heuristic search helps PAT to find the solution faster and able to work with the relatively large state space.

3.3 Editor Interface

This is the user interface layer, which provides the functionality for users to specify their requirements, such as to select the inputs which users can supply; select the desired outputs; enter the constraints on the QoS properties (soft constraints); and enter the LTL expressions described the temporal relationship between web services in the composition. Another function of the user interface layer is to return the composition result, as well as the system information related to the composition process, such as processing time, the used system memory, the number of expanded states, the number of visited states.

4 Functionality Comparisons

Table 3 shows the comparison of WSCOVER with some of the existing tools. These tools either compose or verify the web services, based on functional or non-functional properties.

One of the latest tools that attempted web service verification is VeriWS [3]. It uses the web service composition schema expressed in BPEL [8], then transforms that schema into a LTS model and uses a model checker to verify (the model). It will return the information to show that whether the composition has satisfied the user requirements or not, both on hard and soft constraints. Although this is the first tool handling combined functional and non-functional properties, it only verifies the web service composition and required a full composition schema to be created in advance, which is a *NP-hard* problem [15].

Table 3. Web service composition and verification tools

Tool	Composition	Verification	Constraint	Input
OWLS-XPlan [10]	✓		Hard	OWL-S (primitive WSs)
PORSCE II [5]	✓		Hard	OWL-S (primitive WSs)
WS-Engineer [4]		✓	Hard	BPEL processes
AgFlow [17]		✓	Soft	Statecharts
VeriWS [3]	✓	✓	Combined	BPEL processes
WSCOVER	✓	✓	Combined	OWL-S (primitive WSs)

PORSCE II [5] is a framework implementing the WSC based on the requirements on functional properties. It performs automatic web service composition exploiting artificial intelligent (AI) techniques, specifically planning. PORSCE II transforms the web service composition problem into a solver-ready planning domain and problem. Web services used in PORSCE II are described by OWL-S language [13]. Therefore, PORSCE II developed the function which transforms web services described in OWL-S into actions described by PDDL [14] and used the planner to do composition. The result of planning is a chain of actions whose result satisfied the user requirement. Finally, the chain of actions is transformed back in web service terms. Although PORSCE II solves the web service composition problem, it only handles the functional properties (hard constraint) and does not solve. Similarly, OWLS-XPlan [10] also uses web services expressed by OWL-S [13] and transforms the problem from WSC domain to planning domain.

WS-Engineer [4] is a typical work for the web service verification based on functional properties. In WS-Engineer, the authors described a model-based approach to analyze web service interactions for web service choreography and their coordinated compositions. The approach employs several formal analysis techniques and perspectives, and applies these to verify the web service composition. The resulted web service composition was described as a BPEL process. Then, this process will be specified using the Finite State Process (FSP) algebra notation. The verification was done on this FSP model by using the Finite State Machine (FSM).

AgFlow [17] is a tool implementing the verification of composite web service based on non-functional properties (soft constraints). To use AgFlow, users have to provide the composite web service described as statecharts. Then, these statecharts are put into the *multidimensional QoS model* to monitor the execution of the composite web service. AgFlow will check whether the composite web service conflicts with the user non-functional constraints or not.

Compared to existing tools, WSCOVER is distinguished by several features. First, WSCOVER performs combined composition and verification at the same time and in on-the-fly manner, supporting functional and non-functional properties, also the temporal relations between component web services. Second, WSCOVER uses component web services as the input, i.e. it does not need to build the composition schema in advance as VeriWS [3], or AgFlow [17]. In addition, WSCOVER has integrated a lot of techniques which help to improve the performance where other tools do not have, such as using the heuristic search based on the characteristics of web services, and using the logic-based clustering the cluster the web service repository.

5 Demonstration and Experiment

5.1 Demonstration

In this section, we show a demonstration to illustrate the function as well as the way of working of WSCOVER. This demonstration comes with a video whose link we mentioned at the end of this paper. We use the Travel Online Booking service (TOB) (extracted from [9]) for our work.

The business situation assumed for our demonstration is as follows. It is supposed that we have a repository of 10 web services as given in Table 1, and a user would like to use those services for his convenience when booking a tour.

Table 4 shows some requirements and its corresponding composition results. For each requirement, we evaluate the performance of our tool based on four scenarios: using neither Logic-based Clustering (LbC) nor Heuristic Search (HS); using HS but not LbC; using LbC but not HS; and using both LbS and HS. Besides the composition results, we also collect the system information to show the effectiveness of each module in the framework. The system information consists of the number of Expanded States (ES), the number of Visited States (VS), the Execution Time (ET) in millisecond, and the Used Memory (UM) in kilobyte. These requirements can be hard constraints, soft constraints or both of them. In addition, some requirements also contain the constraints on temporal relationship between web services.

Table 4. The demonstration requirements (LbC: Logic-based clustering, HS: Heuristic search, ES: Expanded states, VS: Visited states, ET: Execution time (ms), UM: Used memory (Kb))

#	User requirement	LbC	HS	ES	VS	ET	UM	Composition result
R1	$TOB \models reach(Date \wedge Sightseeing \rightarrow Price \wedge HotelReservation)$			251	26	30	8,693	$SCH \bullet HR \bullet HP$
			✓	121	13	21	8,675	
		✓		101	26	37	8,642	
		✓	✓	33	9	184	7,319	
R2	$TOB \models reach((Date \wedge Sightseeing \rightarrow Price \wedge HotelReservation) \wedge (respTime \leq 30))$			3,371	338	35	9,801	$SC \bullet CH \bullet HR \bullet HP$
			✓	341	35	25	9,214	
		✓		1,405	352	33	8,055	
		✓	✓	61	16	126	7,326	
R3	$TOB \models \Box((Dates \wedge Sightseeing \rightarrow Price \wedge HotelReservation) \wedge (\neg HotelReservation \cup Price))$			881	89	82	8,979	$SCH \bullet HP \bullet HR$
			✓	131	14	20	8,958	
		✓		237	60	30	7,257	
		✓	✓	37	10	11	7,185	
R4	$TOB \models \Box((Dates \wedge Sightseeing \rightarrow Price \wedge HotelReservation) \wedge (respTime \leq 30) \wedge (\neg HotelReservation \cup Price))$			10,911	1,092	89	11,471	$SC \bullet CH \bullet HP \bullet HR$
			✓	341	35	12	10,796	
		✓		3,029	758	91	10,452	
		✓	✓	113	29	8	8,957	

5.2 Experiments

In this section, we present the experimental results of WSCOVER. Our experiments work on the real datasets obtained from the project OWLS-TC [9] with over 1000 web services classified into different domains and described by OWL-S [13]. In this dataset, we select five sub-datasets with the number of web services is varied 20 to 1000 services as shown in Table 5.

Table 5. The experimentation datasets

Dataset	No. of web service	Description
Travel Booking (TB)	20	Provide information to serve the travel booking
Medical Services (MS)	50	Support to look up hospital, treatment, medicine, etc.
Education Services (EDS)	100	Supply education services as scholarship, courses
Economy Services (ECS)	200	Provide information about goods, restaurant, etc.
Global	1000	1000 random services from OWLS-TC [9]

We conduct the experiment scenarios based on three approaches. These include *Full verification* (i.e. non-heuristic); *Heuristic*; and *Logic-based clustering*. The experiments were executed on a PC with core i5-5200 processor (4×2.7 GHz), 8.0 GB of RAM, running on Windows 7 64-bit. The result of experiments is evaluated in three aspects: the number of expanded states, the number of visited states, and the execution time. The experiment results are analyzed statistically in Table 6.

The experimental results confirm our hypothesis that clustering helps in reducing the number of expanded states. Note that, the heuristic algorithm in the WSCOVER always chooses the best way to travel in the state space. The number of visited states shows the number of best states that have been chosen during the search. In our approach, when the contradictory web services cannot be in the same cluster, then the number of "best" states are smaller resulting in the number of visited states obviously smaller.

6 Conclusion

This paper presents WSCOVER, a tool for automatic composition and verification of web services. WSCOVER allows for verification of both hard constraints and soft constraints of the resultant composite web services, as well as the temporal relations over those constraints. This tool also adopts heuristic strategy

Table 6. Experimentation results

Dataset	Approach	Expanded states	Visited states	Execution time (s)
TB (20)	Full verification	1,100	56	0.250
	Heuristic	825	56	0.210
	Logic-based clustering	316	35	0.155
MS (50)	Full verification	6,251	125	1.008
	Heuristic	4,689	125	0.820
	Logic-based clustering	1,126	75	0.614
EDS (100)	Full verification	27,400	275	5.570
	Heuristic	20,824	275	4.579
	Logic-based clustering	3,021	151	3.294
ECS (200)	Full verification	102,800	515	28.198
	Heuristic	76,586	515	23.030
	Logic-based clustering	8,324	287	18.273
Global (1000)	Full verification	-	-	-
	Heuristic	-	-	-
	Logic-based clustering	91,457	1,429	597.228

and logic-based clustering to reduce the state space. Experimental results show that when the number of web services becomes large (e.g. around 1000 web services), only the logic-based clustering approach can manage to return the composite web services and their verification results.

The tool WSCOVER, its guidelines and all experimental datasets can be downloaded from http://cse.hcmut.edu.vn/~save/project/wscover/start. The video for demonstration is available at https://www.youtube.com/watch?v=yPbA5J5Jsrc, and the online version of WSCOVER is available at http://www.cse.hcmut.edu.vn/wscover.

Acknowledgment. This research is funded by Vietnam National Foundation for Science and Technology Development (NAFOSTED) under grant number 102.01-2015.16.

References

1. AllamehAmiri, M., Derhami, V., Ghasemzadeh, M.: QoS-based web service composition based on genetic algorithm. J. AI Data Min. **1**(2), 63–73 (2013)
2. Aznag, M., Quafafou, M., Jarir, Z.: Leveraging formal concept analysis with topic correlation for service clustering and discovery. In: 2014 IEEE International Conference on Web Services (ICWS), pp. 153–160. IEEE (2014)
3. Chen, M., Tan, T.H., Sun, J., Liu, Y., Dong, J.S.: VeriWS: a tool for verification of combined functional and non-functional requirements of web service composition. In: Proceedings of the 36th International Conference on Software Engineering, pp. 564–567. ACM (2014)

4. Foster, H., Uchitel, S., Magee, J., Kramer, J.: WS-engineer: a model-based approach to engineering web service compositions and choreography. In: Baresi, L., Di Nitto, E. (eds.) Test and Analysis of Web Services, pp. 87–119. Springer, Heidelberg (2007). doi:10.1007/978-3-540-72912-9_4

5. Hatzi, O., Vrakas, D., Bassiliades, N., Vlahavas, I.: The PORSCE II framework: using AI planning for automated semantic web service composition. Knowl. Eng. Rev. **28**(02), 137–156 (2013)

6. Huynh, K.T., Quan, T.T., Bui, T.H.: Fast and formalized: heuristics-based on-the-fly web service composition and verification. In: 2nd National Foundation for Science and Technology Development Conference on, pp. 174–179. IEEE (2015)

7. Huynh, K.T., Quan, T.T., Bui, T.H.: Smaller to sharper: efficient web service composition and verification using on-the-fly model checking and logic-based clustering. In: Proceedings of The 8th International Symposium on Software Engineering Processes and Applications (SEPA: 2016), Singapore (2016)

8. Jordan, D., Evdemon, J., Alves, A., et al.: Web services business process execution language version 2.0. In: OASIS standard 11, 10 (2007)

9. Klusch, M.: OWLS-TC: OWL-S service retrieval test collection, version 2.1. http://projects.semwebcentral.org/projects/owls-tc/

10. Klusch, M., Gerber, A., Schmidt, M.: Semantic web service composition planning with OWLS-Xplan. In: Proceedings of the AAAI Fall Symposium on Semantic Web and Agents. AAAI Press, Arlington (2005)

11. Kumara, B.T., Paik, I., Chen, W., Ryu, K.H.: Web service clustering using a hybrid term-similarity measure with ontology learning. Int. J. Web Serv. Res. (IJWSR) **11**(2), 24–45 (2014)

12. Liu, Y., Sun, J., Dong, J.S.: Developing model checkers using PAT. In: Bouajjani, A., Chin, W.-N. (eds.) ATVA 2010. LNCS, vol. 6252, pp. 371–377. Springer, Heidelberg (2010). doi:10.1007/978-3-642-15643-4_30

13. Martin, D., Burstein, M., Hobbs, J., Lassila, O., McDermott, D., McIlraith, S., Narayanan, S., Paolucci, M., Parsia, B., Payne, T., et al.: OWL-S: semantic markup for web services. W3C member submission 22, 2007-04 (2004)

14. McDermott, D., Ghallab, M., Howe, A., Knoblock, C., Ram, A., Veloso, M., Weld, D., Wilkins, D.: PDDL-the planning domain definition language. Technical report (1998)

15. Pejman, E., Rastegari, Y., Esfahani, P.M., Salajegheh, A.: Web service composition methods: a survey. In: Proceedings of the International MultiConference of Engineers and Computer Scientists, vol. 1 (2012)

16. Xie, L.L., Chen, F.Z., Kou, J.S.: Ontology-based semantic web services clustering. In: 2011 IEEE 18th International Conference on Industrial Engineering and Engineering Management (IE&EM), pp. 2075–2079. IEEE (2011)

17. Zeng, L., Benatallah, B., Ngu, A.H., Dumas, M., Kalagnanam, J., Chang, H.: QoS-aware middleware for web services composition. IEEE Trans. Softw. Eng. **30**(5), 311–327 (2004)

An Improvement of Pattern-Based Information Extraction Using Intuitionistic Fuzzy Sets

Peerasak Intarapaiboon[1(✉)] and Thanaruk Theeramunkong[2]

[1] Deportment of Mathematics and Statistics, Faculty of Science and Technology,
Thammasat University, Pathum Thani, Thailand
peerasak@mathstat.tu.ac.th
[2] School of Information and Computer Technology,
Sirindhorn International Institute of Technology, Thammasat University,
Pathum Thani, Thailand
thanaruk@siit.tu.ac.th

Abstract. Multi-slot information extraction (IE) is a task that identify several related entities simultaneously. Most researches on this task are concerned with applying IE patterns (rules) to extract related entities from unstructured documents. An important obstacle for the success in this task is unknowingness where text portions containing interested information are. This problem is more complicated when involving languages with sentence boundary ambiguity, e.g. the Thai language. Applying IE rules to all reasonable text portions can degrade the effect of the obstacle, but it raises another problem that is incorrect (unwanted) extractions. This paper aims to present a method for removing incorrect extractions. In the method, extractions are represented as intuitionistic fuzzy sets (IFSs), and a similarity measure for IFSs is used to calculate distance between IFS of an unclassified extraction and that of each already-classified extraction. The concept of k nearest neighbor is adopted to design whether the unclassified extraction is correct of not. From the preliminary experiment on a medical domain, the proposed technique improves extraction precision while satisfactorily preserving recall.

1 Introduction

Information extraction (IE) is a process of identifying and extracting desired pieces of information. Multi-slot IE is a special task of IE that extract related pieces of information simultaneously and connecting them in a form of multiple-field relational records. Most IE systems usually involve rule-based approaches, which an IE rule is often represented in terms of a regular expression, e.g., WI [1], CRYSTAL [2], LIEP [3], and WHISK [4]. Applying IE rules to documents with unknown target-phrase locations tends to make false positives (incorrect extractions), since these rules probably match with text portions that do not convey information of interest. As such, several IE frameworks come up with components to alleviate the detriment suffered by the aforementioned issue. One approach to overcome the problem is removing inefficient rules [5,6]. An alternative approach uses the all IE rules and then eliminates unwanted extractions [7–10].

© Springer International Publishing AG 2016
C. Sombattheera et al. (Eds.): MIWAI 2016, LNAI 10053, pp. 63–75, 2016.
DOI: 10.1007/978-3-319-49397-8_6

Recently, intuitionistic fuzzy set (IFS) [11] has been much explored in both theory and application. Differing from representation of a fuzzy set (FS) [12], an IFS considers both the membership and non-membership of elements belonging or not belonging to such a set. IFS is therefore more flexible to handle the uncertainty than FS. Measuring similarity and distance between IFSs is one of most research areas to which many researchers have focused. After Dengfeng and Chuntian [13] gave the axiomatic definition of similarity measures between IFSs, various similarity measures have been proposed continuously [14–19]. One of most applications of IFS similarity measures is classification problems. Khatibi and Montazer [18] conducted experiments for bacterial classification using similarity measures for FSs and IFSs. The results indicated that each measure for IFSs outperformed that for FSs. Ye [19] cosine and weighted cosine similarity measures for IFSs were proposed and applied to a small medical diagnosis problem.

By the success of research in IFS, especially similarity measurement, it is anticipated that IFS technologies will contribute to improve performance of an IE framework. This work presents an IFS-based method aimed to eliminate incorrect extractions. The main contribution of this work is twofold: (i) how to represent an extracted frame in terms of an IFS and (ii) how to apply a similarity measure between IFSs for removing incorrect extraction.

The remainder of the paper proceeds as follows: Section 2 provides a literature review about information extraction with incorrect extraction removal. Section 3 explains a pattern-based IE framework from Thai texts. Section 4 reviews IFS and similarity measures for IFSs. Section 5 presents our filtering method, then the experiments is detailed in Sect. 6. Finally, Sect. 7 gives conclusions and outlines future works.

2   Related Works

From a machine-learning viewpoint, the task of detecting false extractions can be reduced to a binary classification problem. A classification can be constructed to predict whether extractions are correct. In [7], biological events, each of which consists of three slots—one interaction type, one effect, and one reactant—were extracted from unstructured texts using a pattern-based strategy. In order to determine whether an extracted event is correct, a maximum entropy classifier is employed to assign one slot type to each slot filer in the event. When the slot type of a slot filler assigned by the classifier is inconsistent with that by the IE pattern the extracted event is discarded. Similarly, Intarapaiboon [8] proposed an pattern-based IE framework to extract multi-slot frames. To improve precision by removing false extraction, two extraction filtering modules were proposed. The first module uses a binary classifier, e.g. naïve bayes and support vector machine, for prediction of rule application across a target-phrase boundary; the second one uses weighted classification confidence to resolve conflicts arising from overlapping extractions. In [9], linguistic patterns were used for extracting medication information, including medical name, dosage, frequency, duration, and reason, from free-text medical records. Occasionally, medical records contain side effects which are out of scope and usually extracted as reasons. A hand-crafted semantic rule set was constructed and used to filter out such side-effect statements.

3   Information Extraction from Thai Texts

This section briefly explains the idea of domain-specific information extraction for Thai unstructured texts using extraction rules.

3.1   Preprocessing

By detecting paragraph breaks, a text document is decomposed into paragraphs, referred to as *information entries*, then word segmentation is applied to all information entries as part of a preprocessing step. A domain-specific ontology, along with a lexicon for concepts in the ontology, is then employed to partially annotate word-segmented phrases with tags denoting the semantic classes of occurring words with respect to the lexicon.

In the medical domain, as an example, suppose we focus on two types of symptom descriptions: one is concerned with abnormal characteristics of some observable entities and the other with human-body locations at which primitive symptoms appear. Figure 1 illustrates a portion of word-segmented and partially annotated information entry describing acute bronchitis, obtained from the text-preprocessing phase, where '|' indicates a word boundary, '∼' signifies a space, and the tags "sec," "col," "sym," "org," and "ptime" denote the semantic classes "Secretion," "Color," "Symptom," "Organ," and "Time period," respectively, in our medical-symptom domain ontology. The portion contains three target symptom phrases, which are underlined in the figure. Figure 2 provides a literal English translation of this text portion; the translations of the three target phrases are also underlined. Figure 3 shows the frame required to be extracted

เป็น|โรค|ที่|พบ|บ่อย|หลัง|จาก|เป็น|ไข้หวัด|∼|ผู้ป่วย|ส่วนใหญ่|มัก|จะ|มี|[sec เสมหะ]|เป็น|[col สีเขียว]|∼|

มี|[sym อาการเจ็บ]|ที่|บริเวณ|[org หน้าอก]|อยู่|เป็น|เวลา|นาน|∼|[ptime 6-12 วัน]|∼|มี|[sym อาการไอ]|จน|

เกิด|[sym อาการเจ็บ]|ที่|[org ชายโครง]|อยู่|นาน|∼|[ptime 3-4 วัน]|∼|ผู้ป่วย|อาจ|มี|สุขภาพ|ทั่วไป|แข็งแรง|...

Fig. 1. A portion of a partially annotated word-segmented information entry

It is a disease that often begins after flu. A patient may have [col *green*] [sec *mucus*], *and may have a* [sym *pain*] *in his* [org *chest*], *which lasts* [ptime *6-12 days*], *and a* [sym *cough*] *that leads to a* [sym *pain*] *in his* [org *lower rib cage*] *lasting* [ptime *3-4 days*]. *A patient may have regular health*...

Fig. 2. A literal English translation of the partially annotated Thai text in Fig. 1

Target phrase: |มี|[sym อาการเจ็บ]|ที่|บริเวณ|[org หน้าอก]|อยู่|เป็น|เวลา|นาน|~|[ptime 6-12 วัน]|

English translation: *have a* [sym *pain*] *in his* [org *chest*], *which lasts* [ptime *6-12 days*]

Extracted frame: {Sym [sym อาการเจ็บ]}{Loc [org หน้าอก]}{Per [ptime 6-12 วัน]}

English translation: {Sym [sym *pain*]}{Loc [org *chest*] }{Per [ptime *6-12 days*]}

Fig. 3. A target phrase and an extracted frame

from the second underlined symptom phrase in Fig. 1. It contains three slots, i.e., Sym, Loc, and Per, which stand for "symptom," "location," and "period," respectively.

3.2   IE Rules and Rule Application

A well-known supervised rule learning algorithm, called WHISK [Sodeland, 1999], is used as the core algorithm for constructing extraction rules. Figure 4 gives a typical example of an IE rule. Its pattern part contains (i) three triggering class tags, i.e., sym, org, and ptime, (ii) four internal wildcards, and (iii) one triggering word (between the last two wildcards). The three triggering class tags also serve as *slot markers*—the terms into which they are instantiated are taken as fillers of their respective slots in the resulting extracted frame. When instantiated into the target phrase in Fig. 3, this rule yields the extracted frame shown in the same figure.

Pattern: *(sym)*(org)*นาน*(ptime)

Output template: {Sym $1}{Loc $2}{Per $3}

Fig. 4. An IE rule example

WHISK rules are usually applied to individual sentences. In the Thai writing system, however, the end point of a sentence is usually not specified. To apply IE rules to free text with unknown boundaries of sentences and potential target text portions, rule application using sliding windows (RAW) is employed. Roughly speaking, by RAW, a particular rule is applied to each k-word portion of an information entry one-by-one sequentially, where the window size, k, is predefined depending on the rule. As shown in Fig. 5, when the rule in Fig. 4 is applied to the information entry in Fig. 1 using a 10-word sliding window, it makes extractions from the [21, 30]-portion, the [33, 42]-portion, and the [34, 43]-portion of the entry. Table 1 shows the resulting extracted frames. Only the extractions made from the first and third portions are correct. When the rule is applied to the second portion, the slot filler taken through the first slot marker of the rule, i.e., "sym," does not belong to the symptom phrase containing the filler taken through the second slot marker of it, i.e., "org," whence an incorrect extraction occurs.

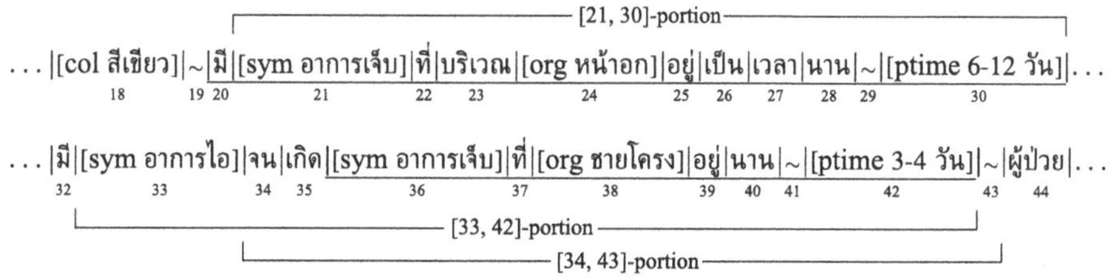

... |[col สีเขียว]|~|มี|[sym อาการเจ็บ]|ที่|บริเวณ|[org หน้าอก]|อยู่|เป็น|เวลา|นาน|~|[ptime 6-12 วัน]|...

... |มี|[sym อาการไอ]|จน|เกิด|[sym อาการเจ็บ]|ที่|[org ชายโครง]|อยู่|นาน|~|[ptime 3-4 วัน]|~|ผู้ป่วย|...

Fig. 5. Text portions from which extractions are made when the rule in Fig. 4 is applied to the information entry in Fig. 1 using a 10-word sliding window

Table 1. Frames extracted from the text portions in Fig. 5 by the rule in Fig. 4

Portion	Extracted frame	Correctness
[21, 30]	{SYM [sym อาการเจ็บ]}{LOC [org หน้าอก]}{PER [ptime 6-12 วัน]}	Correct
[33, 42]	{SYM [sym อาการไอ]}{LOC [org ชายโครง]}{PER [ptime 3-4 วัน]}	Incorrect
[34, 43]	{SYM [sym อาการเจ็บ]}{LOC [org ชายโครง]}{PER [ptime 3-4 วัน]}	Correct

4 Intuitionistic Fuzzy Sets and Their Similarity Measures

In this section, some basic concepts for IFSs and their similarity measures are presented. For the convenience of explanation, the following notations are used hereinafter: $X = \{x_1, x_2, \ldots, x_h\}$ is a discrete universe of discourse and $IFS(X)$ is the class of all IFSs of X. Atanassov [11] defined an intuitionistic fuzzy set A in $IFS(X)$ as follows:

$$A = \{\langle x, \mu_A(x), \nu_A(x) \rangle | x \in X\} \tag{1}$$

which is characterized by a membership function $\mu_A(x)$ and a non-membership function $\nu_A(x)$. The two functions are defined as:

$$\mu_A : X \to [0, 1], \tag{2}$$
$$\nu_A : X \to [0, 1], \tag{3}$$

such that

$$0 \le \mu_A(x) + \nu_A(x) \le 1, \forall x \in X. \tag{4}$$

In the IFS theory, the hesitancy degree of x belonging to A is also defined by:

$$\pi_A(x) = 1 - \mu_A(x) - \nu_A(x). \tag{5}$$

Definition 1 [15]. *A similarity measure S for $IFS(X)$ is a real function $S : IFS(X) \times IFS(X) \to [0, 1]$, which satisfies the following properties:*

P1: $0 \le S(A, B) \le 1$,
P2: $S(A, B) = S(B, A), \forall A, B \in IFS(X)$,

Table 2. Some similarity measures between IFSs.

Author	Expression
Dengfeng and Chuntian [13]	$S_d^p(A,B) = 1 - \frac{1}{\sqrt[p]{h}}\sqrt[p]{\sum_{i=1}^{h}|\varphi_A(i) - \varphi_B(i)|^p}$ where $\varphi_k(i) = (\mu_k(x_i) + 1 - \nu_k(x_i))/2, k = \{A,B\}$, and $p = 1,2,3,\ldots$
Mitchell [15]	$S_m^p(A,B) = \frac{1}{2}(\rho_\mu(A,B) + \rho_f(A,B))$ where $\rho_\mu(A,B) = S_d^p(\mu_A(x_i), \mu_B(x_i))$ and $\rho_f(A,B) = S_d^p(1 - \nu_A(x_i), 1 - \nu_B(x_i))$
Ye [19]	$S_C(A,B) = \frac{1}{h}\sum_{i=1}^{h}\frac{\mu_A(x_i)\mu_B(x_i) + \nu_A(x_i)\nu_B(x_i)}{\sqrt{\mu_A^2(x_i) + \nu_A^2(x_i)}\sqrt{\mu_B^2(x_i) + \nu_B^2(x_i)}}$

P3: $S(A,B) = 1$ iff $A = B$,
P4: If $A \subseteq B \subseteq C$, then $S(A,C) \leq S(A,B)$ and
$\quad S(A,C) \leq S(B,C)$, for all A, B, and $C \in IFS(X)$.

Let $A = \{\langle x_i, \mu_A(x_i), \nu_A(x_i)\rangle | x_i \in X\}$ and $B = \{\langle x_i, \mu_B(x_i), \nu_B(x_i)\rangle | x_i \in X\}$ be in $IFS(X)$, Table 2 highlights some similarity measures between IFSs.

5   The Proposed Technique—*IFS-Based Extraction Filtering*

As the example shown in Sect. 3, RAW probably produces false extractions. Hence, to improve the extraction accuracy, a method for removing unwanted extractions is necessary. The idea behind our method for removing incorrect extractions is based on the fact that if an internal wildcard[1] of a rule is instantiated across a target-phrase boundary, then an incorrect extraction is made. Predicting whether an internal wildcard is instantiated across a target-phrase boundary can be regarded as a binary classification problem.

In our technique, an intuitionistic fuzzy set will be generated for each extracted frame. Like k-NN, to determine whether an extraction E is correct or not, a majority vote among the k nearest neighbors of the IFS corresponding to E is applied, where a distance is calculated by an IFS similarity measure. Given an IE rule r with n internal wildcards, the precise steps of the proposed method are detailed as follows:

5.1   Preprocessing

Vector-Based Document Representation.

(a1) Apply the rule r into all information entries in the training corpus, whence semantic frames are obtained.

---

[1] A wildcard occurs between the first and the last slot markers of a rule, called an *internal wildcard*.

(a2) For each internal wildcard, observe plain words in which the wildcard instantiates during Step (a1). These words are separated to 2 sets: one containing different words only when correct extractions are made; and the other containing those only when incorrect ones are made. For convenience, W_{cor}^k and W_{inc}^k are referred to the former set and the latter set, respectively, of the k-th internal wildcard.

(a3) Construct a feature vector corresponding to each extracted frame. Denoted by

$$\boldsymbol{V}_i = \boldsymbol{v}_i^1 \parallel \boldsymbol{v}_i^2 \parallel \cdots \parallel \boldsymbol{v}_i^n,$$

a feature vector observed when the i-th frame is extracted where $\boldsymbol{v}_i^k$ is a 4-dimensional feature vector corresponding to the instantiation of the k-th internal wildcard in the rule pattern, and '$\parallel$' refers to vector concatenation. A feature vector of k-th internal wildcard is defined as:

$$\boldsymbol{v}_i^k = [f_{i,1}^k, f_{i,2}^k, f_{i,3}^k, f_{i,4}^k],$$

where $f_{i,1}^k$, $f_{i,2}^k$, $f_{i,3}^k$, and $f_{i,4}^k$ are the length of tokens[2], the number of spaces, the number of plain words in W_{cor}^k, and the number of plain words in W_{inc}^k observed from the text portion into which the wildcard is instantiated.

IFS-Based Document Representation. To convert a feature vector for an extraction to an IFS, we propose one method which its conceptual idea is explained as follows: Suppose $A_i = \{\langle HF_j^k, \mu_i(HF_j^k), \nu_i(HF_j^k)\rangle, \}$ is an IFS for the vector V_i, when j and k are indexes for feature types and internal wildcards, respectively. In this work, $\mu_i(HF_j^k)$ presents a confidential level to say that $f_{i,j}^k$ in the feature vector of the i-th extraction is relatively high comparing to those values of the same feature type, j, and the same wildcard, k, in the other feature vectors. In contrast, $\nu_i(HF_j^k)$ does a confidential level to say that $f_{i,j}^k$ in the i-th feature vector is not relatively high. The next example gives more details.

Example 1. Assume a considered rule has two internal wildcard and there are three extractions made by the rule. Let the feature vectors for these extractions be

$$\boldsymbol{V}_1 = [5, 2, 1, 3, 1, 1, 0, 0], \quad \boldsymbol{V}_2 = [1, 0, 1, 0, 1, 0, 1, 0], \quad \boldsymbol{V}_3 = [2, 1, 0, 1, 3, 1, 0, 1].$$

To interpret this situation, for the first extraction, the first internal wildcard instantiates into a five-token-long text portion in which two tokens are white spaces, one token is in W_{cor}^1, three tokens are in W_{inc}^1. It is worthy to note that the other token in the portion is in either $W_{cor}^1 \cap W_{inc}^1$ or $(W_{cor}^1 \cup W_{inc}^1)^c$. Since $f_{1,1}^1 > f_{3,1}^1 > f_{2,1}^1$, the confidential level to say that the first internal wildcard matches with a longer text portion for the first extraction than those for the rest extractions. Hence, $\mu_1(HF_1^1) > \mu_3(HF_1^1) > \mu_2(HF_1^1)$ and $\nu_1(HF_1^1) < \nu_3(HF_1^1) < \nu_2(HF_1^1)$. □

---

[2] A token might be a word, a white space, or a symbol.

Based on the idea discussed above, the process of transformation will be formally explained. Given the universe of discourse

$$X = \{HF_1^1, HF_2^1, HF_3^1, HF_4^1 \ldots, HF_1^n, HF_2^n, HF_3^n, HF_4^n\}.$$

Every value $f_{i,j}^k$ in the vector-based representation of the i-th extraction is then converted in terms of the three degrees of HF_j^k as the following steps:

(b1) $f_{i,j}^k$ is normalized by:

$$z_{i,j}^k = \frac{f_{i,j}^k - \overline{X}_j^k}{s_j^k}, \tag{6}$$

where $\overline{X}_j^k$ and s_j^k are the mean and the standard deviation, respectively, of the feature type j for the wildcard k over extractions. More precisely, if E is the set of extractions made by the r,

$$\overline{X}_j^k = \frac{\sum_{i=1}^{|E|} f_{i,j}^k}{|E|}, \tag{7}$$

and

$$s_j^k = \left(\frac{\sum_{i=1}^{|E|} (f_{i,j}^k - \overline{X}_j^k)^2}{|E|}\right)^{1/2}. \tag{8}$$

(b2) Denoted by $\mu_i(HF_j^k)$, a membership degree of HF_j^k with respect to the extraction i and the wildcard k is determined by a weighted sigmoid function:

$$\mu_i(HF_j^k) = r_j^k \frac{1}{1 + e^{-z_{i,j}^k}}, \tag{9}$$

where $0 < r_j^k \leq 1$ is a weight for HF_j.

(b3) Denoted by $\mu_i(HF_j^k)$, a non-membership degree of HF_j^k with respect to the extraction i and the wildcard k is determined by a weighted sigmoid function:

$$\nu_i(HF_j^k) = \bar{r}_j^k \frac{1}{1 + e^{-z_{i,j}^k}}, \tag{10}$$

where $0 < \bar{r}_j^k \leq 1$ is a weight for HF_j.

(b4) Denoted by $\pi_i(HF_j^k)$, the hesitancy degree of the document i with respect to HF_j^k is calculated by (5), i.e.,

$$\pi_i(HF_j^k) = 1 - \mu_i(HF_j^k) - \nu_i(HF_j^k).$$

Example 2. Assume a considered rule has two internal wildcard and there are only three extractions made, i.e., V_1, V_2, and V_3 as shown in Example 1. For convenience, the extractions are gathered and represented in terms of the matrix as below:

$$E = \begin{bmatrix} 5\,2\,1\,3\,1\,1\,0\,0 \\ 1\,0\,1\,0\,1\,0\,1\,0 \\ 2\,1\,0\,1\,3\,1\,0\,1 \end{bmatrix}.$$

Next, we compute the mean and the standard deviation for each feature type of each wildcard, then the results are presented as the row matrices M and SD:

$$M = \begin{bmatrix} 2.67 \ 1.00 \ 0.67 \ 1.33 \ 1.33 \ 0.67 \ 0.33 \ 0.33 \end{bmatrix},$$
$$SD = \begin{bmatrix} 2.08 \ 1.00 \ 0.58 \ 1.53 \ 1.53 \ 0.58 \ 0.58 \ 0.58 \end{bmatrix}.$$

More precisely, each entry of M and SD is obtained by columnwise computation of E, e.g. the first entry of M is the average of the first column of E. By the step (b1), we have the matrix Z containing the normalizing values:

$$Z = \begin{bmatrix} 1.12 & 1.00 & 0.58 & 1.09 & -0.58 & 0.58 & -0.58 & -0.58 \\ -0.80 & -1.00 & 0.58 & -0.87 & -0.58 & -1.15 & 1.15 & -0.58 \\ -0.32 & 0.00 & -1.15 & -0.22 & 1.15 & 0.58 & -0.58 & 1.15 \end{bmatrix}.$$

Suppose that the weights r_j^k and $\bar{r}_j^k$ are equal to 0.8 and 0.9, respectively, after applying (b2) and (b3), we have the membership and non-membership degrees which are represented as the following two matrices, respectively:

$$D_\mu = \begin{bmatrix} 0.60 \ 0.58 \ 0.51 \ 0.60 \ 0.29 \ 0.51 \ 0.29 \ 0.29 \\ 0.25 \ 0.22 \ 0.51 \ 0.24 \ 0.29 \ 0.19 \ 0.61 \ 0.29 \\ 0.34 \ 0.40 \ 0.19 \ 0.36 \ 0.61 \ 0.51 \ 0.29 \ 0.61 \end{bmatrix},$$

$$D_\nu = \begin{bmatrix} 0.22 \ 0.24 \ 0.32 \ 0.23 \ 0.58 \ 0.32 \ 0.58 \ 0.58 \\ 0.62 \ 0.66 \ 0.32 \ 0.63 \ 0.58 \ 0.68 \ 0.22 \ 0.58 \\ 0.52 \ 0.45 \ 0.68 \ 0.50 \ 0.22 \ 0.32 \ 0.58 \ 0.22 \end{bmatrix}.$$

Finally, we can convert the feature vectors V_1, V_2, and V_3 to IFSs by using D_μ, and D_ν. For instance, gathering the first row of the matrices, we can form an IFS, namely IFS_1 corresponding to V_1:

$$IFS_1 = \{\langle HF_1^1, 0.60, 0.22\rangle \langle HF_2^1, 0.58, 0.24\rangle, \langle HF_3^1, 0.51, 0.32\rangle, \langle HF_4^1, 0.60, 0.23\rangle,$$
$$\langle HF_1^2, 0.29, 0.58\rangle, \langle HF_2^2, 0.51, 0.32\rangle, \langle HF_3^2, 0.29, 0.58\rangle, \langle HF_4^2, 0.29, 0.58\rangle\}.$$

□

5.2   Extraction Classification

Recalling again that E is the set of all extractions—no matter whether each of them is correct or not—when apply the rule r into the training corpus, by the pre-process, we then have IFSs for those extractions. Let us refer them as IFS_1, IFS_2, ..., $IFS_{|E|}$.

To determine whether an extraction e_t made by the rule r is correct or not, it begins with representing e_t in terms of an IFS by the same values of parameters, i.e., means, standard deviations, and weights, used in the training process. The IFS representation of e_t here is referred to as IFS_t. Like the concept of k-nearest neighbor classification, the extraction e_t is classified by assigning the label which is most frequent among the k IFSs corresponding to extractions in E nearest to IFS_t, where a distance is measured by an IFS similarity measure.

6   Experimental Results

6.1   Data Sets and Output Templates

Information Entries. We constructed the corpus by gathering medicinal and pharmaceutical web sites from 2759 URLs. The obtained data covers 474 diseases and 770 medicinal chemical substances, with approximately 6600 and 3350 information entries, respectively. Disease information entries were divided into 3 data sets, i.e., D1, D2, and D3, based on their disease groups. D1 comprises distinct information entries obtained from 5 disease groups, i.e., the circulatory system, the urology system, the reproductive system, the eye system, and the ear system; D2 from 6 groups, i.e., the skin/dermal system, the skeletal system, the endocrine system, the nervous system, parasitic diseases, and venereal diseases; D3 from 4 groups, i.e., the respiratory system, the gastrointestinal tract system, infectious diseases, and accidental diseases. The collected information entries were preprocessed using a word segmentation program, called CTTEX, developed by NECTEC, and were then partially annotated with semantic class tags using a predefined ontology lexicon. Table 3 summarizes the characteristics of the three data sets. The second column shows the number of information entries in each data set. It is followed by a column group showing the maximum number, the average number, and the minimum number of words per information entry in each data set. The last two column groups of this table characterize the three data sets in terms of the number of symptom phrases and their occurrences.

Table 3. Data set characteristics

Data set	No. of info. entries	No. of words per info. entry			No. of distinct symptom phrases		No. of symptom phrase occurrences	
		Max.	Avg.	Min.	MD1	MD2	MD1	MD2
D1	59	130	44	9	179	77	213	84
D2	56	146	45	7	136	66	160	69
D3	58	140	55	8	161	65	210	73

Symptom Phrases and Output Templates. A collected information entry typically contains several symptom phrases, which provide several kinds of symptom-related information. Two basic types of symptom phrases, referred to

Table 4. Output templates and their meanings.

Type	Output template	Meaning
MD1	{OBS O}{ATTR A}{PER T}	An abnormal characteristic A is found at an observed entity O for a time period T
MD2	{SYM S}{LOC P}{PER T}	A primitive named symptom S appears at a human-body part P for a time period T

Table 5. Target phrase information.

Type	Data set	No. of distinct target phrases	Target-phrase length			No. of target phrases per info. entry		
			Max.	Avg.	Min.	Max.	Avg.	Min.
MD1	D1	90	11	3.5	2	7	3.6	1
MD1	D2	136	11	3.4	2	11	2.9	1
MD1	D3	160	14	3.3	2	11	3.6	0
MD2	D1	80	15	4.1	2	3	1.4	0
MD2	D2	66	13	4.3	2	5	1.2	0
MD2	D3	65	13	3.8	2	5	1.3	0

as *MD1* and *MD2*, are considered in our experiments. Table 4 gives the output-template forms for the two types along with their intended meanings. The slot PER in the MD1 template is optional. One of the slots LOC and PER, but not both, may be omitted in the MD2 template. Table 5 provides some key characteristics of each template type in the data sets, e.g., the number of distinct symptom phrases, target-phrase lengths (in words), and target-phrase density.

6.2 Experimental Results

D1 was used as training set. All MD1 and MD2 symptom phrases occurring in D1 were manually tagged with desired output frames and were used for rule learning. The length of the longest symptom phrase observed when a rule yields correct extractions on the training set is taken as the window size for the rule. By applying the obtained rules to the information entries in D1 using RAW, an IFS-based representation for each extraction was constructed when r_j^k and $\bar{r}_j^k$ in Eqs. (9) and (10) were set based on statistical characteristics of the corresponding wildcard instantiation by:

$$r_j^k = \bar{r}_j^k = \left| \frac{1 - s_j^k}{1 + s_j^k} \right|,$$

where s_j^k is the standard deviation for the feature type j of the wildcard k (see Eq. (8)).

The proposed framework was evaluated on D2 and D3. Recall and precision are used as performance measures, where the former is the proportion of correct extractions to relevant symptom phrases and the latter is the proportion of correct extractions to all obtained extractions. Table 6 shows the evaluation results obtained from using RAW without any extraction filtering and RAW with the proposed filtering method using the similarity measures in Table 2, i.e., $\text{RAW} + S_d^p$, $\text{RAW} + S_m^p$, and $\text{RAW} + S_C$. In the table, 'R' and 'P' stand for recall and precision, which are given in percentage. Compared to the results obtained using RAW alone, regardless of which similarity measure is used, the IFS-based

Table 6. Evaluation results

Type	Data set	RAW		RAW $+ S_d^p$		RAW $+ S_m^p$		RAW $+ S_C$	
		R	P	R	P	R	P	R	P
MD1	D2	88.1	60.3	88.1	93.4	86.3	93.9	86.9	97.9
	D3	89.0	55.3	89.0	93.0	88.6	94.9	88.6	96.9
MD2	D2	100.0	37.5	98.6	86.1	98.6	84.4	97.1	86.4
	D3	98.6	31.9	98.6	83.2	94.5	83.6	97.3	87.6

filtering module improves precision while satisfactorily preserving recall for all template types and all test sets. Among the three measures, it is clear that S_C outperforms the others. On further analysis, we found that the precision values for MD2 are significantly lower than those for MD1 because the variety of the structures for the MD2 template is more than that for the MD1 template. There are two optional slots for MD2, but only one for MD1, see their descriptions in Sect. 6.1.

7   Conclusions and Future Works

From a set of manually collected target phrases, IE rules are created using WHISK. To apply the obtained rules to unstructured text without predetermining target-phrase boundaries, rule application using sliding windows is introduced. An IFS-based filtering technique is proposed for removal of false positives resulting from rule application across target-phrase boundaries. The experimental results show that the technique improves extraction precision while satisfactorily preserving recall. Further works include extension of the types of target phrases and empirical investigation of framework application in different data domains as well as different similarity measures.

Acknowledgement. This work has been supported by the Thailand Research Fund (TRF), under Grant No. MRG5980067.

References

1. Kushmerick, N., Weld, D.S., Doorenbos, R.: Wrapper induction for information extraction. In: Proceedings of International Joint Conferences on Artificial Intelligence, Nagoya, Japan, pp. 729–737 (1997)
2. Soderland, S.: Learning to extract text-based information from the world wide web. In: Proceedings of the 3rd International Conference on Knowledge Discovery and Data Mining, pp. 251–254 (1997)
3. Huffman, S.B.: Learning information extraction patterns from examples. In: Wermter, S., Riloff, E., Scheler, G. (eds.) IJCAI 1995. LNCS, vol. 1040, pp. 246–260. Springer, Heidelberg (1996). doi:10.1007/3-540-60925-3_51
4. Soderland, S.: Learning information extraction rules for semi-structured and free text. Mach. Learn. **34**(1–3), 233–272 (1999)

5. Nguyen, Q.L., Tikk, D., Leser, U.: Simple tricks for improving pattern-based information extraction from the biomedical literature. J. Biomed. Semant. **1**(9), 1–17 (2010)
6. Liua, Q., Gaoa, Z., Liuc, B., Zhang, Y.: Automated rule selection for opinion target extraction. Knowl.-Based Syst. **104**, 74–88 (2016)
7. Kim, E., Song, Y., Lee, C., Kim, K., Lee, G., Yi, B.-K.: Two-phase learning for biological event extraction and verification. ACM T. Asian Lang. Inf. Process. **5**(1), 61–73 (2006)
8. Intarapaiboon, P., Nantajeewarawat, E., Theeramunkong, T.: Extracting semantic frames from Thai medical-symptom unstructured text with unknown target-phrase boundaries. IEICE Trans. Inf. Syst. **E94.D**(3), 465–478 (2012)
9. Spasić, I., Sarafraz, F., Keane, J.A., Nenadic, G.: Medication information extraction with linguistic pattern matching and semantic rules. J. Am. Med. Inform. Assoc. **17**(5), 532–535 (2010)
10. Zhang, J., El-Gohary, N.: Semantic NLP-based information extraction from construction regulatory documents for automated compliance checking. J. Comput. Civ. Eng. **30**(2), 1–14 (2014)
11. Atanassov, K.: Intuitionistic fuzzy sets. Fuzzy Set Syst. **20**, 87–96 (1986)
12. Zadeh, L.A.: Fuzzy sets. Inf. Control **8**, 338–353 (1965)
13. Dengfeng, L., Chuntian, C.: New similarity measures of intuitionistic fuzzy sets and application to pattern recognition. Pattern Recogn. Lett. **23**, 221–225 (2002)
14. Liang, Z., Shi, P.: Similarity measures on intuitionistic fuzzy sets. Pattern Recogn. Lett. **24**, 2687–2693 (2003)
15. Mitchell, H.B.: On the Dengfeng-Chuntian similarity measure and its application to pattern recognition. Pattern Recogn. Lett. **24**, 3101–3104 (2003)
16. Hung, W.-L., Yang, M.-S.: Similarity measures of intuitionistic fuzzy sets based on Hausdorff distance. Pattern Recogn. Lett. **25**, 1603–1611 (2004)
17. Xu, Z.: Some similarity measures of intuitionistic fuzzy sets and their applications to multiple attribute decision making. Fuzzy Optim. Decis. Making **6**, 109–121 (2007)
18. Khatibi, V., Montazer, G.A.: Intuitionistic fuzzy set vs. fuzzy set application in medical pattern recognition. Artif. Intell. Med. **47**, 43–52 (2009)
19. Ye, J.: Cosine similarity measures for intuitionistic fuzzy sets and their applications. Math. Comput. Model. **53**, 91–97 (2011)

Shape Optimization in Product Design Using Interactive Genetic Algorithm Integrated with Multi-objective Optimization

Somlak Wannarumon Kielarova[1(✉)] and Sunisa Sansri[2]

[1] iD3 -Industrial Design, Decision and Development Research Unit,
Faculty of Engineering, Naresuan University, Phitsanulok, Thailand
somlakw@nu.ac.th
[2] Department of Industrial Engineering, Faculty of Engineering,
Naresuan University, Phitsanulok, Thailand
sunisasan.nu@gmail.com

Abstract. This paper proposes an interactive genetic algorithm (IGA) integrated with multi-objective genetic algorithm (MOGA) in development of a generative design system. IGA is used in initializing and handling single dimensionally qualitative objectives. MOGA is used in optimizing two quantitative objectives. Qualitative factors are considered as design objectives to be optimized together with quantitative criteria. The multi-objective optimization is regarded to concurrent handling of two quantitative criteria. Shape of product is modeled by parametric modeling with Rhinoceros and Grasshopper. IGA is processed using Galapagos in Grasshopper. Shape optimization of the product is processed by using MOGA in MATLAB and linked to Grasshopper. Pareto-optimal front is generated to show the optimal solutions, which is able to support designers in decision making. The perfume bottle design is used as an illustration of the proposed framework, but the framework is applicable to other design problems.

Keywords: Generative design · Interactive genetic algorithm · Multi-objective genetic algorithm · Qualitative · Multi-objective optimization · Pareto-optimal front · Bottle design

1 Introduction

At present, computer-aided art and design tools play important role in the entire design and development process. These tools are able to support artists and designers from initial conceptual ideas, through optimization of design parameters and aesthetic considerations [1–3]. Conceptual design is considered as a process in which designers generate broad and various alternatives [4], consequently, in this stage, designers usually deal with the activities such as generating and recording ideas, and deciding to continue to generate more ideas or desire to explore the possibilities of the existing ones [5]. Product design problems then cove with both qualitative and quantitative criteria, which can be considered as multi-objective optimization.

© Springer International Publishing AG 2016
C. Sombattheera et al. (Eds.): MIWAI 2016, LNAI 10053, pp. 76–86, 2016.
DOI: 10.1007/978-3-319-49397-8_7

Interactive Evolutionary Computation (IEC) becomes more important in design process that is directly related with human factors such as emotion, preference, feeling, etc. [6].

Designers desire the tools that can effectively support their activities, decision making, and allow them to easily collaborate with the tools since the conceptual design phase. Therefore, it is highly motivating and constructive to develop an interactive generative design system with optimization of multiple objectives for supporting designers during generating their ideas in the beginning of design process. The mentioned system is supposed to support designers who are not familiar with CAD systems. This paper aims to develop an interactive genetic algorithm working with multi-objective optimization, which considers qualitative and quantitative design constraints.

The paper is organized in five main sections. The related works such as interactive evolutionary computation and multi-objective optimization with genetic algorithms are described in Sect. 2. Section 3 introduces the proposed interactive genetic algorithm with multiple objectives. Section 4 provides the illustration of the case study of perfume bottle design, experimental results and discussion. Lastly, Sect. 5 provides the conclusions of the research and the future works.

2 Related Works

2.1 Interactive Evolutionary Computation

Interactive Evolutionary Computation (IEC) is an optimization technique that based on Evolutionary Computation (EC) in which fitness function replaced with human subjective evaluation [6]. Therefore, IEC system optimizes the target system to achieve the desired outputs based on the user's evaluation. Takagi [6] said that the IEC is considered as an approach that embeds human emotion, preference, intuition or named kansei into the target system. IEC is applied in various applications such as artistic image creations, product design, and engineering. In this paper, we focus on the applications of product design and industrial design.

Brintrup et al. [7] developed an interactive genetic algorithm (IGA) for designing ergonomic chairs with qualitative and quantitative fitness. They compared different IGA types in several criteria. Hu et al. [8] developed an interactive co-evolutionary CAD system used for garment pattern design. The system core is based on inspired co-evolutionary algorithm for working with human experts. Sun et al. [9] developed an interactive genetic algorithm for designing sunglass lens. The algorithm is able to work with large population using semi-supervised learning. Lu et al. [10] proposed an interactive evolutionary design to create marble-like textile patterns. Dou et al. [11] proposed a multi-stage interactive genetic algorithm (MS-IGA). It divides the large population of the traditional interactive genetic algorithm into many stages in relation to different functional requirements. They applied the MS-IGA to the car console conceptual design system. It is aimed to capture the knowledge of users' personalized requirements and to achieve the product design.

2.2   Multi-objective Optimization with Genetic Algorithm

The applications of EC techniques in multi-objective optimization have become popular. Genetic Algorithms (GAs) is a subset of EC. They are suitable for multiple-objective optimization with several reasons described in [12].

Shibuya *et al.* [13] integrated multi-objective optimization to interactive genetic algorithms in animation design application. Their system allows users to assign subjective ratings from each objective's point of view. Brintrup *et al.* [14] proposed a framework of multi-objective optimization with interactive genetic algorithm for ergonomic chair design problem. Zhang *et al.* [15] proposed a multi-objective genetic algorithm (MOGA) to optimize free-form shape of building in the severe cold zones of China. The proposed MOGA is to accomplish the three objectives: to maximize solar radiation gain, to maximize space efficiency, and to minimize the shape coefficient. A Pareto frontier is created to present the optimal solutions and to support designers in final decision making. It seems that these soft computing based frameworks are successfully handling qualitativeness in the design systems.

3   Interactive Genetic Algorithm with Multi-objective Optimization in Product Design: A Framework

The proposed interactive genetic algorithm copes with multiple objectives to optimize the qualitative and quantitative features in product design application.

Most real life design problems concern with multiple objectives with qualitative and quantitative features. Unlike quantitative criteria, qualitative criteria mostly involve human feelings or emotions that are very difficult to derive in terms of mathematic models. Therefore, fitness function for evaluating qualitative criteria is usually done by human-based fitness evaluation. The qualitative objective is user preference of product's shape. While the quantitative objectives are such as weight, size, volume, and other computable features.

This paper proposes the framework of IGA with multi-objective optimization for shape optimization in product design application.

Interactive Genetic Algorithm (IGA) is used in initializing and handling single dimensionally qualitative objectives.

Multi-objective Genetic Algorithm (MOGA) is used in optimizing two quantitative objectives.

After the optimization process is performed, the results are generated. A common method used to evaluate the results of MOGA in this work is Pareto optimization. It intends to achieve a balance between two objectives and to discover non-dominated solutions called the Pareto-optimal solution, which is not dominated by any other solution in the solution space. The Pareto-optimal front is formed by all non-dominated solutions [15, 16].

We construct a local Pareto-optimal front for two objective functions using genetic algorithm. The objective functions are real-valued functions with seven decision variables and their bound constraints. The Pareto front is plotted in every generation. In this work, the Pareto-optimal front is disconnected displaying two completing objectives.

Firstly, the initial population is randomly generated. The next generation is then formed using non-dominated rank and distance measure of the individuals in the current generation. Each individual is assigned the non-dominated rank using the relative fitness. For distance measure, it is used to compare individuals with equal rank. It is a measure of distance between an individual and the other individuals with the same rank.

It is very important to maintain diversity of the population for convergence to an optimal Pareto-optimal front [16]. The proposed MOGA uses a controlled elitist GA to increase the diversity of the population. To control elitism, two strategies are used: number of individuals or elite members on the Pareto-optimal front is limited and the distance measure by preferring individuals that are relatively far away on the front.

The termination criteria used in this work are the maximum number of generations is reached or the average change in the spread of the Pareto-optimal front over the stall generations.

In the proposed generative design system, designer firstly chooses the preferred product shape from a set of the generated shapes, and then the system optimizes the pre-defined volumes and predefined height of the preferred shape. The system process is illustrated in Fig. 1.

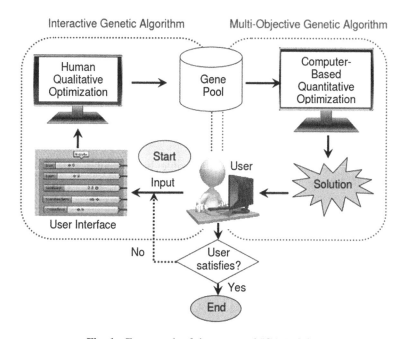

Fig. 1. Framework of the proposed IGA-MOGA

4   Illustration of the Case Study: Bottle Design

4.1   Problem Definition

The IGA and MOGA modules of the proposed generative design system framework were applied to the bottle design.

In the case study of bottle design, the qualitative fitness is user preference of shape of bottle, while the quantitative fitness is determined volume and height of the designing bottle. The goal of the system is to design the perfume bottle or to optimize shape of the perfume bottle with the commercial constraints of its volume is targeted to 100 ml. and its height is targeted to 10 cm, while to maximize subjective user evaluation obtained by IGA.

The perfume bottle is divided into three main sections as shown in Fig. 2. The first section includes the parameters that form shape of base such as base's radius and base's height. The second section forms the shape of bottle's body that includes the parameters such as radius of cross-sections in lower and upper parts, and height of body section. The third section covers the neck part, which is shaped by radius and height of bottle neck. The mentioned shape parameters are illustrated in Fig. 3.

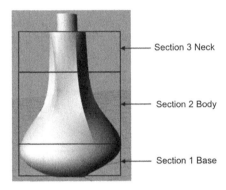

Fig. 2. Schematic outline of the designing perfume bottle.

In IGA, user can change shape of the bottle by changing shape parameters. Therefore, in IGA there are nine parameters for generating shape of perfume bottle.

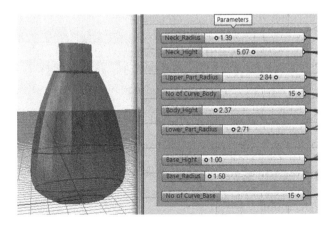

Fig. 3. Illustration of shape parameters of the designing perfume bottle.

In this work, the seven control parameters that control volume and height of the bottle are encoded in real-valued chromosome as shown in Fig. 4.

Base		Body			Neck	
Base Radius	Base Height	Lower Part: Cross-section Radius	Upper Part: Cross-section Radius	Body Height	Neck Radius	Neck Height
1.50-2.50	1.00-4.00	2.50-5.00	1.50-3.50	2.00-6.00	1.30-2.50	2.00-8.-00

Fig. 4. Illustration of chromosome for controlling volume and height of the perfume bottle

Two objective functions are derived to measure volume and height of the designed perfume bottle. Considering volume of the bottle, the shape of the bottle is combined from three pieces of the frustum of cones. Therefore, volume of base section (V_{base}) is calculated from

$$V_{base} = \frac{\pi h_{base}}{12} \left(d_{base}^2 + 2 d_{base} d_{lower\_body} + d_{lower\_body}^2 \right) \tag{1}$$

when h_{base} is height of base section, d_{base} is diameter of base and d_{lower_body} is diameter of lower part of body.

Volume of body section (V_{body}) is calculated from

$$V_{body} = \frac{\pi h_{body}}{12} \left(d_{lower\_body}^2 + 2 d_{lower\_body} d_{upper\_body} + d_{upper\_body}^2 \right) \tag{2}$$

when h_{body} is height of body section and d_{upper_body} is diameter of upper part of body.

Volume of neck section (V_{neck}) is calculated from

$$V_{neck} = \frac{\pi h_{neck}}{12} \left(d_{upper\_body}^2 + 2 d_{upper\_body} d_{neck} + d_{neck}^2 \right) \tag{3}$$

when h_{neck} is height of neck section and d_{neck} is diameter of neck part.

The volume of the bottle is calculated by combining these three sections. Therefore the volume of the bottle is computed from

$$V_{bottle} = V_{base} + V_{body} + V_{neck} \tag{4}$$

The second objective function is to measure the height of the bottle (H_{bottle}) from

$$H_{bottle} = h_{base} + h_{body} + h_{neck} + h_{cap} \tag{5}$$

The bound constraints of the seven decision variables are as follows

$$1.00 \leq h_{base} \leq 4.00$$
$$2.00 \leq h_{body} \leq 6.00$$
$$2.00 \leq h_{neck} \leq 8.00$$
$$3.00 \leq d_{base} \leq 5.00$$
$$5.00 \leq d_{lower\_body} \leq 10.00$$
$$3.00 \leq d_{upper\_body} \leq 7.00$$
$$2.60 \leq d_{neck} \leq 5.00$$

The targets of the optimization of two objective functions are obtaining the volume of the bottle 100 ml. and height of the bottle 10 cm.

4.2   Setup and Experiments

The prototype system was developed using Grasshopper in Rhinoceros 5.0 and MATLAB on a computer workstation with Intel Xeon CPU Processor 1.8 GHz Dual and 4.0 GB of RAM, working on 64 bit.

We apply Galapagos plug-in in Grasshopper [17] to build an IGA. The generative design system starts with user inputs or random generation of a set of the perfume bottles. Number of generated bottles depends on user determined. Then user chooses the preferred bottles to the gene pool. MOGA works for optimization of the shape parameters to optimize the volume and height of the bottle 100 ml. and 10 cm. respectively. The generative design system is outlined on Fig. 5.

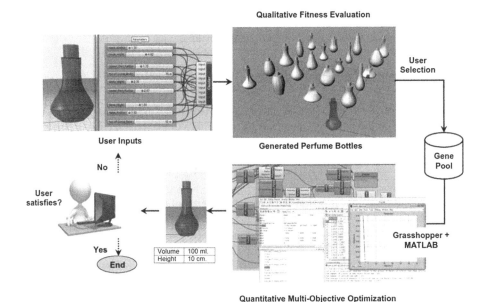

Fig. 5. Outline of the proposed generative design system based on IGA and MOGA

We use gamultiobj solver in MATLAB [18] to develop a MOGA. The experiments of parameter study are organized. The important parameters are such as population size, Pareto fraction, number of individuals on the Pareto-Optimal front, stall generation limits and function tolerance. The levels of the parameter study are listed in Table 1. The termination criteria used are maximum number of generations (1,400 generations) is reached or the average change in the spread of the Pareto-optimal front over the stall generations.

4.3 Experimental Results and Discussions

The experimental results consist of two sections. The first section involves the results of IGA development. From the initial set-up parameters, the experimental results of the proposed generative design system based on IGA show the convergence starts at the sixth generation in average to acceptable results within the practical number of generations of 43 generations. The parameter studies show that when crossover rate is increased to 0.95, the convergence starts at the fifth generation and the average number of generations of 28 generations can reduce the computational time, but increase diversity of the solutions. With 0.95 of crossover rate and 0.10 of mutation rate, the result shows that the system provides higher diversity in the convergence starts at the fifth generation within average number of generations of 60. Therefore, the increase of mutation rate causes the longer computational time. With the set of parameters 0.95 of crossover rate, 0.05 of mutation rate and double population size to be 40, the results indicate that the evolutionary patterns is different from the above cases. The convergence of the fitness is longer; therefore the creativity and diversity are increased, but the computational time is reduced since the average number of generation is 32 generations.

Table 1. List of parameters and their levels in MOGA

Parameters	Levels
Population size	100, 200, 300, 500, 1000
Pareto fraction	0.25, 0.50, 0.75
Stall generation limits	100, 200, 300
Function tolerance	10^{-3}, 10^{-6}

Therefore the suitable population size is 40, mutation rate of 0.05 and crossover rate of 0.95 to be used in the proposed generative design system based on IGA.

In the point of view of user consistency, when using IGA we cannot guarantee a consistency of user during working with the system (Fig. 6).

The second section, for the proposed MOGA, we have studied on the parameters such as population size, number of individuals on the Pareto-optimal front, stall generation limits and function tolerance. From the experiments, we found that the larger population size such as 1,000 leads to expensive computational time, but the results are not different from the smaller one. The function tolerance at 10^{-3} works well at smaller stall generation limits, while the tolerance at 10^{-6} works better at larger stall generation limits.

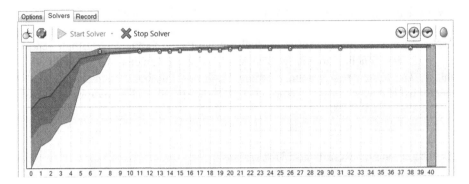

Fig. 6. Some results of IGA using Galapagos in grasshopper

From the experimental design, the best set of parameters is population size = 200, Pareto fraction = 0.50, stall generation limits = 300, function tolerance = 10^{-6}, which generates number of individuals on the Pareto-optimal front = 100. The resulting Pareto-optimal front is shown in Fig. 7.

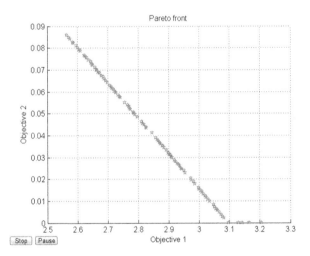

Fig. 7. Pareto-optimal front of the optimization from gamultiobj solver in MATLAB

The satisfaction of three objectives of different nature in a single framework is achieved with five users participating in the test. It is necessary to further test the proposed system with higher number of users and experts.

5   Conclusions and Future Works

This paper proposes the framework of IGA with multi-objective optimization for shape optimization in product design application. The proposed framework has been tested using a perfume bottle design. The system considers both qualitative and quantitative objectives in the single framework. The framework can be used with multi-objective optimization to make real-life experimentation. IGA is used for work with user with qualitative aspects, while MOGA is used to optimize multi-quantitative aspects. Shape of the perfume bottle is modeled by parametric modeling with Rhinoceros and Grasshopper. IGA is processed using Galapagos in Grasshopper. Shape of bottle is optimized by using MOGA in MATLAB and linked to Grasshopper. Pareto-optimal front is generated to show the optimal solutions, which is able to support designers in decision making. The proposed algorithm provides the reasonable computational time with diversity of the solutions.

The future research will cover the test of the proposed system with higher number of users and experts. The framework can be applied to other design problems. As well as hybrid algorithms will be studied to improve the system.

Acknowledgement. The research has been carried out as part of the research projects funded by National Research Council of Thailand and Naresaun University with Contract No. R2560B005. The author would like to gratefully thank all participants for their collaborations in this research.

References

1. Sequin, C.H.: Virtual prototyping of scherk-collins saddle rings. Leonardo **30**, 89–96 (1997)
2. Sequin, C.H.: CAD tools for aesthetic engineering. Comput. Aided Des. **37**, 737–750 (2005)
3. Sequin, C.H.: Computer-aided design and realization of geometrical sculptures. Comput.-Aided Des. Appl. Spec. Issue CAD Arts **4**, 671–681 (2007)
4. French, M.J.: Conceptual Design for Engineers. Springer, London (1999)
5. Kolli, R., Pasman, G.J., Hennessey, J.M.: Some considerations for designing a user environment for creative ideation. In: Proceedings of INTERFACE 1993, Human Factors and Ergonomics Society, Santa Monica, USA, pp. 72–77 (1993)
6. Takagi, H.: Interactive evolutionary computation: fusion of the capabilities of EC optimization and human evaluation. Proc. IEEE **89**, 1275–1296 (2001)
7. Brintrup, A.M., Ramsden, J., Tiwari, A.: An interactive genetic algorithm-based framework for handling qualitative criteria in design optimization. Comput. Ind. **58**, 279–291 (2007)
8. Hu, Z.-H., Ding, Y.-S., Zhang, W.-B., Yan, Q.: An interactive co-evolutionary CAD system for garment pattern design. Comput. Aided Des. **40**, 1094–1104 (2008)
9. Sun, X., Gong, D., Zhang, W.: Interactive genetic algorithms with large population and semi-supervised learning. Appl. Soft Comput. **12**, 3004–3013 (2012)
10. Lu, S., Mok, P.Y., Jin, X.: From design methodology to evolutionary design: An interactive creation of marble-like textile patterns. Eng. Appl. Artif. Intell. **32**, 124–135 (2014)
11. Dou, R., Zong, C., Nan, G.: Multi-stage interactive genetic algorithm for collaborative product customization. Knowl.-Based Syst. **92**, 43–54 (2016)
12. Deb, K.: Multi-objective optimization using evolutionary algorithms: an introduction. KanGAL Report Number 2011003 (2011)
13. Shibuya, M., Kita, H., Kobayashi, S.: Integration of multi-objective and interactive genetic algorithms and its application to animation design. In: IEEE International Conference on Systems, Man, and Cybernetics, pp. 646–651. IEEE Press, New York (1999)

14. Brintrup, A.M., Ramsden, J., Takagi, H., Tiwari, A.: Ergonomic chair design by fusing qualitative and quantitative criteria using interactive genetic algorithms. IEEE Trans. Evol. Comput. **12**, 343–354 (2008)
15. Zhang, L., Zhang, L., Wang, Y.: Shape optimization of free-form buildings based on solar radiation gain and space efficiency using a multi-objective genetic algorithm in the severe cold zones of China. Sol. Energy **132**, 38–50 (2016)
16. Deb, K., Kalyanmoy, D.: Multi-objective Optimization Using Evolutionary Algorithms. Wiley, New York (2001)
17. Grasshopper http://www.grasshopper3d.com/group/galapagos
18. MathWorks https://www.mathworks.com/help/gads/gamultiobj.html

A Hierarchical Learning Approach for Finding Multiple Vehicle Number Plate Under Cluttered Background

Kirti[✉], C. Raghavendra Rao, Rajeev Wankar, and Arun Agarwal

School of Computer and Information Sciences, University of Hyderabad,
Hyderabad, India
kirti05ai@gmail.com, {crrcs,wankarcs,aruncs}@uohyd.ernet.in

Abstract. Traffic control is one of the biggest problems faced by the surveillance department in every country. Manual surveillance supported by well-defined rules and effective equipment is a common solution to this problem. But, often traffic personnel fail to isolate and recognize the number plate and hence to penalize the owner of the vehicle who violated the traffic rules due to the speeding vehicle and mounting of number plates at arbitrary parts of the vehicle. Hence, his inability renders the traffic surveillance a challenging research area. The automatic vehicle number plate recognition system provides one of the solutions to this problem but constrained by various limitations. The most challenging aspect is to detect the number plates itself present in an image. The presence of multiple vehicle number plates and cluttered background makes our work different from earlier approaches. A single step process may not be able to detect all the number plates in an image, hence we propose Hierarchical Filtering (HF) approach which employs several transformation functions on the input image. The proposed HF models the characteristics of vehicle number plates and label them by fitting the Logistic Regression model. The proposed method is able to detect all the number plates present in an image but at the expense of some non-number plate regions in the form of a minimum bounding rectangle. The proposed model is tested on a wide range of inputs, and the results are profiled in terms of precision and recall measures.

1 Introduction

Population growth, the rise in living standard, an increase in the number of cars in a family, the cheaper rate of investment or loans on cars, etc., increases the number of vehicles and its users. Thus, managing traffic and tracking of vehicles is a troublesome task and utmost important to deal with. The image-processing technique is used to identify vehicles by their vehicle number plate. Number plate detection is considered as the most crucial stage in the number plate recognition system. For detecting different shapes under cluttered background [1], the black and white image is considered. There are many algorithms for detecting number plates, such as edge detection [2], corner detection, template matching [3],

© Springer International Publishing AG 2016
C. Sombattheera et al. (Eds.): MIWAI 2016, LNAI 10053, pp. 87–98, 2016.
DOI: 10.1007/978-3-319-49397-8_8

histogram analysis, morphological operations, threshold based techniques [4], wavelet transformation method [5], coarse-to-fine strategy [6] and color based approaches [5]. But, all these approaches are constrained by various factors. This paper proposes a method of detecting multiple number plates from cluttered scenes in various environmental conditions.

Automatic Vehicle Identification (AVI) system has a camera monitoring the roads and captures video. The study confines to segmentation process, region discovery, and region labeling in a hierarchical manner. Since the objective of this study is to detect multiple vehicle number plate from an image having cluttered background so, the domain knowledge pertains to vehicle number plate and its characteristics which are provided in Sect. 2. The captured frames i.e. image is subjected to image analysis for identifying particular rectangles which represent the number plate. The process of hierarchical filtering of rectangles is discussed in Sect. 3. Identified rectangles are then labeled using Logistic Regression (LR) modeling. This LR classifier is built based on a training & testing set which is discussed in Sect. 4. Section 5 discusses the results of experimentation on several images in traffic prone area. Observations are provided in Sect. 6, and it also concludes the study.

2 Vehicle Number Plate Characteristics

There are two types of standard vehicle number plates available, as shown in Fig. 1(a), (b), (c) and (d). The format of the standard number plate is described in Fig. 2.

(a) Single Line Black Background (b)Single Line Yellow Background

(c) Double Line Black Background (d) Double Line Yellow Background

Fig. 1. Types of number plates (Color figure online)

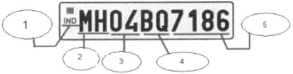

Fig. 2. Standard vehicle plate format

1. Country Code
2. State Code
3. District code
4. Type of Vehicle (Two wheeler, four wheeler, commercial etc.)
5. Actual Registration Number

According to Central Motor Vehicle rules, 1989, and Central Motor Vehicles (Amendment) Rules,1993, various standards for size of number plate are as listed below-

– For two and three wheelers: Front: 285×45 mm, Rear: 200×100 mm.
– For light motor vehicles/passenger car: 340×200 mm or 500×120 mm.
– For medium/heavy commercial vehicles and trailer/combination: 340×200 mm
– For agricultural tractors: Front: 285×45 mm, Rear: 200×100 mm.

The size of letters and numerals of the number plate have also been standardized. The dimension of letters and figures of the registration mark and the space between different letters and numerals and edges of the plain surface shall not be less than indicated dimensions as shown in Table 1 (Substituted by G.S.R. 338(E), dated 26-3-1993 (w.e.f. 26-3-1993)):

Table 1. Size of letters and numerals for the standard number plate

S.No	Class of vehicles	Letters \ numerals	Height (mm)	Thickness (mm)	Space between (mm)
1	All motor cycles and three wheeled invalid carriages	Rear-letters	35	7	5
2	All motor cycles and three wheeled invalid carriages	Rear-numerals	40	7	5
3	Motor cycles with engine capacity less then 70cc	Front letters & numerals	15	2.5	2.5
4	Other motor cycles	Front letters & numerals	30	5	5
5	Three wheeler of engine capacity not exceeding 500cc	Rear & front letters & numerals	35	7	5
6	Three wheeler of engine capacity exceeding 500cc	Rear & front letters & numerals	40	7	5
7	All other motor vehicles	Rear & front letters & numerals	65	10	10

Several variations such as environmental effects, illumination change, blurring, and reflection, make single feature unable to detect multiple number plate in an image. We can get high success rate for number plate detection by considering multiple characteristic features of the number plate. These features of number plate can be given as:

1. Aspect ratio, i.e. width to height ratio which is ranges from [1, 3.5].
2. Black to white pixel density and black to yellow pixel density.
3. Location of a number plate should not be more than 1 m above the ground.
4. Area of standard vehicle number plate.
5. Number of connected objects in a number plate.

3   Image Analysis and Filtering

The suggested method is a hierarchical solution containing three main stages, (1) extraction of the possible region of interest i.e. number plate by various filtering techniques, (2) collection of statistical information of extracted components, and (3) fitting of the logistic regression model for labeling of extracting candidate components. The proposed technique can also detect multiple number plates with different orientations and sizes (front and rear), present in an image. Figure 3 shows the proposed architecture of hierarchical filtering approach.

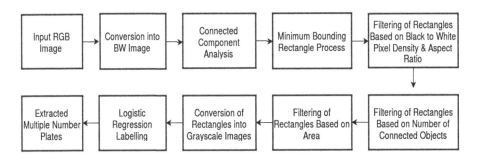

Fig. 3. Proposed architecture

3.1   Filtering Based on Characteristic Properties of a Vehicle Number Plate

The proposed method used characteristic features of a vehicle number plate, as discussed in Sect. 2 for detecting multiple vehicle number plate in an image. We performed filtering based on different characteristics of a number plate such as black to white pixel density, aspect ratio, the number of connected objects and area.

Initially, Connected Component Analysis (CCA) is performed on input images. CCA is a flood fill algorithm to label all the pixels of the image into connected components. Then Minimum Bounding Rectangle (MBR) is applied on each connected component of the image. This process is able to form rectangles on each connected components which further can be filtered, based on the presence of number plate or non-number plate. The output of an input image after performing CCA and applying MBR is shown in Fig. 4.

(a) Input RGB Image   (b)Converted BW Image   (c)Processed MBR Image

Fig. 4. Input image (Color figure online)

Filtering techniques are applied after getting MBRs. The process is described below:

Filtering Based on Black to White Pixel Density and Aspect Ratio.
The range is given for applying filtering based on black to white pixel density (bw_den), and aspect ratio (asp_ratio) is bw_den $>= 0.25$ and asp_ratio is [1, 3.5]. The minimum rectangles are stored in variable min_rect. 'N' is the size of the total number of input images. The algorithm for applying filtering based on black to white pixel density and aspect ratio [7] is given by Algorithm 1.

Algorithm 1. Algorithm for Applying Filtering Based on Black to White Pixel Density & Aspect Ratio

---

Require: Input RGB Image I
Ensure: Extracted rectangles P
  for $i = 1 : N$ **do**
    Convert I into BW image
    Do CCA
    For each component, apply MBR
    if $bw\_density(min\_rect) >= 0.25$ &&$1 <= asp\_ratio <= 3.5$ **then**
      P=Extract min_rect
    else
      delete min_rect
    end
  end

Further filtering is required as some charts or stickers containing alphabets, adhere on vehicles or signboards, in the images that taken at traffic signals makes the above range of values fit for them. So, some non-number plate rectangles are still present in extracted rectangles from filtering. We put number of connected objects in extracted rectangle from the previous filtering, as our next filter.

Filtering Based on Number of Connected Objects. Since every number plate possess a format given in Fig. 2. By this format, every number plate should have ten connected components as they have total ten alphanumeric characters. The algorithm for applying this filtering technique is as follows:

Algorithm 2. Algorithm for Applying Filtering Based on Number of Connected Objects

---

Require: P (Extracted rectangles from Algorithm 1), P=1,2,3,...m
Ensure: Extracted Rectangles R
  for $i = 1 : P$ **do**
    Convert P into BW image
    Do CCA
    if $6 <= Num\_objects(P) <= 25$ **then**
      R=Extract min_rect
    else
      delete min_rect
    end
  end

But this number varies due to various reasons such as: (1) blurring, (2) varying distance of the image from a camera, (3) more than one character can be taken as a single component when the distance between vehicle and camera increases, (4) auxiliary characters are written on the number plate and, (5) orientation of number plate. Still, some extracted rectangles containing non-number plate lies within this range. So further analysis is required.

Filtering Based on Area. Further, the extracted rectangles from Algorithm 2 are filtered by area of number plate, as standard number plate has predefined area as discussed above. Area based filtering technique is described by Algorithm 3.

Algorithm 3. Algorithm for Applying Filtering Based on Area

Require: R (Extracted rectangles from Algorithm 2), R=1,2,3,...l
Ensure: Extracted Rectangles S
  for $i = 1 : R$ **do**
    Convert R into BW image
    if $600 <= Area(R) <= 10000$ **then**
      S=Extract min_rect
    else
      delete min_rect
    end
  end

This filter drastically reduces the number of extracted rectangles containing the non-number plates. After using this filter, some non-number plate rectangles remain that fit into the given range of area due to different orientations, noise, the variation of the distance between vehicle number plate and camera, and blurring in the image. For further refining our search result of detecting vehicle plate, LR based predictor is used. This LR model predicts the candidate regions from the extracted rectangular components that remain after above filtering techniques.

4   LR Modeling

LR is one of the popular statistical regression models and is often used as supervised Machine Learning approach for classification. Mathematically, LR can be given by Eq. 1:

$$E(Y \mid X) = \frac{1}{1 + exp(-G(X))} \tag{1}$$

where Y is the dichotomous dependent variable and X is a vector of independent variables. We are given training data of size N, $[X_i Y_i]$ for i = 1, 2, 3, ... N. $Y_i's$ will be 0 or 1 while $X_i's$ are p-dimensional real vector.

One can build the model by considering different structures for $G(X)$. The following are the two instances of $G(X)$ by considering p = 3, which is known as linear and quadratic and are given by Eqs. 2 and 3 respectively.

$$G(X) = a + b_1 x_1 + b_1 x_1 + b_2 x_2 + b_3 x_3 \tag{2}$$
$$G(X) = a + b_1 x_1 + b_2 x_2 + b_3 x_3 c_1 x_1^2 + c_2 x_2^2 + c_3 x_3^3 + c_4 x_1 x_2 + c_5 x_1 x_3 + c_6 x_2 x_3 \tag{3}$$

Here a, b and c are the parameter of G(X). The values of parameter a and $b_i's$ in the case of linear and a, $b_i's$ and $c_i's$ in case of quadratic are estimated using the training dataset.

4.1   Dataset Creation

A training dataset is created for providing training to LR model, and then testing dataset is created for labeling of extracted rectangles into a number plate or a non-number plate. We collected statistical properties of extracted rectangles that include mean, standard deviation and mixed proportions.

Calculation of Mixed Proportions, Mean and Standard Deviation. All input images, each of size (m×n) are compiled as input data and is used to train Expectation Maximization (EM) algorithm [8]. It is an unsupervised learning approach and of maximal likelihood nature, whereas LR is supervised learning and uses the least square method. It is applied to grayscale images. There are two steps in EM algorithm i.e. Estimation step and Maximising Likelihood step. These steps are performed iteratively for finding values of statistical properties. EM algorithm use empirical distribution of the histogram of an image (number of peaks in image's histogram) for deciding the number of iteration of EM algorithm. Here mean and sigma is distribution parameter and mixed proportions are the mixed parameter and represented by mu, sigma, and alpha respectively. After knowing the value of mixed proportions, we can analyze following property of an image:

1. with alpha factor = 0.5; the image is mixed equally in foreground and background pixels.
2. with alpha factor < 0.5; the background pixels are contributing more to the image.
3. with factor > 0.5; the foreground pixels are contributing more to the image.

Statistical parameter values of standard number plate, as shown in Fig. 5 are, mixed proportions is 0.5952, mean = 217, and standard deviation = 98.1688.

(a) RGB Color Image   (b) GrayScale Image

Fig. 5. Standard number plate ("Image Source: http://www.plateshack.com/y2k/India/indiapl8.jpg)" (Color figure online)

The steps taken for creating training and testing dataset are given by Algorithms 4 and 5, and generated sample datasets are given in Tables 2 and 3 respectively, where D_mu, D_alpha, and D_sigma are euclidean distance between standard number plate statistical properties values and the extracted rectangles statistical properties values.

Algorithm 4. Steps for Training Dataset Creation

Require: Extracted Rectangles S, where S=1,2,....,g
Ensure: Training Dataset.
  for $i = 1 : S$ do
    Calculate mu, alpha, sigma;
    D_mu=std_mu - mu;
    D_alpha=std_alpha - alpha;
    D_sigma=std_sigma - sigma;
    if $Ex\_rect == Num\_plate$ then
        Set Flag=1;
    else
        Set Flag=0;
    end
  end

Fitting of LR Model. After creating the training and testing dataset, we fit LR model on the extracted set of rectangles from area filtering. This model label the rectangles into number plate or non number plate based on the training set provided to it. The result of the labeled rectangles after fitting logistic regression model on provided training set is shown in Table 4.

Table 2. Sample training dataset of extracted rectangles

D_alpha	D_MU	D_sigma	FLAG	Extracted Rectangle
0.7965012705	7744	357555.013428493	0	
0.8053971877	3136	94476.2525981768	1	
0.8053249689	3844	330665.613725272	0	
0.7955967311	4489	1204.5077071723	0	
0.8159570339	9409	322540.9387098	0	
0.7771706676	3844	59.9804198516	1	
0.7888001965	400	211834.530954226	1	

Algorithm 5. Steps for Testing Dataset Creation

Require: Extracted Rectangles S for labelling, where S=1,2,....h
Ensure: Testing Dataset
  for $i = 1 : S$ **do**
    Calculate mu, alpha, sigma;
    D_mu=std_mu - mu;
    D_alpha=std_alpha - alpha;
    D_sigma=std_sigma - sigma;
  end

Table 3. Sample testing dataset of extracted rectangles

D_alpha	D_mu	D_sigma
0.8053249689	3844	330665.613725272
0.885320435	2601	5436160.96920542
0.7771706676	3844	59.9804198516
0.7955967311	4489	1204.5077071723
0.7987699866	3844	1223.5068472127
0.7911521241	1764	163602.492412574

Table 4. Results of labeling using logistic regression model

D_alpha	D_mu	D_sigma	Flag	Expected Flag
0.8053249689	3844	330665.613725272	1	1
0.885320435	2601	5436160.96920542	0	0
0.7771706676	3844	59.9804198516	1	1
0.7955967311	4489	1204.5077071723	0	1
0.7987699866	3844	1223.5068472127	0	0
0.7911521241	1764	163602.492412574	1	1

The overall algorithm for the proposed approach can be given by Algorithm 6. The experimental results after applying this algorithm are shown in next section.

Algorithm 6. Algorithm for Proposed Method

Require: Input RGB Image I, I=1,2,3...N
Ensure: Extracted Rectangles
  for $i = 1 : N$ **do**
    Convert i into BW image
    Do CCA
    DO MBR for each connected component
    Apply Algorithm 1
    Apply Algorithm 2
    Apply Algorithm 3
    Apply Algorithm 4
    Apply Algorithm 5
    Fit LR Model
  end

---

5   Experimental Result and Analysis

Several images were taken from immediate traffic area and were analyzed thoroughly. The proposed method is applied on 50 images and found that all the number plates present in the images with cluttered background are detected in the extracted rectangles with the expense of few non-number plates which later can be identified by doing the structural analysis on images. Figure 6 presents the process flow of hierarchical learning approach and a sample of resulting images at every stage. Figure 6, shows that proposed method is able to detect all the three number plates, which are present in the given input image with some non-number plate regions.

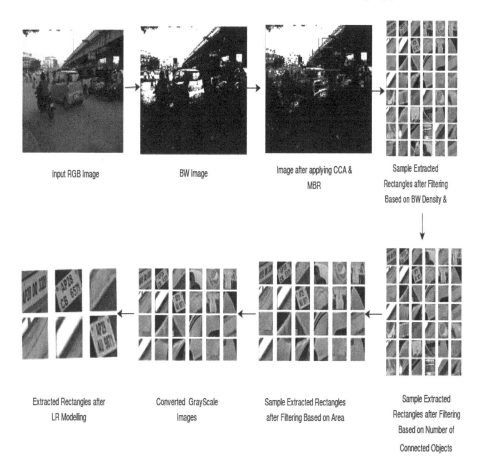

Input RGB Image   BW Image   Image after applying CCA &
MBR   Sample Extracted
Rectangles after Filtering
Based on BW Density &

Extracted Rectangles after
LR Modeling Converted GrayScale
Images Sample Extracted Rectangles
after Filtering Based on Area Sample Extracted
Rectangles after Filtering
Based on Number of
Connected Objects

Fig. 6. Results of hierarchical filtering (Color figure online)

6 Observations and Conclusions

The proposed approach follows hierarchical method to solve the problem of detection of multiple number plates as single process may not be able to achieve this purpose.

One should pay particular attention while creating training dataset for modeling of logistic regression. Training dataset should be a good representative set so that it can properly label the testing dataset.

The experimental results show that our approach has a significant effect in application. So, no matter how the environment change, all the vehicle number plates present in an image are always detected. The calculated average precision of the proposed method for 50 input images is 36.5 % and the average recall is 100 %. The reason of moderate precision in some cases can be broadly classified into three major categories.

- The distance of camera and moving vehicle is more so that taken image is not clear enough.
- Fast moving vehicles.
- Multiple sizes & shape of number plate within one image.

References

1. Gao, D.-S., Zhou, J.: Car license plates detection from complex scene. In: 5th International Conference on Signal Processing Proceedings, 2000. WCCC-ICSP 2000, vol. 2, pp. 1409–1414. IEEE (2000)
2. Lalimi, M.A., Ghofrani, S., McLernon, D.: A vehicle license plate detection method using region and edge based methods. Comput. Electr. Eng. **39**(3), 834–845 (2013)
3. Chen, S., Cremers, D., Radke, R.J.: Image segmentation with one shape prior-a template-based formulation. Image Vis. Comput. **30**(12), 1032–1042 (2012)
4. Chang, S.-L., Chen, L.-S., Chung, Y.-C., Chen, S.-W.: Automatic license plate recognition. IEEE Trans. Intell. Transp. Syst. **5**(1), 42–53 (2004)
5. Sengur, A., Guo, Y.: Color texture image segmentation based on neutrosophic set and wavelet transformation. Comput. Vis. Image Underst. **115**(8), 1134–1144 (2011)
6. Amit, Y., Geman, D., Fan, X.: A coarse-to-fine strategy for multiclass shape detection. IEEE Trans. Pattern Anal. Mach. Intell. **26**(12), 1606–1621 (2004)
7. Naito, T., Tsukada, T., Yamada, K., Kozuka, K., Yamamoto, S.: Robust license-plate recognition method for passing vehicles under outside environment. IEEE Trans. Veh. Technol. **49**(6), 2309–2319 (2000)
8. Chillarige, A., Kamakshi Prasad, V.: Gray image segmentation using expectation maximization approach. In: ICRAMAV - 2012: International Conference on Recent Advances in Design, Development and Operation of Micro Air Vehicles, pp. 359–371. BS Publications (2012)

Finding Risk-Averse Shortest Path with Time-Dependent Stochastic Costs

Dajian Li[1], Paul Weng[2,3($\boxtimes$)], and Orkun Karabasoglu[2,3]

[1] Carnegie Mellon University, Pittsburgh, USA
`dajianl@andrew.cmu.edu`
[2] School of Electronics and Information Technology,
SYSU-CMU Joint Institute of Engineering, Sun Yat-sen University,
Guangzhou, China
`{paweng,karabasoglu}@cmu.edu`
[3] SYSU-CMU Shunde Joint Research Institute, Shunde, China
`http://weng.fr`
`http://karabasoglu.com`

Abstract. In this paper, we tackle the problem of risk-averse route planning in a transportation network with time-dependent and stochastic costs. To solve this problem, we propose an adaptation of the A* algorithm that accommodates any risk measure or decision criterion that is monotonic with first-order stochastic dominance. We also present a case study of our algorithm on the Manhattan, NYC, transportation network.

Keywords: Route planning · Shortest path · Risk-averse decision-making · Conditional Value-at-Risk · Time-dependent stochastic costs

1 Introduction

Shortest path problems have been extensively studied as they are canonical problems that appear in many domains, for instance transportation [2,7], artificial intelligence [28] or circuit design [25] to cite a few. The standard version of this problem can easily be solved with classic shortest path algorithms such as the Dijkstra algorithm [8] or the A* algorithm [16].

In this paper, we focus more particularly on route planning in transportation networks. While classically route planning operates with deterministic information (e.g., expected travel duration), with the advent of intelligent transportation systems that provide real-time and historical traffic data, it becomes possible to design route planning approaches that take into account the stochastic and time-dependent nature of traffic condition. Indeed, as more and more cities open the access to historical traffic data, it is now possible to estimate a probability distribution over durations for each street at different times of the day. Such information can then serve as input to determine "shortest" paths that takes into account the variability of durations.

More specifically, in this paper we focus on building a risk-averse route planning system for drivers in networks with stochastic and time-dependent costs.

© Springer International Publishing AG 2016
C. Sombattheera et al. (Eds.): MIWAI 2016, LNAI 10053, pp. 99–111, 2016.
DOI: 10.1007/978-3-319-49397-8_9

For a given origin and destination positions, it determines a shortest risk-averse path with respect to a pre-specified risk measure or decision criterion. With this system, a driver could not only plan their trip in advance, but also avoid possible congestions. Consequently, this system could help reduce in particular travel time, traffic congestion and as a consequence exhaust emissions.

The contributions of this paper are twofold. First, we propose an adaption of the A* algorithm, which extends and unifies previous algorithms [6,24] for computing a risk-averse shortest path in transportation networks where costs are stochastic and time-dependent. Our approach can accommodate any risk measure or decision criterion that is monotonic with respect to first-order stochastic dominance. Second, we demonstrate our proposition in the Manhattan, NYC transportation network with Conditional Value-at-Risk as a risk measure.

The paper is structured as follows. The next section discusses the related work. Section 3 recalls the standard shortest path problem and the A* algorithm. Section 4 defines the time-dependent stochastic-cost shortest path problem tackled in this paper. Section 5 presents an adapted version of the A* algorithm to solve our problem. Section 6 demonstrates our solution algorithm to route planning in Manhattan, NYC. Finally, we conclude in Sect. 7.

2   Related Work

Over the past decades, much effort has been devoted to the solution of the shortest path problem and its many variants.

Classic shortest-path algorithms such as the Bellman-Ford algorithm [4,11, 19], the Dijkstra algorithm [8] or the A* algorithm [16] have been proposed before 1970s to solve the static version of the problem where edge costs are scalar and constant. However, in route planning, drivers usually value more travel times than distances, which has several implications. Edge costs are generally nonstationary, that is they are a function of time (e.g., driving the same street during peak hours or during normal hours lead to different durations). They also tend to be random, depending on traffic conditions and other drivers. For these reasons, those classic algorithms for static shortest-path problems need to be adapted to this more general setting.

On the one hand, many studies have considered the non-stationary case, i.e., time-dependent shortest path problem (TDSPP). Dreyfus [9] extended the Dijkstra algorithm to TDSPP and Goldberg and Harrelson [15] solved TDSPP with a variant of the A* algorithm. TDSPP has been proven to be solvable in polynomial time under the First In First Out (FIFO) property (i.e., which forbids an earlier arrival time while traversing an edge at a later time) [18] while it reveals to be an NP-hard problem without the FIFO property [23].

On the other hand, since Frank [12] studied stochastic-cost shortest path problem (SSPP), extensive work has been done on this problem (e.g., [14,21, 24,31]). Bertsekas and Tsitsiklis [5] considered an even more general class of stochastic shortest path problems (where node transitions are stochastic) and modeled them as a Markov Decision Problem [26]. In SSPP, some researchers, such as Nie and Wu [21], aimed at determining a shortest path guaranteeing a given probability of arriving on time. Recently, Gavriel et al. [14] and Parmentier and Meunier [24] investigated risk-averse versions of SSPP by considering different risk measures, such as conditional Value-at-Risk (CVaR) [10]. However, neither of them considered the case where costs are time-dependent.

Besides, some work also tackles problems where edge costs are both non-stationary and random. Fu and Rilett [13] considered the problem of expected shortest paths in dynamic and stochastic traffic networks. Chen et al. [6] studied time-dependent stochastic shortest path problems and proposed an adapted A* algorithm with first-order stochastic dominance in order to compute a reliable shortest path. However, the risk measure they use is Value-at-Risk (VaR) [17], which may not always be the most suitable measure. In this paper, we propose a practical algorithm for any risk measure that is monotonic with respect to first-order stochastic dominance (as in [24]) and test it with CVaR, which may be considered a better criterion than VaR as it has better properties [1] and takes into account not only VaR but also the tail distribution.

3   Background

We first recall the definition of the classic shortest path problem. Let $G = (V, E)$ be a directed graph (e.g., corresponding to a transportation network) where V is a set of nodes (e.g., intersections and landmarks in a city) and $E \subset V^2$ is a set of directed edges (e.g., lanes of streets). The set of successors of a node n is denoted $E^+(n)$, i.e., $E^+(n) = \{n' \in V \mid (n, n') \in E\}$. A path π of length k in G is a sequence of k edges in E: $(n_1, n_2), (n_2, n_3), \ldots, (n_k, n_{k+1})$. For convenience, we write $\pi = (n_1, n_2, \ldots, n_{k+1})$. A subpath of a path is a consecutive subsequence of edges of that path.

The edges of graph G are assumed to be valued by a cost function $c : E \to \mathbb{R}$ (e.g., representing the distance or duration of travel in an edge). We assume that costs are non-negative. By extension, the cost of a (sub)path π, denoted $c(\pi)$, is defined as the sum of the costs of the edges in that (sub)path. Let π_{on} be a subpath from node o to node n and π_{nd} a subpath from node n to node d. We denote $\pi_{on} \oplus \pi_{nd}$ the path obtained from the concatenation of the two subpaths. Obviously, $c(\pi_{on} \oplus \pi_{nd}) = c(\pi_{on}) + c(\pi_{nd})$.

Let $o \in V$ (resp. $d \in V$) be an origin (resp. destination) node. The shortest path problem consists in searching for the path starting from node o and ending in node d that has the lowest cost. Many efficient algorithms, such as the Ford-Bellman algorithm [4,11] or the Dijkstra algorithm [8], have been proposed to solve this problem. In the case of transportation networks where the number of nodes may be large, those algorithms, even though polynomial in the size of graph G may become impractical. In that case, the A* algorithm may help to determine a shortest path faster.

The A* algorithm, proposed by Hart et al. [16], has been widely accepted as an efficient algorithm to solve the shortest path problem. As it is well-known, we only recall its principle and not its pseudo-code for space reasons. In this algorithm, an extra heuristic information is assumed to be given: for any node n, an estimation $h(n)$ of the cost of the shortest path from node n to destination node d is available. For instance, in transportation networks, where path distances are minimized, $h(n)$ can be defined as the Euclidean distance from node n to destination node d. The A* algorithm finds a path from origin node o to destination node d by exploring a tree of (sub)paths following a best-first-search strategy. In order to choose the best subpath to extend, the A* algorithm usually maintains a priority queue $\mathcal{O}$ of nodes representing subpaths ending in those nodes. The priority $f(n)$ of a subpath π_{on} ending in a node n is defined as the sum of the cost cumulated so far and the heuristic estimation, i.e., $f(n) = g(n) + h(n)$ where $f(n)$ represents an estimation of the cost of of a path to node d whose subpath is π_{on}, $g(n)$ is the cost of π_{on} and $h(n)$ is the heuristic estimation of the cost of a subpath from node n to node d.

Heuristic function $h(n)$ plays a significant role in the A* algorithm by influencing the number of (sub)paths A* algorithm will examine. Besides, whether the A* algorithm can eventually find the shortest path in the graph depends on the selection of the heuristic function $h(n)$. In order to guarantee the soundness of the A* algorithm, $h(n)$ should satisfy the following inequality: $\forall n \in V$,

$$h(n) \leq \min_{\pi_{nd}} c(\pi_{nd}) \tag{1}$$

where π_{nd} represents a subpath from node n to node d. This property means that the heuristic information provided by $h(n)$ is a lower bound to the best possible cost to reach node d from node n. For instance, the heuristic function defined as the Euclidean distance is admissible. A heuristic function that satisfies inequality (1) is called an *admissible* heuristic function.

4   Problem Statement

We start with some notations. For any random variable X, we denote P_X its probability dension function (pdf), F_X its cumulative distribution (i.e., $F_X(c) = \int_{-\infty}^{c} P_X(x)dx$) and F_X^{-1} the (pseudo)inverse of F_X (i.e., $F_X^{-1}(\alpha) = \inf\{c \in \mathbb{R} \mid F_X(c) \geq \alpha\}$).

In a real transportation network, the duration for traversing an edge (i.e., portion of a street) is stochastic and dynamic. Such a network can be represented as a directed graph $G = (V, E)$ as before, however, edge costs are now time-dependent real random variables. For an edge (n, n'), random variable $C_t(n, n')$ denotes the random cost of traversing that edge at time t. We assume random costs take non-negative values (representing durations) and S-FIFO[1] (Stochastic FIFO) [21], which is a natural property in transportation networks, holds.

---

[1] The SFIFO property states that for any confidence level α, leaving later cannot lead to an earlier arrival time: $t \leq t' \implies t + F^{-1}{}_{C_t}(\alpha) \leq t' + F^{-1}{}_{C_{t'}}(\alpha)$ where t, t' are departure times, $C_t, C_{t'}$ random costs of an edge and $\alpha \in [0, 1]$.

For a path $\pi = (n_1, n_2, \ldots, n_{k+1})$, its cost $C_t(\pi)$ for a departure time t is also a random variable defined as the sum of the random costs of its edges. It can be written recursively as follows:

$$C_t(\pi) = C_t(\pi') + C_{t+C_t(\pi')}(n_k, n_{k+1}) \tag{2}$$

where $\pi' = (n_1, n_2, \ldots, n_k)$. In a similar fashion, the pdf of $C_t(\pi)$ can be written:

$$P_{C_t(\pi)}(c) = \int_{-\infty}^{+\infty} P_{C_t(\pi')}(x) P_{C_{t+x}(n_k, n_{k+1})}(c - x) dx \tag{3}$$

The problem we tackle in this paper can then be formulated: given a risk-averse criterion or risk measure $\rho : \mathcal{X} \to \mathbb{R}$ (with $\mathcal{X}$ the set of real random variables), we search for the ρ-minimum path π^* for a departure time t, i.e.,

$$\rho(C_t(\pi^*)) = \min_\pi \rho(C_t(\pi)) \tag{4}$$

We call π^* a risk-averse shortest path. We assume that criterion ρ satisfies a consistency property that relates ρ to the first-order stochastic dominance, which is a partial order defined over probability distributions [30].

Definition 1. First-order stochastic dominance (FSD) is defined as follows: Let F_1, F_2 be two cumulative distributions, F_1 (weakly) first-order-stochastically dominates (or FSD-dominates) F_2, denoted $F_1 \succsim_{FSD} F_2$, iff $\forall x, F_1(x) \leq F_2(x)$. An illustration of FSD is shown in Fig. 1.

The consistency property that we assume states that ρ is monotonic with respect to first-order stochastic dominance:

FSD $F_X \succsim_{FSD} F_Y \Rightarrow \rho(X) \geq \rho(Y)$

where X and Y are two real random variables and F_X and F_Y are their respective cumulative distributions. This property is important because it will allow us to prune in the adapted A* algorithm.

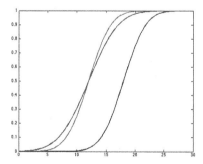

Fig. 1. Illustration of first-order stochastic dominance: the green cumulative distributions FSD-dominates the blue and red ones, while the latter two are incomparable.

Let us introduce another property that states that ρ is increasing with the addition of a non-negative random variable:

INC $\rho(X) \le \rho(X + C)$

where X and C are two real random variables and C takes non-negative values. In our setting, this is a natural property as random variables represent durations.

To prove that FSD implies INC, we first introduce a lemma[2]:

Lemma 1. *Let X be a real random variable and C be a non-negative real random variable. Then, $F_{X+C} \succsim_{FSD} F_X$.*

Then, as a direct consequence of Lemma 1, we obtain:

Proposition 1. *If ρ satisfies FSD, ρ also satisfies INC.*

As illustrations of ρ, we present three examples, Value-at-Risk, Conditional Value-at-Risk and Expected Utility, which all satisfy FSD (and therefore INC).

Example 1. Value-at-Risk (VaR) [17] is a widely-used risk measure in finance. For a fixed $\alpha \in [0, 1]$, it represents the threshold loss value, such that the probability the loss on an investment exceeds this value is α. Formally, in our context, it is defined by: $VaR_\alpha(X) = F_X^{-1}(\alpha) = \inf\{x \in \mathbb{R} \mid F_X(x) \ge \alpha\}$ In other terms, $VaR_\alpha(X)$ is defined in our context as the α-quantile of random variable X. It is well-known that VaR satisfies FSD [3].

Example 2. Conditional Value-at-Risk (CVaR) [10], also called *Expected Shortfall* is a risk measure that refines VaR. Because VaR is a threshold value (for a single fixed probability α), it neglects the risk at the tail of the distribution. CVaR remedies this shortcoming of VaR by measuring the expected loss at the tail above VaR. CVaR is mathematically defined by: $CVaR_\alpha(X) = \mathbb{E}[X \mid X \ge VaR_\alpha(X)]$ where X is a real random variable. The benefit of using CVaR instead of VaR is that CVaR takes into account not only the VaR value but also the tail information of a distribution. CVaR is known to satisfy FSD [3].

Example 3. Expected Utility (EU) is a well-known decision criterion in decision under risk [20] and decision under uncertainty [29], which is known to satisfy FSD [3]. It is defined as follows: $EU(X) = \mathbb{E}(u(X))$ where X is a real random variable and $u : \mathbb{R} \to \mathbb{R}$ is a so-called von Neumann-Morgenstern utility function. The utility value $u(x)$ represents how much x is valuable. For this reason, function u is assumed to be monotonic (i.e., in our settings, $x \le y \Rightarrow u(x) \ge u(y)$). In decision theory, it is well-known that a concave (resp. convex) function u leads to a risk-averse (resp. risk-seeking) decision criterion. Although we focus on risk-averse criteria in this paper (as it is most people's concern in transportation), note that our approach could also tackle the risk-seeking case.

There are many other possible examples of ρ that satisfies property FSD: for instance, semideviations [22], rank-dependent utility [27], Yaari's dual model [32]... Our solution algorithm covers all those cases.

---

[2] For space reasons, we do not include the proofs.

Algorithm 1. Proposed adapted A* algorithm

Data: graph $G = (V, E)$, random costs C_t, heuristic h, criterion ρ, upperbound UB, origin node o, destination node d, departure time t

Result: risk-averse shortest path

```
1  begin
2  |    O ← {(o, 0)}
3  |    while O ≠ ∅ do
4  |    |    (n, C) ← highest priority pair in O
5  |    |    if n = d then return corresponding path
6  |    |    remove (n, C) from O
7  |    |    for n′ ∈ E⁺(n) do
8  |    |    |    C′ ← C + C_{t+C}(n, n′)      ▷initial time t + C selects the edge cost
9  |    |    |    f(n′, C′) ← ρ(C′ + h(n′))
10 |    |    |    if n′ = d and f(n′, C′) < UB then
11 |    |    |    |    UB ← f(n′, C′)
12 |    |    |    else
13 |    |    |    |    if f(n′, C′) ≥ UB then continue
14 |    |    |    if n′ ∉ O then
15 |    |    |    |    add (n′, C′) in O
16 |    |    |    else
17 |    |    |    |    if C′ not FSD-dominating any (n′, C″) ∈ O then
18 |    |    |    |    |    add (n′, C′) in O and remove FSD-dominating (n′, C″) ∈ O
19 |    |    |    |    else
20 |    |    |    |    |    continue
21 |    return ∅
```

---

5 Solution Algorithm

We propose an algorithm that is an adapted version of the standard A* algorithm to solve the proposed risk-averse shortest path problem using time-dependent stochastic costs. It generalizes the algorithm proposed by Chen et al. [6] to general ρ measures that satisfies FSD and extends the algorithm proposed by Parmentier and Meunier [24] to the time-dependent cost setting.

The proposed algorithm keeps the basic features of the standard A* algorithm, for example, an open set $\mathcal{O}$ while adding new features such as labeling with random variables and path pruning using FSD dominance. We now explain why the notions of label (for evaluating the value of a subpath ending in node n) and priority (for guiding the order the subpaths are examined) need to be redefined in our setting and how they can be redefined. In the standard A* algorithm for computing a shortest path, a node n in the priority queue $\mathcal{O}$ is the end node of a subpath for which only one label (i.e., cumulated cost $g(n) = c(\pi_{on})$) needs to be stored. This is possible because we have $\forall n \in V$:

$$c(\pi_{on}) \leq c(\pi'_{on}) \implies c(\pi_{on} \oplus \pi_{nd}) \leq c(\pi'_{on} \oplus \pi_{nd})$$

Table 1. Cumulative distributions.

x	0	1	2	3	4
$F_{C_t(\pi_{on})}$	0	0.95	1	1	1
$F_{C_t(\pi'_{on})}$	0.9	0.9	1	1	1
$F_{C_t(\pi_{nd})}$	0.8	0.9	1	1	1
$F_{C_t(\pi_{on}\oplus\pi_{nd})}$	0	0.76	0.895	0.995	1
$F_{C_t(\pi'_{on}\oplus\pi_{nd})}$	0.72	0.81	0.98	0.99	1

where π_{on} and π'_{on} are two paths from node o to node n and π_{nd} is a path from n to d. Unfortunately, in our setting, a counterpart of these inequalities with respect to ρ does not hold, due to the possible non-linearity of criterion ρ:

$$\rho(C_t(\pi_{on})) \leq \rho(C_t(\pi'_{on})) \;\not\Rightarrow\; \rho(C_t(\pi_{on} \oplus \pi_{nd})) \leq \rho(C_t(\pi'_{on} \oplus \pi_{nd})) \qquad (5)$$

In words, a dominated subpath can become non-dominated when extended.

Example 4. We give an example for the case when ρ is VaR with $\alpha = 95\,\%$. Assume the probability distributions are given in Table 1. One can check that:

$$VaR(C_t(\pi_{on})) = 1 < 2 = VaR(C_t(\pi'_{on})) \text{ and}$$
$$VaR(C_t(\pi_{on} \oplus \pi_{nd})) = 3 > 2 = VaR(C_t(\pi'_{on} \oplus \pi_{nd}))$$

As a consequence of (5), labels have a more complex form. Following previous related work [6,21,24], the label of node n is defined as $C_t(\pi_{on})$ instead of $\rho(C_t(\pi_{on}))$. For a given node, two labels can be compared with FSD-dominance (thanks to Corollary 1). As it is a partial order, a node can then receive several labels. For this reason, elements of $\mathcal{O}$ are pairs (n, C) where n is a node and C is a random variable representing the cost of a subpath from node o to node n.

The priority of a pair (n, C) in $\mathcal{O}$ is defined as $f(n, C) = \rho(C + h(n))$ where $h(n)$ is a known heuristic evaluation of a subpath from node n to node o. We assume that $h(n)$ is FSD-dominated by the random cost of any subpath from node n to node d. Heuristic $h(n)$ can be a deterministic value [6] as usual or more generally a random variable [24].

Defining the label as such and comparing them with FSD dominance are justified because of the following lemma [6,21,24] and corollary, written for X, Y, Z three real random variables.

Lemma 2. *If* $F_X \succsim_{FSD} F_Y$, *then* $F_{X+Z} \succsim_{FSD} F_{Y+Z}$.

This lemma can be interpreted in our context as follows: If the label (i.e., random variable or its associated probability distribution more exactly) of a subpath π ending in n FSD-dominates the label of another subpath also ending in n, then any extension of those two subpaths will keep the direction of the dominance.

As a direct consequence of Lemma 2 and FSD, we have:

Corollary 1. *If $F_X \succsim_{FSD} F_Y$, then $\rho(X + Z) \geq \rho(Y + Z)$.*

This corollary states that if a given node n has two labels, one FSD-dominating the other, the former label can be pruned as it will lead to a higher ρ value.

Thanks to INC, the following proposition explains why it is sound to end the algorithm as soon as node d is examined (Line 5 of Algorithm 1).

Proposition 2. *When node d is chosen, the corresponding path is ρ-minimum.*

Property INC was not considered in Parmentier and et al.'s work [24]. Contrary to their algorithm, ours can stop as soon as a path to node d is found.

In order to avoid generating too many subpaths, we use an upperbound UB on the best ρ value known so far. When starting Algorithm 1, we can use $UB = +\infty$ or better compute a standard shortest path and use its ρ value as an upperbound. Then, UB can be updated each time a path to d is found (Line 10). Besides, this algorithm can be sped up by pruning with any other known lower bound to the ρ value (see the case study where we use the expected duration).

Note that in general, Line 5 may be hard to compute. In our case study, we assume the time is discretized into equal-length intervals on which probability distributions are assumed to be constant. Moreover, we also assume all distributions are discretized. In the next section, we explain this in more details.

6   Case Study

We demonstrate our algorithm with ρ chosen as the conditional value-at-risk (CVaR) with $\alpha = 90\%$. This seems to be a better choice than VaR, which was used in Chen et al.'s work [6], because it not only takes into account the VaR threshold, but also the tail distribution. Besides, being a coherent risk measure [1], it enjoys nicer properties than VaR. We implemented our adapted A* algorithm in OpenTripPlanner[3], an open-source platform for route planning, which offers a map-based web interface and standard shortest path algorithms.

In order to work with real traffic data, we estimated the dynamic random costs from taxi trip data[4] released by the New York City TLC (Taxi and Limousine Commission). We first explain how a probability distribution for the random duration of an edge was estimated and then present an illustration of results that can be obtained thanks to our algorithm.

Data Cleaning and Estimation. The dataset contains records of taxi trips in Manhattan from 2009 to 2015. We only used the data from 2011 to 2015, as the data size was large and we preferred focusing on the most recent records. The dataset contains trip information including pick-up/drop-off locations and

---

[3] http://www.opentripplanner.org.
[4] http://www.nyc.gov/html/tlc/html/about/trip_record_data.shtml.

pick-up/drop-off times. We only took into account trips inside the Manhattan area, which represents a network of 5,111 nodes and 16,396 edges. During the data cleaning phase, we filtered out trips that had a pick-up or drop-off location outside Manhattan. We also removed abnormal trips, which may be due to incorrect GPS readings.

Because the actual path of a trip and detailed times at each intersection of a trip were not provided, we had to make two assumptions to extract random duration $C_t(e)$ of an edge $e \in E$ from the dataset:

A1 A trip follows the shortest path from origin to destination.
A2 The driver maintains the same speed along the trip.

Given the nature of the dataset, the assumptions seem reasonable enough. A1 leads to a small overestimation of travel durations in each edge. For our risk-averse route planning problem, overestimation is better than underestimation. A2 is a simplifying assumption, which neglects the effects of traffic lights, intersections, turns... We do not think it has a too big impact for our application, especially given that we have already overestimated the durations.

Based on A1, for each trip, we computed its shortest path from its origin to its destination using standard A* in terms of duration, where the duration of an edge (i.e., portion of a street) equals to the length (i.e., distance) of an edge divided by the maximum speed limit allowed in that edge. Then, given the computed shortest path π, we could generate a duration sample for each of its edge based on A2 with $c_e = c_\pi \times \frac{l_e}{l_\pi}$ where c_e is the duration of an edge e of π, c_π the total duration of the trip, l_e the length of edge e and l_π the length of π.

Samples c_e's were then collected and used to estimate $P_{C_t(e)}$. As we expect different traffic patterns on weekdays and during weekends, we divided the days of a week into two classes: $Weekdays = \{Mon., Tues., Wed., Thur., Fri.\}$ and $Weekends = \{Sat., Sun.\}$. We divided a day into 24 bins of 1 h. For a specific edge e, we obtained 24 distributions $P_{C_t(e)}$ (one for each hour) and assume the distribution was constant during an interval of one hour. Moreover, we assume those distributions are discrete and defined over 100 bins of 6 s. Durations that exceeds 600 s were counted as 600 s.

Experimental Results. With the adapted A* algorithm described before, we can find the risk-averse path between any pair of origin and destination. For this case study, following Chen et al.'s work [6], we define the heuristic function used in the implemented risk-averse path finding system as $h(n) = \frac{d(n)}{v_{max}}$ where $d(n)$ is the shortest length of a path from node n to node d and v_{max} is the maximum travel speed in the network. Therefore, $h(n)$ is the shortest possible duration to go from n to d.

Besides, it is known that $CVaR_\alpha$ is increasing with α and $CVaR_{0\%}$ is the expectation. Therefore, we also maintained an expected duration of a subpath π_{on} and estimated a lowerbound of the expected duration of an extension of π_{on} to node d (as in standard A*). This lower bound can be used to prune subpaths by comparing it with the upperbound UB.

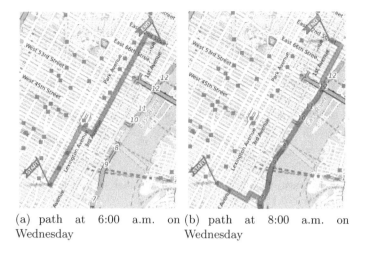

(a) path at 6:00 a.m. on Wednesday   (b) path at 8:00 a.m. on Wednesday

Fig. 2. Examples of risk-averse paths.

To illustrate our system, we present one example where for the same pair of origin and destination nodes, CVaR yields different risk-averse paths depending on the departure time. As depicted in Fig. 2, at 6:00 a.m. on Wednesday, before rush hours, the risk-averse path (with a CVaR of 24 mins) is very similar to the shortest distance path because there is little risk of congestion. Its CVaR can be interpreted as follows: In the worst 10 % of the case, the average duration of the trip will be 24 mins. And, in most cases, the observed travel duration would be much less than 24 min. In contrast, at 8:00 a.m. on the same day during rush hour, the risk-averse path is no longer the shortest distance path, but a path (with a CVaR of 31 mins) that passes via a highway, which has less probability of congestion. Although the risk-averse path may be a longer path to drive, it is a less risky path in terms of CVaR.

The computation times depend on the origin and destination nodes. By averaging over 100 runs where those pairs where selected randomly, the average computation time was less than one second (976.2 millisecs) using a computer equipped with an *Intel Xeon E31225 @ 3.10GHz*. To make this system usable in a real application, the computation time could be further improved. We expect this could be achieved with different optimization techniques: e.g., memoization, better heuristics, fitting duration samples to continuous distributions... As we wanted to demonstrate the feasibility of our approach, we leave this as future work.

7 Conclusion

In this paper, we proposed an adapted A* algorithm, which accommodates any risk measure or decision criterion that is monotonic with first-order stochastic dominance, to find a risk-averse shortest path in a transportation network with time-dependent stochastic costs. Besides, we demonstrated our algorithm on a case study with NYC taxi data and obtained reasonable results.

As future work, we plan to improve the computational efficiency of our method, taking inspiration from the techniques developed for standard short-est path problems [2,7]. Moreover, we would like to test our system on more accurate historical traffic data. Finally, we plan to extend the approach to take into account other kinds of costs, such as power consumption.

References

1. Artzner, P., Delbaen, F., Eber, J., Heath, D.: Coherent measures of risk. Mathe. Finan. **9**(3), 203–228 (1999)
2. Bast, H., Delling, D., Goldberg, A., Müller-Hannemann, M., Pajor, T., Sanders, P., Wagner, D., Werneck, R.: Route planning in transportation networks (2015). arXiv:1504.05140v1
3. Bäuerle, N., Müller, A.: Stochastic orders and risk measures: Consistency and bounds. Math. Econ. **38**, 132–148 (2006)
4. Bellman, R.: On a routing problem. Q. Appl. Math. **16**, 87–90 (1958)
5. Bertsekas, D., Tsitsiklis, J.: An analysis of stochastic shortest paths problems. Math. Oper. Res. **16**, 580–595 (1991)
6. Chen, B.Y., Lam, W.H.K., Sumalee, A., Li, Q., Tam, M.L.: Reliable shortest path problems in stochastic time-dependent networks. J. Intell. Transp. Syst. **18**(2), 177–189 (2014)
7. Delling, D., Sanders, P., Schultes, D., Wagner, D.: Engineering route planning algorithms. In: Lerner, J., Wagner, D., Zweig, K.A. (eds.) Algorithmics of Large and Complex Networks. LNCS, vol. 5515, pp. 117–139. Springer, Heidelberg (2009). doi:10.1007/978-3-642-02094-0_7
8. Dijkstra, E.: A note on two problems in connexion with graphs. Numer. Math. **1**, 269–271 (1959)
9. Dreyfus, S.: An appraisal of some shortest-path algorithms. Oper. Res. **17**(3), 395–412 (1969)
10. Embrechts, P., Kluppelberg, C., Mikosch, T.: Modelling Extremal Events for Insur-ance and Finance. Springer, Berlin (1997)
11. Ford, L.J.: Network flow theory. Technical report, Rand Corporation (1956)
12. Frank, H.: Shortest paths in probabilistic graphs. Oper. Res. **17**(4), 583–599 (1969)
13. Fu, L., Rilett, L.: Expected shortest paths in dynamic and stochastic traffic net-works. Transp. Res. Part B: Methodol. **32**(7), 499–516 (1998)
14. Gavriel, C., Hanasusanto, G., Kuhn, D.: Risk-averse shortest path problems. In: IEEE 51st Annual Conference on Decision and Control, pp. 2533–2538 (2012)
15. Goldberg, A., Harrelson, C.: Computing the shortest path: A* meets graph theory. In: SODA, pp. 156–165 (2005)
16. Hart, P.E., Nilsson, N.J., Raphael, B.: A formal basis for the heuristic determina-tion of minimum cost paths. IEEE Trans. Syst. Cybern. **4**(2), 100–107 (1968)
17. Jorion, P.: Value-at-Risk: The New Benchmark for Managing Financial Risk. McGraw-Hill, New York (2006)
18. Kaufman, D., Smith, R.: Fastest paths in time-dependent networks for intelligent vehicle-highway systems application. J. Intell. Transp. Syst. **1**(1), 1–11 (1993)
19. Moore, E.F.: The shortest path through a maze. In: Proceedings of the Interna-tional Symposium on the Theory of Switching, pp. 285–292 (1959)

20. von Neumann, J., Morgenstern, O.: Theory of Games and Economic Behavior. Princeton University Press, Princeton (1944)
21. Nie, Y., Wu, X.: Shortest path problem considering on-time arrival probability. Transp. Res. Part B: Methodol. **43**(6), 597–613 (2009)
22. Ogryczak, W., Ruszczynski, A.: From stochastic dominance to mean-risk models: semideviations as risk measures. Eur. J. Oper. Res. **116**, 33–50 (1999)
23. Orda, A., Rom, R.: Shortest-path and minimum delay algorithms in networks with time-dependent edge-length. J. ACM **37**(3), 607–625 (1990)
24. Parmentier, A., Meunier, F.: Stochastic shortest paths and risk measures. In: arXiv preprint (2014)
25. Peyer, S., RautenBach, D., Vygen, J.: A generalization of Dijkstra's shortest path algorithm with applications to VLSI routing. J. Discret. Algorithms **7**(4), 377–390 (2009)
26. Puterman, M.: Markov Decision Processes: Discrete Stochastic Dynamic Programming. Wiley, Hoboken (1994)
27. Quiggin, J.: Generalized Expected Utility Theory: The Rank-dependent Model. Kluwer Academic Publishers, Berlin (1993)
28. Russell, S., Norvig, P.: Artificial Intelligence: A Modern Approach, 2nd edn. Prentice-Hall, Upper Saddle River (2003)
29. Savage, L.: The Foundations of Statistics. Wiley, Hoboken (1954)
30. Shaked, M., Shanthikumar, J.: Stochastic Orders and Their Applications. Academic Press, New York (1994)
31. Sigal, C., Pritsker, A., Solberg, J.: The stochastic shortest route problem. Oper. Res. **28**, 1122–1129 (1980)
32. Yaari, M.: The dual theory of choice under risk. Econometrica **55**, 95–115 (1987)

Causal Basis for Probabilistic Belief Change: Distance *vs.* Closeness

Seemran Mishra[1] and Abhaya Nayak[2]([⊠])

[1] National Institute of Technology, Rourkela, India
seemran.mishra1996@gmail.com
[2] Macquarie University, Sydney, Australia
abhaya.nayak@mq.edu.au

Abstract. In probabilistic accounts of belief change, traditionally Bayesian conditioning is employed when the received information is consistent with the current knowledge, and imaging is used otherwise. It is well recognised that imaging can be used even if the received information is consistent with the current knowledge. Imaging assumes, *inter alia*, a relational measure of similarity among worlds. In a recent work, Rens and Meyer have argued that when, in light of new evidence, we no longer consider a world ω to be a serious possibility, worlds more similar to it should be considered relatively less plausible, and hence more dissimilar (distant) a world is from ω, the larger should be its share in the original probability mass of ω. In this paper we argue that this approach leads to results that revolt against our causal intuition, and propose a converse account where a larger share of ω's mass move to worlds that are more similar (closer) to it instead.

1 Introduction

Our knowledge is often fallible, uncertain and incomplete. How this knowledge should evolve in light of observations made and evidence acquired is a subject of much research. Research in this area can be divided into two broad approaches. In the first, "deterministic approach", knowledge is assumed to be certain but fallible, and the crux of the problem is to devise a rational model for modifying it in light of evidence that contravenes this knowledge. Literature in this approach is directly or indirectly inspired by the AGM paradigm [1] and deals with issues such as the problem of repeated belief change [10,11], the problem of belief change when knowledge is finitely represented [5] and the problem of belief modification when the world described by the knowledge is dynamic [7]. A knowledge state (or belief state) in such approaches is represented as a set of sentences together with a mechanism to capture the firmness of beliefs, semantically underpinned by a plausibility ranking of worlds. In the second, "probabilistic approach", a knowledge state is often represented as a probability function,

This research has been partially supported by the Australian Research Council (ARC), Discovery Project: DP150104133.

C. Sombattheera et al. (Eds.): MIWAI 2016, LNAI 10053, pp. 112–125, 2016.
DOI: 10.1007/978-3-319-49397-8_10

with beliefs being interpreted as "full beliefs", meaning propositions assigned probability 1. Evidence with non-zero prior is then processed using standard Bayesian conditioning; however belief-contravening evidence, that is evidence with zero prior needs special treatment using "imaging" first introduced by David Lewis in the context of analysing conditionals [9].

It has been proposed that distance between worlds captured by a pseudo-distance function can be used to "implement" the notion of imaging and provide a construction of belief-contravening operations in a probabilistic framework. The idea is that if a world ω has to lose a portion (or all) of the probability apportioned to it, rationality dictates that this unaccounted for probability should be gained by the world that is closest to ω among the set of potentially deserving worlds.[1] In a recent work it has been suggested that such bias in the movement of probability may not be appropriate [13]. The rationale behind this suggestion is that a world that is losing probability is doing so because it is no longer deemed plausible, and hence, worlds similar to it would also suffer from a reduction in plausibility. Increasing probability of such worlds runs counter to this intuition, and so the probability salvaged should be distributed in proportion to their distance from the worlds losing that probability.

Out primary objective is to examine this suggestion. First of all, the suggestion that salvaged probabilities should be mostly contributed to worlds farthest from the "victim worlds" is based on abstract intuitions which are not much better than the counter-intuition that such probabilities should mostly be contributed to wolds closer to them. Consider for instance the counterfactual, *If Oswald had not killed Kennedy, someone else would have.* In order to evaluate such a counterfactual, it is natural to consider worlds that are very similar to the "real world" except that in them Oswald didn't kill Kennedy (but yet, it was the post-Cuban crisis period, JFK was a Democrat visiting Dallas on a reelection campaign, bullets are designed to kill people, and so on). If we take the suggestion in [13] seriously, we should instead consider worlds that are diametrically opposite to the real world in which, not only that Oswald didn't kill Kennedy, but also, presumably, JFK was a Republican visiting Alaska, bullets have salutary effects on humans, and so on which would make a complete mess of our understanding of counterfactuals. Secondly, while our intuitions about counterfactuals may not be very reliable,[2] we have quite strong causal intuitions. In the next section we look at a simple cause-effect scenario and examine what happens if, instead of doing Bayesian conditioning using the ratio principle we distribute probability along the way suggested by [13] and observe that it leads to counterintuitive consequences.

This leads us to explore the converse approach in causal domains. In Sect. 3 we define *closeness* between worlds based on the *distance* between them, and examine some properties of this measure. This notion of closeness is then used to develop an account of probabilistic reasoning, particularly probabilistic expansion, and

---

[1] Assuming such a unique closest world exists. Short of it, an appropriate distribution mechanism should be employed.

[2] Recall the standard refrain of the politicians, *I don't answer hypothetical questions.*

study its behaviour in our chosen example. Finally, in Sect. 4, we briefly outline
the implications and limitations of our proposal, and our future research work.

2   Farther is not Better

We assume a propositional language and classical logic with standard notation.
A world is defined to be a unique assignment of truth values $\{0,1\}$ to all the
atoms in the propositional language. The set of all the worlds is denoted by
Ω. An agent's body of beliefs, denoted b, contains information as to different
subjective probability it assigns to different worlds, and the set of all possible
belief sets is denoted by B.

Definition 2.1. *A belief set b is the set of pairs $\bigcup_{\omega_i \in \Omega}\{(\omega_i, p_i)\}$ where world
ω_i is allotted probability $p_i = P(\omega_i)$ such that $\Sigma_i p_i = 1$.*

It is understood that the probability that the agent assigns to a proposition Φ
is given by $P(\Phi) = \Sigma_i\{p_i \mid \omega_i \models \Phi\}$.

The belief state of an individual is rationally modified in the light of new
information Ψ that the agent receives. Received wisdom has it that the nature
of this modification is sensitive to whether the knowledge domain is static or
dynamic [7] and whether the new information contravenes current knowledge or
not [4]. In this paper we will assume that the knowledge domain is static, and the
received information is not knowledge contravening. In such case, as advocated
in [4], the belief modification should be carried out using Bayesian conditioning,
in other words, $b + \Psi = \bigcup_{\omega_i \in \Omega}\{(\omega_i, p'_i)\}$ where $p'_i = \frac{p_i}{P(\Psi)}$ if $\omega_i \models \Psi$ and 0
otherwise. Note that for this purpose no other machinery such as a plausibility
ranking of worlds is used – if there is any implied notion of plausibility at all, it
is presumably captured by probability.

Intuitively, similar scenarios are similarly plausible – that, at least, is the
picture portrayed by plausibility rankings. However, this nice picture does not
extend to probability calculation, as illustrated by the following example, and
hence, arguably, *plausibility* and *probability* may not be reducible to each other.

Example 1. Three coins are tossed. We know that they are similarly biased:
the odds of getting head are *same for each of the three coins* – either 9:1 or
1:9 – but we don't know which. If the bias favours heads, probabilities of the
events $(HHH, HHT, HTH, HTT, THH, THT, TTH, TTT)$ are respectively $(0.729, 0.081, 0.081, 0.009, 0.081, 0.009, 0.009, 0.001)$. The respective
probabilities are $(0.001, 0.009, 0.009, 0.081, 0.009, 0.081, 0.081, 0.729)$ on
the other hand if the bias favours tails. Appealing to the principle of indif-
ference we obtain $(0.365, 0.45, 0.045, 0.045, 0.45, 0.045, 0.045, 0.365)$ as the
final probabilities. Clearly, the outcome HHH is more similar to HHT
(and HTH and THH), and less similar to the outcome TTT. Yet, the
outcomes HHH and TTT are equally probable, and that probability is
very different from the probabilities of HHT!

Indeed, probability has been supplemented by an extraneous notion of plausibility in an account of probabilistic belief change advocated in [3]. In this approach it has been argued that *imaging* [9] which is designed to deal with knowledge update in dynamic worlds can also be used to capture probabilistic belief change in static worlds, and they use (pseudo-)distance between worlds [8] to compute the image of a world among a given set of worlds for this purpose. This approach can be used irrespective of whether the new information is belief-contravening or not. Nonetheless, this approach is rather restrictive in that given a world ω and a set X of worlds, the former has a unique image in the latter, that is, there is a unique world in X that is closer or more similar to ω than any other world in X. In a subsequent work [13] Rens and Meyer have developed a more general method that effectively allows multiple images of a world. We will now briefly outline how they deal with probabilistic belief change (called belief expansion) when the evidence does not contravene current knowledge before examining it. We assume *here onwards* that the new evidence is not belief-contravening.

In general, when evidence Ψ does not contravene existing knowledge, there are at least some worlds $\omega \models \Psi$ with non-zero prior. Evidence Ψ suggests that any world $\omega' \not\models \Psi$ is a scenario not longer deemed seriously possible and must be eliminated from the hypotheses space, that is it must be assigned a (posterior) probability 0, and the probability thus salvaged must be distributed among worlds $\omega \models \Psi$. For simplicity let us use the following notation:

Notation 1. *Removal set R and Acceptance set S.*

1. $R = \{\omega_{\in \Omega} \not\models \Psi\}$ *is the set of worlds that conflict with the evidence and should receive zero posterior.*
2. $S = \Omega \setminus R = \{\omega_{\in \Omega} \models \Psi\}$ *is the set of worlds consistent with the evidence and the probability salvaged from members of R should be distributed among its members.*

In the Bayesian conditioning, so to speak, the probabilities salvaged from members of R are all pooled together and then distributed among those in S in proportion to their prior probabilities. The worlds in S with zero prior will continue have zero posterior. In the approach described in [13], instead of using the priors in S as the basis of distribution, the distance between an individual world $\omega^{\times} \in R$ and the worlds $\omega \in S$ is used to determine the share of ω in the probability of $\omega^{\times}$. This distance between worlds is captured by a pseudo-distance function d.

Definition 2.2 *[8]. The pseudo-distance $d : \Omega \times \Omega \to Z$ between two worlds satisfies:*

1. $d(\omega, \omega') \geq 0$ *(Non-negativity)*
2. $d(\omega, \omega) = 0$ *(Identity)*
3. $d(\omega, \omega') = d(\omega', \omega)$ *(Symmetry)*
4. $d(\omega, \omega') + d(\omega', \omega'') \geq d(\omega, \omega'')$ *(Triangle Inequality)*

for all worlds ω, ω' and $\omega'' \in \Omega$,

This distance function is then used to determine the different *weights* that members of $\omega \in S$ will carry while receiving their share of probability from any $\omega^{\times} \in R$. This weight, $\delta^{rem}(\omega^{\times}, \omega, S)$ is the distance $d(\omega^{\times}, \omega)$ normalised over the total distance of different worlds in S from $\omega^{\times}$.

Definition 2.3. *[13]* $\delta^{rem}(\omega^{\times}, \omega, S) = \dfrac{d(\omega^{\times}, \omega)}{\sum_{\omega' \in S} d(\omega^{\times}, \omega')}$

The weight $\delta^{rem}(\omega^{\times}, \omega, S)$ is used to compute total share of any $\omega \in S$ in the probability salvaged from R, denoted $\sigma(\omega, S, R)$:

Definition 2.4. $\sigma(\omega, S, R) = \sum_{\omega^{\times} \in R} P(\omega^{\times}) * \delta^{rem}(\omega^{\times}, \omega, S)$.

We note that the probability function P in Definition 2.4 is sensitive to the contextually fixed belief set b. The probability $P(\omega^{\times})$ is the $p^{\times}$ extracted from the pair $(\omega,^{\times} p^{\times}) \in b$.

Finally, an operation $\langle prem \rangle : B \times 2^{\Omega} \to B$ is used to determine the result of modifying a belief set b in light of evidence Ψ by topping up the existing probability of each world in $\omega \in S$ by its share $\sigma(\omega, S, R)$.

Definition 2.5. $b\langle prem \rangle R = \bigcup_{\omega \in S}\{(\omega, p') \mid (\omega, p) \in b \ and \ p' = p + \sigma(\omega, S, R)\}$.

Let us illustrate the application of the proposed probabilistic reasoning using a simple example that also brings to the surface a problem with this approach.

Example 2. The major causes of asthma are polluted air and stress. **R** is a factory town with bad air pollution. On a given day, the chance of **R** having high pollution index is 60 %. On the other hand, children in **R** have easy, stress-free life, and the probability that a child in this town suffers from stress is negligible, say 5 %. Both pollution and stress are equally efficacious in causing asthma, with a probability of 10 % each. In presence of both pollution and stress, there is a multiplier effect and the chance of asthma attack goes up to 25 %. On a particular day a little child who lives in **R** suffered from an asthma attack. Between pollution and stress, which factor should we blame?

This scenario is compactly represented as a Bayesian Network [12] as depicted in Fig. 1 below.

Intuitively, since pollution and stress are equally efficacious in causing asthma attack, and the prior of pollution is a lot higher, around 12 fold, than the prior of stress, pollution is more likely the cause of the asthma attack. Indeed this is indicated by Bayesian updating: the posterior probabilities, $P'(P) = P(P|A) = 0.91$ and $P'(S) = P(S|A) = 0.13$. Both pollution and stress contributed to the attack and accordingly both of their probabilities hiked.

Let us now consider what happens if, instead of Bayesian Conditioning, we employ the distance based approach outlined above. There are eight worlds here that we denote as APS, $AP\overline{S}\ldots\overline{APS}$ along expected lines. The A-worlds here represent the *accepted set* S and $\overline{A}$-worlds the *removal set* R. We use *Hamming Distance* as the pseudo-distance function d between different worlds.

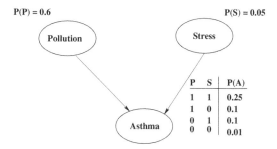

Fig. 1. A simple Bayesian Network depicting pollution and stress as the causes of asthma, with conditional probability tables (CPTs) given for each node. Pollution, stress and asthma are represented by P, S and A respectively.

Table 1 provides the distance between different worlds,[3] the weight of different A-worlds ω with respect to different $\overline{A}$-worlds $\omega^\times$,[4] and accordingly the share of ω in the probability of $\omega^\times$. The original probability of different $\overline{A}$-worlds is conveniently given beside their names in parentheses. For instance, the entry on top left says that $d(APS, \overline{A}PS) = 1$, since the total distance from $\overline{A}PS$ to different A-worlds is 8, the weight of APS with respect to $\overline{A}PS$ is $\frac{1}{8}$, and hence the share APS will claim in the probability of $\overline{A}PS$ (0.0225) is $\frac{0.0225}{8} \approx 0.0028$.

Table 1. Hamming distance, weight and share of an A-world with respect to different $\overline{A}$-worlds.

	$\overline{A}PS$ (0.0225)	$\overline{A}P\overline{S}$ (0.513)	$\overline{A}\overline{P}S$ (0.018)	$\overline{A}\,\overline{P}\,\overline{S}$ (0.3762)
APS	(1, 1/8, 0.0028)	(2, 2/8, 0.1282)	(2, 2/8, 0.0045)	(3, 3/8, 0.141)
$AP\overline{S}$	(2, 2/8, 0.0056)	(1, 1/8, 0.0641)	(3, 3/8, 0.0066)	(2, 2/8, 0.094)
$A\overline{P}S$	(2, 2/8, 0.0056)	(3, 3/8, 0.1923)	(1, 1/8, 0.0022)	(2, 2/8, 0.094)
$A\overline{P}\,\overline{S}$	(3, 3/8, 0.0084)	(2, 2/8, 0.1282)	(2, 2/8, 0.0044)	(1, 1/8, 0.047)

Now, the total probability share that an A-world receives from $\overline{A}$-worlds can be computed by simply adding up the third figures for each entry in a row. For instance, the total share that the world APS receives is $0.0028 + 0.1282 + 0.0045 + 0.141 \approx 0.2766$. These are shown in the third column in Table 2. It is easily noted that among the $\overline{A}$-worlds, the worlds $\overline{A}P\overline{S}$ and $\overline{A}\,\overline{P}\,\overline{S}$ account for most of the initial probability, and consequently the two A-worlds APS and $A\overline{P}S$ benefit the most from the probabilities initially allotted to the $\overline{A}$-worlds since they are *both* relatively "far away" from *each* of $\overline{A}P\overline{S}$ and $\overline{A}\,\overline{P}\,\overline{S}$.

---

[3] The distance between different A-worlds, or between different $\overline{A}$-worlds is not shown since they will not be used in the calculation.

[4] The more distant/different a world is from a target world, the higher is its relative weight.

The result of topping up the old probabilities by these shares gives the new probability of the A-worlds, as shown in the fourth column. Now, to compute the posterior probability of *Pollution*, say $P''(P)$, following this distance based approach, we simply add up the new probabilities of APS and $AP\overline{S}$, and get $0.2841 + 0.2275 \equiv 0.51$.[5] Similarly, the posterior $P''(S) = 0.2841 + 0.2963 \equiv 0.58$. In other words, stress is the more likely cause of the asthma attack than pollution. This is rather unexpected! Pollution and stress are equally effective in causing an asthma attack, the chance of pollution is very high, and the presence stress was assessed to be unlikely. And yet, when an asthma attack happens we conclude that stress was the likely cause of the attack. This is akin to believing in miracles.

Table 2. Old probabilities of A-worlds, their total probability share from $\overline{A}$-worlds, and the new probabilities.

	Old probability	Share from $\overline{A}$	New probability
APS	0.0075	0.2766	0.2841
$AP\overline{S}$	0.0570	0.1705	0.2275
$A\overline{P}S$	0.0020	0.2943	0.2963
$A\overline{P}\overline{S}$	0.0038	0.1882	0.1920

So we consider closeness rather than distance as the basis for probabilistic reasoning instead.

3   From Distance to Closeness

In this section we will develop an account of probabilistic belief change that uses a closeness measure between worlds. Intuitively, the smaller the distance between two worlds is, the closer they are, and this measure will exploit this conviction. In many ways it will mimic the approach in [13] but trail imaging by moving the probability of a "discredited" world to similar worlds. This process makes use of the pseudo-distance function but dispenses with the unique closest world assumption, and in that it generalises the approach developed in [3].

We denote by $c(\omega, \omega')$ the closeness or similarity between two worlds ω and ω', and let the minimum closeness between two worlds to be 0 and the maximum to be 1. Intuitively, $c(\omega, \omega')$ attains the value 0 when ω and ω' are at farthest distance from one another. On the other hand, it is standard practice in the modal semantics to assume that each world is most similar to itself. This is indeed indicated by the fact that $d(\omega, \omega) = 0$ in case of pseudo-distance. Hence we would want that the closeness between any world and itself should be maximum

---

[5] We need not worry about adding the new probabilities of relevant $\overline{A}$-worlds since they are all zero.

and equal to 1, that is $c(\omega, \omega) = 1$. We will also assume that the closeness or similarity is symmetric, that is $c(\omega, \omega') = c(\omega', \omega)$ for all worlds ω and ω'.[6]

These properties are very similar to those of a pseudo-distance function. Where these two functions diverge is the *Triangle Inequality* which is satisfied by the pseudo-distance function but not satisfied by closeness function. It is easily seen that this is not a desirable property for closeness function. For instance, consider the major cities in Australia. Sydney is not very close to Perth, say the $c(Sydney, Perth) \approx 0$, and hence by *Symmetry* we also have $c(Perth, Sydney) \approx 0$. On the other hand, by *Identity* we have $c(Sydney, Sydney) = 1$. However, *Triangle Inequality* mandates that $c(Sydney, Perth) + c(Perth, Sydney) \approx 0 \geq c(Sydney, Sydney) = 1$.

In order to limit the range of $c(\omega, \omega')$ to $[0, 1]$, apart from the pseudo distance, we will need the maximum distance between any two worlds in any set X that we call its diameter $\Delta(X)$.

Definition 3.1. $\Delta(X) = max\{d(\omega, \omega') \mid \omega, \omega' \in X\}$ *for any set X of worlds.*

Closeness, being the complementary concept of distance, it is natural to assume that the closeness between two worlds would correspond to the gap between the distance between them and the maximum distance possible between any two worlds. We define the closeness function $c : \Omega \times \Omega \rightarrow [0, 1]$, parametrised to a relevant set $X \subseteq \Omega$ as:

Definition 3.2. $c_X(\omega, \omega') = \dfrac{\Delta(X) - d(\omega, \omega')}{\Delta(X)}$ *for ω and $\omega' \in X \subseteq \Omega$.*

Note that when we set X to be Ω in this definition, we get the absolute closeness between two worlds, and in that case we drop the subscript Ω in $c_\Omega(\cdot, \cdot)$. In Fig. 2 below we graphically illustrate the notion of this parametrised closeness (or comparative similarity). Two sets of worlds are represented as two spheres with diameters X and X'. We want to capture the comparative similarity between two pairs of worlds, (a, b) in the first set and (a', b') in the second. The actual distances between the two pairs of worlds is given by $d(a, b) = x$ and $d(a', b') = x'$. The "raw similarity" between the pairs is given by y and y'. The corresponding degrees of comparative closeness are given by $\frac{y}{x+y}$ and $\frac{y'}{x'+y'}$. Assuming $\frac{y}{x+y} < \frac{y'}{x'+y'}$, the worlds a' and b' are comparatively closer to each other in comparison to the worlds a and b.

It is easily shown that the closeness measure as defined in Definition 3.2 has the desirable properties we discussed earlier.

Observation 1. *The closeness function c (appropriately parametrised to a set X) satisfies the following conditions:*

1. $0 \leq c(\omega, \omega) \leq 1$ (Range)

---

[6] Arguably the use of similarity in common parlance is non-symmetric. For instance, if John is non-violent, we would say *John is like Gandhi*. But saying *Gandhi is like John* would mean a very different thing. Capturing such asymmetry in our simple framework may not be quite feasible.

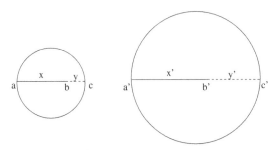

Fig. 2. Worlds a' and b' are closer to each other than a and b are since $\dfrac{y'}{x'+y'} > \dfrac{y}{x+y}$.

2. $c(w, w) = 1$ *(Identity)*
3. $c(w, w') = c(w', w)$ *(Symmetry)*

Now, let us see how we can distribute the probability of a world $w^\times \in R$ among worlds $w \in S$ with $R, S \subseteq \Omega$. One way would be to use the absolute closeness $c(w^\times, w)$ for different $w \in S$ and then compute the weight of different worlds in S with respect to $w^\times$ based on it analogous to the approach in [13]. This would entail that *all* worlds in S except the farthest ones will receive some non-zero share from the probability of $w^\times$. It would appear rather ad hoc since, if every other world in S benefits from the probability of $w^\times$, there is no reason why the least close ones should be deprived of this benefit. Hence we suggest that only those worlds in S that are in the "neighbourhood" of $w^\times$ should receive a portion of the latter's probability. We view *neighbourhood* as a very flexible concept – ranging from a very tight neighbourhood encompassing only those worlds in S closest to $w^\times$, to a very liberal one that includes almost the whole of S – and it can be adapted to suit the needs as necessary. (It is easily seen that classical imaging can be thus modeled by suitably choosing the distance function and required closeness of neighbours.) Here we provide only a particular interpretation of neighbourhood via what we call *mean proximity* below. Also, we use absolute closeness for convenience since it will not make any difference at the end. The mean proximity of a world $w^\times \in R$ with respect to set S is given as:

Definition 3.3. *Mean proximity of $w^\times$ wrt S:* $\pi(w^\times, S) = \dfrac{\sum_{w' \in S} c(w^\times, w')}{|S|}$

Intuitively, mean proximity gives us the boundary of the neighbourhood of $w^\times$ in which its probability will be distributed. We define the neighbourhood of a world $w^\times$ in a set S, denoted by $\nu(w^\times, S)$, as those worlds in S that are at most mean-proximity away from $w^\times$.

Definition 3.4. $\nu(w^\times, S) = \{w \in S \mid c(w^\times, w) \geq \pi(w^\times, S)\}$

Now we define the closeness-based weights δ^{cl} of the worlds in the neighbourhood of $w^\times$ along the expected lines. The δ^{cl} of w to $w^\times$ is its closeness to $w^\times$ appropriately normalised. The closeness-weight of w wrt $w^\times$ is given by:

Definition 3.5. *For all* $\omega^\times \in R, \omega' \in \nu(\omega^\times, S)$,

$$
\delta^{cl}(\omega^\times, \omega, S) = \begin{cases} \dfrac{c(\omega^\times, \omega)}{\sum_{\omega' \in \nu(\omega^\times, S)} c(\omega^\times, \omega')} & \text{if} \quad \omega \in \nu(\omega^\times, S) \\ 0 & \text{otherwise.} \end{cases}
$$

Now, analogous to Definition 2.4, we use the weights $\delta^{cl}(\omega^\times, \omega, S)$ to compute the total share of any $\omega \in S$ in the overall probability salvaged from R, denoted $\sigma^{cl}(\omega, S, R)$:

Definition 3.6. $\sigma^{cl}(\omega, S, R) = \sum_{\omega^\times \in R} P(\omega^\times) * \delta^{cl}(\omega^\times, \omega, S)$.

Subsequently, we use the operation $\langle prem^{cl} \rangle : B \times 2^\Omega \to B$ to determine the result of modifying a belief set b by new evidence Ψ through supplementing the existing probability of each world in $\omega \in S$ by its share $\sigma^{cl}(\omega, S, R)$:

Definition 3.7. $b\langle prem^{cl}\rangle R = \bigcup_{\omega \in S}\{(\omega, p'') \mid (\omega, p) \in b,\ p'' = p + \sigma^{cl}(\omega, S, R)\}$.

The closeness based probabilistic belief expansion operation $\langle prem^{cl} \rangle$ described above may be operationalised by the algorithm displayed below.

Algorithm: Closeness based Expansion

Input: b: belief-state, R: set of worlds that conflict with evidence
Output: new belief-state b'; R has total probability 0

1. **foreach** $\omega^\times \in R$ **do**
2.   **foreach** $\omega \in S$ **do**
3.   **Calculate** closeness c(ω, ω')
4.   **endfor**
5.   **Calculate** mean proximity $\pi(\omega^\times, S)$
6.   **Determine** neighborhood $\nu(\omega^\times, S)$
7.   **foreach** $\omega' \in \nu(\omega^\times, S)$ **do**
8. $\delta^{cl}(\omega^\times, \omega', S) \leftarrow \dfrac{c(\omega^\times, \omega')}{\sum_{\omega' \in \nu(\omega^\times, S)} c(\omega^\times, \omega')}$
9. $p_{new} \leftarrow p_{old} + p^\times * \delta^{cl}(\omega^\times, \omega, S)$
10.   **endfor**
11. **endfor**

Let us now see how our proposal fares *vis-a-vis* Example 2. As before, we use the Hamming distance displayed in Table 1 as the pseudo-distance. As the parameter X for computing closeness we will use the space Ω, and hence, for this purpose, use as diameter $\Delta(\Omega) = 3$. Closeness function is availed to calculate the closeness between pairs of worlds as shown in Table 3. For instance,

$$c(APS, \overline{A}PS) = \frac{\Delta(\Omega) - d(APS, \overline{A}PS)}{\Delta(\Omega)} = \frac{3 - 1}{3} \approx 0.67.$$

Table 3. Closeness between A-worlds and $\overline{A}$-worlds.

	$\overline{A}PS$	$\overline{A}P\overline{S}$	$\overline{A}\,\overline{P}S$	$\overline{A}\,\overline{P}\,\overline{S}$
APS	0.67	0.33	0.33	0
$AP\overline{S}$	0.33	0.67	0	0.33
$A\overline{P}S$	0.33	0	0.67	0.33
$A\overline{P}\,\overline{S}$	0	0.33	0.33	0.67

Now, when we learn A: *the child has had an asthma attack*, the $\overline{A}$-worlds will lose their probabilities which would be distributed in their respective neighbourhoods. We provide in Table 4 below for each $\overline{A}$-world its mean-proximity to the set of A-worlds, its neighbourhood among A-worlds, and the respective weight of each such neighbour. As in the case of distance based approach, we can compute the total probability share that an A-world receives from $\overline{A}$-worlds by adding up the weighted probabilities of each $\overline{A}$-world in whose neighbourhood it resides. For instance, APS is in the neighbourhoods of three $\overline{A}$-worlds: $\overline{A}PS$, $\overline{A}P\overline{S}$, and $\overline{A}\,\overline{P}S$. Furthermore, its claim to the probabilities of these $\overline{A}$-worlds receives the respective weights of 0.5, 0.25 and 0.25 (the first entries in the last column for the three relevant rows of Table 4). The total share of APS in the probabilities of $\overline{A}$-worlds is then the weighted sum of their probabilities, that is, $0.0225 * 0.5 + 0.513 * 0.25 + 0.018 * 0.25 \approx 0.144$. These total shares that different A-worlds receive from $\overline{A}$-worlds are shown in the third column in Table 5. The result of topping up the A-worlds' old probabilities by these shares gives the new probability of the A-worlds, as shown in the fourth column in Table 5.

Table 4. Mean proximity and neighbourhood of an $\overline{A}$-world and the respective weights of its neighbours. Probabilities of the $\overline{A}$-worlds are also provided for convenience.

$\overline{A}$-worlds	Mean proximity	Neighborhood	Weights
$\overline{A}PS$ (0.0225)	0.33	$\{APS, AP\overline{S}, A\overline{P}S\}$	(0.5, 0.25, 0.25)
$\overline{A}P\overline{S}$ (0.513)	0.33	$\{APS, AP\overline{S}, A\overline{P}\,\overline{S}\}$	(0.25, 0.5, 0.25)
$\overline{A}\,\overline{P}S$ (0.018)	0.33	$\{APS, A\overline{P}S, A\overline{P}\,\overline{S}\}$	(0.25, 0.5, 0.25)
$\overline{A}\,\overline{P}\,\overline{S}$ (0.3762)	0.33	$\{AP\overline{S}, A\overline{P}S, A\overline{P}\,\overline{S}\}$	(0.25, 0.25, 0.5)

Now, we compute the posterior probability of *Pollution*, say $P''_{cl}(P)$, following this closeness based approach by adding up the new probabilities of

APS and $AP\overline{S}$, and get $0.1515 + 0.4131 \equiv 0.564$. Similarly, the posterior $P''_{cl}(S) = 0.1515 + 0.1106 \equiv 0.262$. The observation of asthma attack has resulted in slight decrease in probability of pollution, and some increase in the probability of stress, and yet pollution remains the major causal contributor. This is more in alignment with our causal intuition.

Table 5. Old probabilities of A-worlds, their total probability share from $\overline{A}$-worlds, and the new probabilities using closeness based approach.

	Old probability	Share from $\overline{A}$	New probability
APS	0.0075	0.144	0.1515
$AP\overline{S}$	0.0570	0.3561	0.4131
$A\overline{P}S$	0.0020	0.1086	0.1106
$A\overline{P}\,\overline{S}$	0.0038	0.3208	0.3246

4   Discussion and Future Work

We looked at a simple scenario from the causal domain, described as Example 2 and sought to examine the propagation of probabilities when an effect is observed. The example was a case of multiple causes with a single effect. The two causes, stress and pollution, are equally efficacious as far as effecting an asthma attack is concerned. However, the pollution is a lot more prevalent than stress. When a child gets an asthma attack, which factor should we causally attribute it to?

Bayesian conditioning leads to results that are pleasing to the intuition. However since Bayesian conditioning cannot handle belief-contravening evidence we sought to explore unified methods that are more comprehensive.

A method advocated by Rens and Meyer [13] is one such more general approach. This approach exploits the distance between worlds to determine how the probabilities of the worlds "eliminated" by evidence should be distributed among those that "survive". It shifts the probability of a world to be eliminated to other worlds in the proportion of their distance from it. The rationale behind this approach is that if a world is considered implausible, then the worlds similar to it should also be considered implausible as well. The major flaw inherent in this method is that, as illustrated in Example 2, it produces counter-intuitive results. Given multiple causes of an effect, if the causes have equal efficacy, intuition demands that updating in light of the effect should not reverse the profile of the causal priors – if the prior of one cause dominates the prior of the other, their posteriors should exhibit that trend as well. In Example 2, both pollution and stress are taken to be equally good in effecting asthma attack, and prior of pollution is a lot higher than the prior of stress. However, the distanced based approach using Hamming distance shows a reversal of this trend – the posterior of stress is higher than the posterior of pollution.

The approach developed in this paper is based on closeness instead of distance. It may be considered to be a generalisation of imaging [9], as well as of an approach proposed in [3], in that it dispenses with the *unique closest world assumption* and distributes the probability of an eliminated world ω among those not eliminated roughly based on their similarity to ω. It is shown that when applied to Example 2, the posteriors of the causes do not reverse the direction of their priors. This indicates that the intuition behind imaging is basically sound, and probabilities of worlds eliminated by observed evidence should be moved to worlds similar to them, not to worlds dissimilar to them.

This paper is based on our preliminary exploration to probabilistic belief change based on closeness. There are a number of issues to be addressed down the road:

1. In this paper the closeness measure we employed is based on Hamming distance. The properties of Hamming distance that primarily contributed to the desirability of the outcome need to be formally captured.
2. This paper is based on a single, simple example. It does not show if the approach will work in other examples. We have tried it with a few other examples, and when there is substantial conflict between the distance based approach and the closeness based approach, the result of the latter appeared more intuitive. That, however, is no substitute for a formal proof.
3. The scope of this paper was restricted to belief non-contravening evidence. It will be interesting to see how the closeness based reasoning behaves when the evidence is belief-contravening.
4. It is likely that in different problem domains different approaches to probabilistic reasoning will be more appropriate. It will be fruitful to comprehensively explore this issue and compile the findings.
5. In the literature on logic of action there have been attempts to marry qualitative and quantitative approaches to causality and diagnosis (see, e.g., [2,6]). It will be interesting to see how our proposal sits in with such approaches.

We seek to address these issues in our future work.

References

1. Alchourrón, C.E., Gärdenfors, P., Makinson, D.: On the logic of theory change: partial meet contraction and revision functions. J. Symb. Log. **50**(2), 510–530 (1985)
2. Baral, C., Tran, N., Tuan, L.: Reasoning about actions in a probabilistic setting. In: Proceedings of the AAAI, pp. 507–512 (2002)
3. Chhogyal, K., Nayak, A., Schwitter, R., Sattar, A.: Probabilistic belief revision via imaging. In: Pham, D.-N., Park, S.-B. (eds.) PRICAI 2014. LNCS(LNAI), vol. 8862, pp. 694–707. Springer, Heidelberg (2014). doi:10.1007/978-3-319-13560-1_55
4. Gärdenfors, P.: Knowledge in Flux. MIT Press, Cambridge (1988)
5. Hansson, S.O.: Knowledge-level analysis of belief base operations. Artif. Intell. **82**(1–2), 215–235 (1996)

6. Iocchi, L., Lukasiewicz, T., Nardi, D., Rosati, R.: Reasoning about actions with sensing under qualitative and probabilistic uncertainty. ACM Trans. Comput. Log. 10(1) (2009). doi:10.1145/1459010.1459015
7. Katsuno, H., Mendelzon, A.O.: On the difference between updating a knowledge base and revising it. In: Proceedings of the KR, pp. 387–394 (1991)
8. Lehmann, D.J., Magidor, M., Schlechta, K.: Distance semantics for belief revision. J. Symb. Log. **66**(1), 295–317 (2001)
9. Lewis, D.: Probabilities of conditionals and conditional probabilities. Philos. Rev. **85**(3), 297–315 (1976)
10. Nayak, A.C.: Iterated belief change based on epistemic entrenchment. Erkenntnis **41**, 353–390 (1994)
11. Nayak, A., Goebel, R., Orgun, M., Pham, T.: Taking LEVI IDENTITY seriously: a plea for iterated belief contraction. In: Lehner, F., Fteimi, N. (eds.) KSEM 2016. LNCS(LNAI), vol. 9983, pp. 305–317. Springer, Heidelberg (2006). doi:10.1007/11811220_26
12. Pearl, J.: Probabilistic Reasoning in Intelligent Systems - Networks of Plausible Inference. Morgan Kaufmann, Burlington (1989)
13. Rens, G.B., Meyer, T.: A new approach to probabilistic belief change. In: Proceedings of the FLAIRS, pp. 581–587 (2015)

A Scalable Spatial Anisotropic Interpolation Approach for Object Removal from Images Using Elastic Net Regularization

M. Raghava[(⊠)], Arun Agarwal, and C. Raghavendra Rao

School of Computer and Information Science,
University of Hyderabad, Hyderabad 500046, India
raghava.m@cvr.ac.in, {aruncs,crrcs}@uohyd.ernet.in

Abstract. Object removal from an image is a novel problem with a lot of applications, in the area of computer vision. The ill-posed nature of the problem and the non-stationary content present in the image render it a complicated task. The diffusion-based and self-similarity based algorithms available in the literature explicitly model either the structures or the textures but not the both. They are good at solving small instances of the problem. However, they tend to produce low fidelity results and turn out to be intractable if the relative size of the object to the input image increases. The moving average based Spatial Anisotropic Interpolation (SAI) for text removal, proposed in our previous work also failed due to its poor extrapolation capability. Thus, it is imperative to develop a sampling scheme which can retain the interpolation feature while showing an apposite concern to the non-stationary features present in the image. The proposed, Design of Computer Experiments (DACE) driven Scalable SAI (SSAI) is a natural extension of SAI in three aspects. Precisely, it extends the Systematic Sampling to 'Not only Symmetric Hierarchical Sampling' (NoSHS), intelligently selects a basis based on Hurst Exponent, and employs Elastic Net regularization of Gaussian regression error for determining the order of the polynomial. Hence, these adaptive features increase the fidelity of the results. This paper elaborates the proposed framework- SSAI and demonstrates its capabilities by comparing the results with the latest hybrid approaches using the PSNR metric.

Keywords: Spatial anisotropic interpolation · DACE · Kriging · Hurst exponent · Elastic net regularization · NoSHS

1 Introduction

In real world situations, an expert artist renovates the damaged wall paintings, manually either by continuing the surrounding information or copying the coherent portions that are present in the vicinity of the target region. Translation of this natural art form into a computer program to process the digital images is a challenge for computer vision community. Researchers from various domains- computational fluid dynamics, mathematics, computer graphics

© Springer International Publishing AG 2016
C. Sombattheera et al. (Eds.): MIWAI 2016, LNAI 10053, pp. 126–140, 2016.
DOI: 10.1007/978-3-319-49397-8_11

and signal processing have been attempting to solve this problem by utilizing the domain-specific tools. The process of restoring the missing portions of an image by using the information available in the neighboring locations, therein the modifications are indiscernible to new observers is called Image Inpainting [1]. The scale of the problem ranges from removal of small scratches to larger objects from images. Let F be the image which is mathematically defined as $F : \Re^m \rightarrow \Re^n$ where $m = 2$ and $n = 3$ for color images; denoting the radiometric color channels Red(R), Green(G) and Blue(B) within RGB model or $n = 1$ for grayscale images. Concerning image inpainting, the input image is assumed to suffer from some degradation process, denoted with an operator T leading to missing of some pixel values. The resulting image I is seen as a composition of two disjoint parts: Ω- represents the locations x at which the pixel values are not known and Φ- corresponds to the locations of which pixel values are known. Ω is considered as the inpaint region and Φ as the source region. Mathematically the degraded/damaged image is expressed as $I = TF$, which is a composition of T and F. The goal of inpainting is to recover the image by estimating the pixel value p_x at all locations $x \in \Omega$ by utilizing the information available in Φ. Hence this problem can be viewed as an inverse problem for which more than one solution may be possible. In literature, such problems are referred to as ill-posed problems [2], and none of the possible solutions is perfect. Object removal is one of the inpainting category problems wherein the region to be filled is large and is of arbitrary shape that renders the task complicated. Figure 1(a) presents an instance of the object removal wherein the man standing to the right is treated as an unwanted object, and the effect of his removal is presented in Fig. 1(b). Overall, an object removal problem is abstracted to ensure that the inpainting model should produce a visually plausible result and retains the overall unity of the image. Now the existing models in the literature are presented briefly.

(a)   (b)

Fig. 1. (a) Image with unwanted man present (b) and unwanted man removed

1.1   Local and Greedy Approaches

The first set of formal models for inpainting is derived from Partial Differential Equations (PDE). These methods just fill the missing parts of the image by propagating the information available in the surrounding regions, thereby fall under the local category. Bertalmio [1] pioneered in this field by developing a non-linear PDE of order three, for controlled transportation of data into the inpaint region, which is named as anisotropic diffusion. Cahn-Hillard equations which are of order four [2] are developed that are capable of interpreting the geometrical structures succinctly. But these methods are less preferred for interactive applications because of their high computational complexity and are best suited for extending small structures only. Another set of models represent the inpaint problem in a variational framework which models the problem as an energy function, and the solutions to minimize it turn out to be ill-posed as the inverse of the energy function is unbounded. A standard way to handle the ill-posed nature is to include some *aprior* information into the model along with the fidelity term, which is referred to as regularization in the literature [2]. The chronological developments in variational models [2] range from simple Total Variation (TV) to Curvature Driven Diffusion (CDD) of information into the inpaint region. These models also come under the local category as they don't involve global features such as large scale textures. The second category of regularization models is based on self-similarity measure which is grossly referred to as exemplar-based methods [3]. These methods iteratively select a patch from the boundary of the inpaint region, analyze its content and search for a similar patch in the source region. Then the best matching patch content is copied into the selected patch. These exemplar-based methods are the examples of non-local methods but work in a greedy manner i.e. they inpaint the hole in a single pass. Though this model seems to be simple, it is necessarily required to answer many design issues that otherwise end up with texture garbages and have a significant bearing on the quality of the result. Even though a lot of variants are derived, the primary features of these methods aim at large-scale texture reproduction but fail to propagate the structures [3].

1.2   Global and Dynamic Approaches

This set of algorithms address the ailments of the local and greedy approaches by defining the inpainting aspect as an *energy* function and evolving an iterative algorithm to minimize it. The energy minimization, for example, is realized by combining multiple patches and verifying the coherence among the neighboring patches. This spatial phenomenon is achieved over a Markov Random Field (MRF) which offers suitable inference algorithms to attain global optimal solutions [4]. The MRF based solutions follow dynamic programming model that strive to achieve spatial coherence [4], among the Exemplar patches while inpainting the image. Authors in [5] observed that the Exemplar patches are separated by the same offset in the image and histogram offer a clue about statistics of dominant offsets. Then the image is shifted based on the dominant offsets, and the resulting images are combined through the MRF framework to achieve better results. But these methods involve manual intervention to separate the texture and structure components and are sensitive to the initial setting.

1.3 Hybrid Approaches

The domain decomposition approaches [6] solve the structure propagation and texture synthesis problems simultaneously by decomposing the image into cartoon and texture components using Morphological Component Analysis (MCA), inpaint both the components independently based on component specific algorithms and combine the partial results to get the overall results. These decomposition based methods are computationally expensive and are capable of filling medium size gaps. Another set of models [7] combine the energy terms related to self-similarity for texture synthesis, diffusion of information for structure Propagation and coherence between inpaint and source region pixels based on the precomputed correspondence map. This method is very sensitive to the initial setting and could solve small scale problems. The latest trend in image inpainting domain is to use non-local statistical information and to combine the Exemplar method with non-local models [8]. These methods are sound in addressing both the functional aspects of inpainting but are applicable for small to medium scale problems. The proposed DACE model attempts to solve the medium to large scale inpainting instance which can address both the texture and structures concurrently without modeling them explicitly.

2 Object Removal as a Spatial Interpolation Problem

For many problems in Computer Vision, it is a formulation necessity to incorporate the notion of representing the gray value or color information of a pixel as a random variable. While imputing the mission values, this association offers a tolerance to a certain degree of variation and is expressed as *error* [9]. The spatial arrangement of random variables $Z(x_i), i = 1 : n$ at locations x_i forms a random field. For example, an image I of size $M \times N$ can represent a random field which is composed of the rectangular arrangement of pixels referring to design sites, and the corresponding pixel values are the responses. In the model building, $X = [x_1, ..., x_m]^T$; $x_i \in \Re^n$ stands for the design sites and $Y = [y_1, ..., y_m]^T$; $y_i \in \Re^q$ represents the corresponding pixel values- the responses. Then an attempt to solve the inpainting problem by casting it as a regular image interpolation problem fails because of its over sensitive nature to the outliers and the absence of error modeling capabilities [9].

2.1 Design and Analysis of Computer Experiments (DACE) Model

DACE [10] is a 'surrogate computer model' which presents the spatial interpolation, namely kriging, in a deterministic way. Kriging model exploits the spatial correlation among the design sites of the underlying random field in contrast to ordinary interpolation which is independent of the location [11]. At a high-level description, kriging model involves the convolution of Polynomial regression with Gaussian regression model. The polynomial regression effectively captures the global patterns (also called as the trend), and the Gaussian regression controls the prediction error incurred over Ω. Universal kriging [9], a variant of kriging, assumes that the design sites exhibit some trend, and it is possible to capture it, through fitting a higher order polynomial F with unknown coefficients β which constitutes the polynomial regression. Subsequently, kriging controls the prediction error in an unbiased and quantitative manner by associating a stationary field $Z(x)$, which is further governed either by a pre-conceived or well-understood correlation function R. This feature of kriging strikes a behavioral advantage over general regression that analyzes prediction errors through white noise [11]. The combined kriging model could predict the response $\hat{y}$ at a design site x as

$$\hat{y}(x) = \sum_{j=1}^{p} \beta_j F_j + z(x). \tag{1}$$

The first term on the right-hand side in Eq. 1, models the trend through a polynomial regression through an assumed basis F and the unknown coefficients β_j. The second term $z(x)$, $x \in \Re^n$ on the right hand side models the residual over a random field $Z(X)$. The residual is assumed to follow the second order stationary property i.e. the error exhibits zero mean and finite covariance and its minimization requires computation of covariance matrix as $\sigma^2 R(\theta, x_i, x_l)$, for $i, l = 1...m$ where R is the correlation kernel with θ as the parameter and σ is the standard deviation. R, inherently offers a quantitative description for the spatial anisotropy. Thus, kriging addresses the modeling of the given trend and the associated correlation structure simultaneously [13] for computing β. The kriging, as a linear regression model, collects the responses Y at the given design sites in $2 - D$ and predicts the response at an unobserved site x as

$$\hat{y}(x) = C^T Y \quad \text{with} \quad C \in \Re^m. \tag{2}$$

Subsequently, kriging computes three terms: $F = \{F(x_1)...F(x_m)\}^T$- the collocation matrix over the selected basis function, $R = \{R(x_i, x_l)\}$- the spatial correlation expressed in terms of the lag between every pair of design sites i, l and $r(x) = [R(x, x_1), ...R(x, x_m)]^T$- the correlation between every design site x_i and an untried location x. Then the Best Linear Unbiased Estimator (BLUE) of kriging predicts the response at x from Eq. 2 as [10]

$$\hat{y}(x) = r^T R^{-1} Y - (F^T R^{-1} r - f)^T (F^T R^{-1} F)^{-1} F^T R^{-1} Y. \tag{3}$$

Now the Generalized Least Square Solution (GLS) for the multivariate polynomial regression [10]

$$F\beta = Y \tag{4}$$

is expressed as

$$\hat{\beta} = (F^T R^{-1} F)^{-1} F^T R^{-1} Y. \tag{5}$$

The overall predictor can be modeled [10] from Eqs. 3 and 5 as

$$\hat{y}(x) = F^T \hat{\beta} + r^T R^{-1} \quad (Y - F\hat{\beta}), \tag{6}$$

with an estimate of error variance

$$\sigma^2(x) = \frac{1}{m}(Y - F\hat{\beta})^T R^{-1}(Y - F\hat{\beta}). \tag{7}$$

The correlation kernel, in case of 2-D exponential model, is defined as the tensor product of the individual correlations along the columns and the rows

$$R(x, y) = \prod_{p=1}^{2} \exp^{(-\theta_p|d_p|)}. \tag{8}$$

It is evident from the correlation kernel, given in Eq. 8, the correlation decreases exponentially with lag d_p, and the parameter θ_p plays a role in model fitting. In DACE, the parameter θ_p is learned through Maximum Likelihood (ML) principle [10]. Hence, the vector θ can induce the anisotropy if the correlation among the pixel locations along columns differs from the rows.

2.2 Not only Symmetric Hierarchical Sampling Scheme

If the number of design sites grow high then the Kriging model, which involves computation of the inverse for a large scale correlation matrix, suffers from memory hungry problem. This aspect is referred to as 'curse of dimensionality' in the literature [14]. Hence, it is essential to select a subset of representatives from the entire set of available design sites. The proposed Not only Symmetric Hierarchical Sampling scheme (NoSHS) is an extension of the sampling schemes widely used in the super resolution analysis [13]. Systematic Sampling addressed in [14] generates only one sub-image from the input image. Where as, NoSHS produces multiple sub-images recursively by probing the entire image with a dummy element of size $l \times k$ for $l, k \in \{2, 3\}$ and arranges each pixel spanned by the probing element spatially along the rows and the columns of the respective sub-image indexed by l and k. For example, if an image I of size $M \times N$ is subjected to NoSHS for once, with $l = 3$ and $k = 2$ using the Algorithm Extract_6, then the sampling by factor 3 along the row and 2 along the column produces six sub-images I_n with $n = 1 : 6$, each of size $\frac{M}{3} \times \frac{N}{2}$. This sampling stands for an asymmetric instance of NoSHS and is chosen if the underlying image exhibits more spatial correlation along the rows than the columns. In contrast, if the data present in the image exhibits isotropic correlation then Extract_4 (which is not presented in this article) employs a probing element of size is 2×2 and the NoSHS generates four sub-images through symmetric sampling along both the dimensions. Figure 2 presents the functionality of NoSHS which takes a synthetic input image of size 4×4 (column 1) as input and produces 4 sub-images I_1 - I_4 (columns 2–4) of size 2×2. On the input image NoSHS is performed recursively until the size of the Ω present in a sub-image reduces to a minimum. This hierarchical sampling honors the underlying stationary property witnessed in the image and improves the fidelity of prediction process as elaborated in Sect. 2.3. The intuition behind introducing the NoSHS is depicted in the Fig. 3.

$$I = \begin{bmatrix} \bullet & \circ & \bullet & \circ \\ \diamond & \star & \diamond & \star \\ \bullet & \circ & \bullet & \circ \\ \diamond & \star & \diamond & \star \end{bmatrix} \quad I_1 = \begin{bmatrix} \bullet & \bullet \\ \bullet & \bullet \end{bmatrix} \quad I_2 = \begin{bmatrix} \circ & \circ \\ \circ & \circ \end{bmatrix} \quad I_3 = \begin{bmatrix} \diamond & \diamond \\ \diamond & \diamond \end{bmatrix} \quad I_4 = \begin{bmatrix} \star & \star \\ \star & \star \end{bmatrix}$$

Fig. 2. The first column presents a synthetic input image I and the following columns present the sub-images I_1–I_4 extracted through symmetric version of NoSHS from the input image

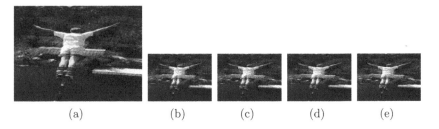

(a)   (b)   (c)   (d)   (e)

Fig. 3. (a) Input image [(b)–(e)] The sub-images extracted through NoSHS

Algorithm 1. Extract_6

Input: Input Image I
Output: Extracted Sub-images $I_i; i = 1 : 6$

1. $l := 1;$ $k := 1;;$ $i_2 := 2;$ $j_2 := 2;$ $j_3 := 3;$
2. for $i_1 := 1 :$ $M - 1 :$ **step by 2**
3. $i_2 := i_1 + 1;$
4. for $j_1 := 1$ $: N - 2 :$ **step by 3**
5. $j_2 := j_1 + 1;$ $j_3 := j_1 + 2;$
6. $I_1(l, k) := I(i_1, j_1);$
7. $I_2(l, k) := I(i_1, j_2);$
8. $I_3(l, k) := I(i_1, j_3);$
9. $I_4(l, k) := I(i_2, j_1);$
10. $I_5(l, k) := I(i_2, j_2);$
11. $I_6(l, k) := I(i_2, j_3);$
12. $k := k + 1;$
13. end
14. $l := l + 1;$ $k := 1;;$ $i_2 := i_2$ $+$ $2;$
15. end

2.3   Higher Order Polynomial Basis as the Trend

The next criticality of kriging is to solve the multivariate polynomial regression problem which models the trend. The Universal kriging fits a non-linear polynomial basis. However, the nature of the basis and the order of the polynomial are not addressed formally and hence assumes immense importance from large scale pattern detection point of view. The authors in [15] developed a Bayesian approach based Blind Kriging to resolve this situation. Blind Kriging is essentially the 'multi kriging' model, with a conservative approach involving the fitting of an increasing sequence of orthogonal polynomials, in search of the best-fit polynomial for the input data. Clearly, this approach turns the prediction process a computationally expensive one. The proposed algorithm SSAI addresses the polynomial basis selection issue by employing a futuristic approach which involves kriging, just for once. Initially, for the given design space the correlation R is expressed as the tensor product of the Correlations: one along the rows and the other along the columns, and they may exhibit anisotropy over $Z(X)$. Subsequently, a higher order polynomial F including the cross product terms is selected to model the global trend. For example, if a polynomial basis of order n is selected then the total number of quotients β_k to be learned is $\frac{n(n+1)}{2}$. As the object removal problem entails with a wide range of design sites when compared to the number of candidate terms of higher order polynomial, the chances of wiggling of the regression due to an overfit is less likely. The regression model given in Eq. 5, is a GLS solution that employs the Mahalanobis distance [9] which is scale invariant. This point substantiates that extraction of sub-images from the input image through the NoSHS will not negatively influence the model fitting. Subsequently, the GLS is appropriately regularized with an intention to annihilate the quotients that are either less important or playing spoilsport in predictions. The literature says such a regularization issue is well addressed through feature selection [16], pattern search, etc. The SSAI utilizes the feature selection approach.

2.4 Feature Selection Using Elastic Net

In the proposed model the feature selection is achieved by applying the Elastic net model, developed by Zou and Hastie [16]. Elastic net regularizes the data term with a convex combination of the L_1 (LASSO) and L_2 (ridge) terms of the regression coefficients as

$$\hat{\beta} = \underset{\beta}{\mathrm{argmin}}(\|Y - X\beta\|^2 + \lambda\|\beta\|^2 + (1 - \lambda)\|\beta\|_1). \tag{9}$$

It inherently replaces the Euclidean distance measure with checkerboard distance which induces sparsity into the solution and the value of λ determines the level of sparsity. According to the literature [16], small values of λ yields a sparse solution. Thus, even if the trend is modeled with higher order basis, the contribution of particular unnecessary basis terms can be, subsequently curtailed by annihilating the corresponding coefficients to zeros and retaining only the remaining. The future experiments are based on the proposed proactive model and is named as Enhanced DACE (EDACE) which is implemented by utilizing the LARS LASSO [16] model.

3   Adaptive Basis Selection

The object removal algorithm SSAI is capable of selecting a basis from the predefined collection of different basis functions in an adaptive manner. From our empirical observations, the nature of the polynomial can be a constant or any higher order polynomial, each of them with varying capabilities while modeling the trend. The subjective issue of basis selection is addressed by extracting the Hurst exponent H [17]. $H \in [0 \quad 1]$, is a self-similarity based measure capable of manifesting the non-stationary behavior present in the content of the image. Signal process community uses H for measuring the strength of singularity [18]. Hurst exponent describes the probability with which the autocorrelation among the pixels of the image changes asymptotically with the lag h as follows:

$$E[\frac{R(h)}{S(h)}] = Ch^H \qquad as \qquad \text{h} \rightarrow \quad \infty, \qquad (10)$$

where R refers to the range of values spanned by h, S represents the standard deviation of the data and C is a constant. H value can be estimated in various ways such as Box Counting, Rescaled Range Analysis [18], Wavelet Spectral Density, etc., and is extended to 2-D images also [15]. To understand Hurst exponent, R/S- one of the estimation methods, is discussed for 1-D data. Given y_i; $i = 1 : n$ the pixel responses at locations x_i; $i = 1 : n$ with mean m, then calculate the deviations d_i of each y_i from m using $d_i = y_i - m$. The series of partial sums p_i over d_i are calculated in a cumulative manner as detailed in Eq. 11.

$$s_k = \sum_{l=1:k} d_l \qquad k = 1 : n. \qquad (11)$$

Then the difference of $max\{s_k\}$ and $min\{s_k\}$ denotes Range R and the slope of 'bestfit' line for $\log(\frac{R}{S})$ vs $\log(n)$ represents the *estimate* of Hurst exponent H. Inherently, it models the Fractal coefficient which is a measure of self-similarity of Brownian motion [17]. The *estimated* Hurst exponent value, can be utilized to characterize the underlying process into three classes: stationary, not-known and non-stationary. The proposed SSAI utilizes the value of H to categorize the images into two classes as follows. If $H \in [0 \quad 0.5]$, then the image is classified to possess second order stationary trend and is modeled by deploying a higher order polynomial. Otherwise, the content of the image is assumed to be smooth, and the trend is represented through a constant. This adaptive feature in SSAI is a novel contribution in inpainting domain and avoids the formulations for texture and structure specific components.

4   The PushBack Operation

As the last step in the proposed framework, all inpainted sub-images belonging to a particular level of the hierarchy required being systematically combined in an anti-recursive manner to get the overall result. It is a PushBack of the chosen hierarchical sampling that merely rearranges the pixels from each inpainted sub-image back into their proper locations until the original size of the input image is reached. Algorithm $PushBack\_4$ takes the 4 inpainted sub-images $I_i, i = 1 : 4$, in the case of symmetric sub-sampling by order 2, and reconstructs the original image I. A similar algorithm can be evolved for Asymmetric Push-Back_6 through a relatively simple effort.

Algorithm 2. PushBack_4

Input: Set of all 4 inpainted sub-images $I_i, i = 1 : 4$ each of size $m \times n$ that are extracted from I

Output: Reconstructed image I

1. $for \quad i := 1 : m : \quad step \quad by \quad 1$
2. $for \quad j := 1 : n : \quad step \quad by \quad 1$
3. $I(2 \times i - 1, 2 \times j - 1) := I_1(i, j);$
4. $I(2 \times i - 1, 2 \times j) := I_2(i, j);$
5. $I(2 \times i, 2 \times j - 1) := I_3(i, j);$
6. $I(2 \times i, 2 \times j) := I_4(i, j);$
7. end
8. end

---

5   The Algorithm: SSAI

Algorithm SSAI, summarizes the steps discussed so far and presents the complete algorithm for object removal from images. If the input image possesses the correlation which is same along the rows and columns then their tensor product gains circular shape (see Fig. 4(a)) and the symmetric kernel of the proposed NoSHS, Extract_4 carries out uniform sampling and extracts four sub-images. In contrast to this, if the correlation is more prominent along the columns than the rows then their tensor product assumes elliptic shape, as shown in Fig. 4(b). Then the asymmetric kernel of NoSHS, Extract_6 performs coarse sampling along the columns and fine-grained sampling along the rows to derive 6 sub-images.

(a)   (b)

Fig. 4. (a) presents the pictorial representation of the Isotropic correlation in which case symmetric kernel for sampling is required and (b) presents the Anisotropic correlation that prompts the selection of asymmetric sampling kernel

Finally, the EDACE model is developed for predicting the pixel values at inpaint locations for each sub-image.

Algorithm 3. SSAI

Data: Input Image I and the associated Mask M
Result: Inpainted image
1. Derive H from I
2. Select the forward sampling kernel Sect. 2.2 under NoSHS
3. Check the size of the inpaint region and determine the levels of hierarchy k
4. Apply the chosen NoSHS scheme on I and M to extract k sub-images
5. Select the sub-images available at bottom-most level of the hierarchy and put them into array A
6. Choose a basis F to represent the trend based on H and the correlation model R.
7. Fit EDACE model for every sub-image on A and predict the responses over Ω
8. Invoke the corresponding PushBack kernel Sect. 4 recursively on inpainted sub-images, about step 2, to produce the inpainted image.

---

6   Experiments and Results

Experiment 1. The fidelity gain through Elastic net regularization-

Here we demonstrated the effect of the Elastic net regularization term introduced in the proposed model. In this experiment, initially a fifth order polynomial with 21 coefficients was selected to represent the trend in the course of polynomial regression. The over sensitive nature of the linear regression to the outliers was addressed by Elastic net regularization while eliminating the non-essential coefficients of the higher order polynomial. This improved the fidelity of the results as shown in Fig. 5- compare Fig. 5(b) with 5(c) and Fig. 5(e) with 5(f).

Experiment 2. Structure and Texture preserving capabilities without modeling them explicitly-

In Experiment 2, images with a large whole marked in white color were taken up. SSAI modeled the trend as a constant and employed the exponential correlation basis for the input images in Fig. 6(a) and (g). We could observe that SSAI was able to fill the holes, see Fig. 6(b) and (h), with a meaningful content on both sides of the linear structures. The anisotropy feature of the interpolation resolved the confusion as to extend the horizontal structure or vertical structure and achieved the better control. In Fig. 6(c) and (d), the staircase was preserved thoroughly by propagating the information along the structures. The image in Fig. 6(e) possesses small H value and is considered to be non-stationary image

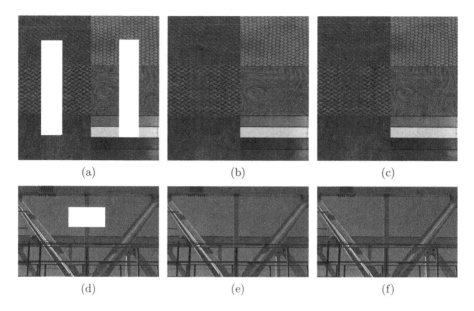

Fig. 5. Column1 presents the input images, column 2 presents the results of the proposed SSAI model and column 3 presents the results in the absence of Elastic net regularization.

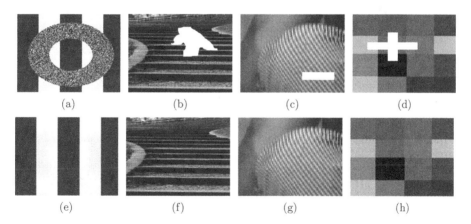

Fig. 6. Row 1 presents the input images. Row 2 presents the corresponding inpainted results of SSAI Model.

and Fig. 6(f) demonstrated the capability of SSAI model to synthesize the fine texture.

Experiment 3. Large Scale object Removal - comparison with Hybrid approach

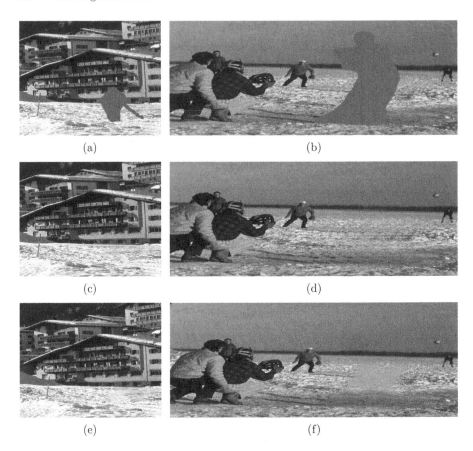

(a)   (b)

(c)   (d)

(e)   (f)

Fig. 7. Row 1- Input images, row 2 presents the results of SSAI and row 3 presents the results of Aria's [7] method.

Table 1. The Hurst exponent based Decision Table and the improvement in PSNR through SSAI

Image	H	Nature	Basis	Before	After
Grid-6(d)	0.89	Stationary	Constant	16.8 db	59.41 db
Trouser-6(c)	0.19	Second-order-stationary	Polynomial	22.6 db	39.76 db
Gantry-5(e)	0.42	Second-order-stationary	Polynomial	16.32 db	46.45 db

In this experiment, Fig. 7, the input images that had cartoon components and texture elements separated by structures were considered. SSAI preserved the structures and interpolated the texture components successfully. Whereas [7], a non-local Total Variational method failed to synthesize the textures clearly, compare Fig. 7(c) with 7(e) and Fig. 7(d) with 7(f). The efficacy of the adaptive basis selection feature of the proposed SSAI model and the quality of results in PSNR values are presented in Table 1.

7   Summary

In this paper, the object removal problem was solved without involving 'locate and copy' or diffusion of information models which were developed around the texture and structure components. In contrast, the proposed SSAI, employed the anisotropy based sampling scheme NoSHS, and addressed the ill-posed nature of the problem, through an adaptive polynomial basis selection based on Hurst exponent. The SSAI enhanced the quality of the solution through Elastic net regularization and overcome the stigma of Blind kriging over DACE. The experiments established the fact that high fidelity results are realizable for large scale inpainting problems through the properly modeled spatial interpolation.

References

1. Bugeau, A., Bertalmio, M., Caselles, V., Sapiro, G.: A comprehensive framework for image inpainting. IEEE Trans. Image Process. **19**(10), 2634–2645 (2010)
2. Guillemot, C., Le Meur, O.: Image inpainting: overview and recent advances. IEEE Signal Process. Mag. **31**, 127–144 (2014)
3. Buyssens, P., David, T., Olivier, L.: Exemplar-based inpainting: technical review and new heuristics for better geometric reconstructions. Trans. IP **24**, 1809–1824 (2015)
4. Komodakis, N., Georgios, T.: Image completion using efficient belief propagation via priority scheduling and dynamic pruning. IEEE Trans. Image Process. **16**, 2649–2661 (2007)
5. He, K., Sun, J.: Statistics of patch offsets for image completion. In: Fitzgibbon, A., Lazebnik, S., Perona, P., Sato, Y., Schmid, C. (eds.) Computer Vision, ECCV 2012. LNCS, vol. 7573, pp. 16–29. Springer, Heidelberg (2012)
6. Elad, M., Starck, J., Querre, P., Donoho, D.L.: Simultaneous cartoon and texture image inpainting using morphological component analysis (MCA). Appl. Comput. Harmon. Anal. **19**, 340–358 (2005)
7. Barnes, C., Shechtman, E., Finkelstein, A., Goldman, D.: Patchmatch: a randomized correspondence algorithm for structural image editing. ACM Trans. Graph. **28**, 24:1–24:11 (2009)
8. Arias, P., Facciolo, G., Caselles, V., Sapiro, G.: A variational framework for exemplar-based image inpainting. Int. J. Comput. Vis. **93**, 319–347 (2011)
9. Cressie, N.: Statistics for Spatial Data, vol. 416. Wiley, Hoboken (1993)
10. Lophaven, N., Nielsen, H.B., Jacob, S.: IMM. Informatics and Mathematical Modeling. Technical University of Denmark, DACE A Matlab Kriging Toolbox (2002)
11. Gentile, M., Courbin, F., Meylan, G.: Interpolating point spread function anisotropy, aap (2013)
12. Miller, A.J.: Subset Selection in Regression. Chapman & Hall, Boca Raton (2002)
13. Miaohui, W., Bo, Y., King, N.: An efficient framework for image, video inpainting. Signal Process.: Image Commun. **28**, 753–762 (2013)
14. Raghava, M., Agarwal, A., Rao, C.R.: Spatial anisotropic interpolation approach for text removal from an image. In: Ramanna, S., Lingras, P., Sombattheera, C., Krishna, A. (eds.) MIWAI 2013. LNCS (LNAI), vol. 8271, pp. 153–164. Springer, Heidelberg (2013). doi:10.1007/978-3-642-44949-9_15
15. Couckuyt, I., Forrester, A., Gorissen, D., De Turck, F., Dhaene, T.: Blind Kriging: implementation and performance analysis. Adv. Eng. Softw. **49**, 1–13 (2012)

16. Zou, H., Hastie, T.: Regularization and variable selection via the elastic net. J. R. Stat. Soc. **67**, 301–320 (2005)
17. Orietta, N., Garutti, C., Vidakovic, B.: 2-D wavelet-based spectra with application in analysis of geophysical images, GIT, February 2006
18. Carbone, A.: Algorithm to estimate the Hurst exponent of high-dimensional fractals. Phys. Rev. E **76** (2007)

Identification of Relevant and Redundant Automatic Metrics for MT Evaluation

Michal Munk[1], Daša Munková[2], and Ľubomír Benko[3(✉)]

[1] Department of Informatics, Faculty of Natural Sciences,
Constantine the Philosopher University in Nitra, Nitra, Slovak Republic
mmunk@ukf.sk
[2] Department of Translation Studies,
Constantine the Philosopher University in Nitra, Nitra, Slovak Republic
dmunkova@ukf.sk
[3] Institute of System Engineering and Informatics,
University of Pardubice, Pardubice, Czech Republic
lubomir.benko@gmail.com

Abstract. The paper is aimed at automatic metrics for translation quality assessment (TQA), specifically at machine translation (MT) output and the metrics for the evaluation of MT output (Precision, Recall, F-measure, BLEU, PER, WER and CDER). We examine their reliability and we determine the metrics which show decreasing reliability of the automatic evaluation of MT output. Besides the traditional measures (Cronbach's alpha and standardized alpha) we use entropy for assessing the reliability of the automatic metrics of MT output. The results were obtained on a dataset covering translation from a low resource language (SK) into English (EN). The main contribution consists of the identification of the redundant automatic MT evaluation metrics.

Keywords: Machine translation · Evaluation · Automatic metrics · Reliability · Entropy · Redundancy

1 Introduction

Machine translation (MT), its specifics and function are still relatively under researched fields, not only in Translation Studies, but also in Natural Language Processing. As its tools are still in their infancy and need a lot of adjustments and improvements to achieve better translation quality in target languages, MT theory itself is in its early stages in the context of low resource languages. This is caused not only by the short and recent development time of MT tools and MT systems, but also because machine translation is an interdisciplinary field comprising various research areas such as Translation studies, Computer engineering or Computational linguistics. Translation studies is focused on the evaluation of machine translation output from the point of view of linguistic phenomena, whereas Computer engineering or Computational linguistics are focused on the determination of the effectiveness of existing MT systems and on the optimization of algorithms implemented in MT systems as well as on the

© Springer International Publishing AG 2016
C. Sombattheera et al. (Eds.): MIWAI 2016, LNAI 10053, pp. 141–152, 2016.
DOI: 10.1007/978-3-319-49397-8_12

performance of MT systems. Progress relies on translation quality assessment through systematic effective evaluation approaches. Better evaluation metrics lead to better machine translation [1]. There are many evaluation approaches used in MT evaluation. Babych et al. [2] examined the effectiveness of the performance of MT system translating from low-resource languages into English via closely-related and well-developed translation resources. Adly and Al Ansary [3] evaluated the effectiveness of MT system based on the Interlingua approach. Vandeghinste et al. [4] evaluated the METIS-II system, Machine Translation System for Low Resource Languages, based on the automatic metrics BLEU, NIST and TER.

In general, there are two main approaches to MT evaluation: Glass box and Black box. We focus on a black box approach to MT evaluation, measuring the performance of a system upon a same test set and within the black box on intrinsic metrics. Intrinsic metrics – manual and automatic - are used to assess the accuracy of MT output. They focus on the quality of MT output and they compare MT output with one or more references (high quality translation, usually done by a human translator). Manual (human) intrinsic metrics assess the quality of Mt output as fluency and adequacy by human, which is not only subjective, but expensive, slow and difficult to standardize. Vilar et al. [5] remarked that the subjectivity of manual evaluation causes a problem in terms of the lack of clear guidelines as to how to assign values to translations. Automatic intrinsic metrics correlate well with human judgements [6–10] using quantitative scores of adequacy and fluency [11]. They compute sentence similarity -matches based on comparisons between a set of references (fixed translations) and the corresponding MT output. Automatic evaluation offers easy, low cost and high speed of evaluation compared to human translation, which is regarded as the most reliable.

Statistical machine translation systems have been the most widely used for many recent surveys and events. Since 2006 the evaluation campaigns, during the Annual workshop on Statistical machine translation (WMT), have been organized by the special interest group of machine translation (SIGMT) focusing on European languages [12–21]. The tested language pairs are divided into two directions from English into other (French, German, Spanish, Czech, Hungarian, Haitian Creole and Russian) and vice versa, i.e. from other languages into English. These campaigns do not cover other European language – Slovak, which is the subject of this paper.

We demonstrate how the analysis of reliability of automatic intrinsic metrics using Cronbach's or Standardized alpha or entropy can help by the identification of the relevant and redundant intrinsic metrics of automatic MT evaluation. Moreover, it can be used as a starting point for the automatic identification of errors and the classification of MT output.

Issues of MT and MT evaluation are more topical since there are not many studies concerning the less resource languages, such as the inflectional Slovak language.

This paper is constructed as follow: Sect. 2 introduces seven automatic intrinsic metrics for MT evaluation, Sect. 3 describes experiment setting, Sect. 4 presents the results of the analyses and finally Sect. 5 consists of conclusions.

2   Intrinsic Automatic Metrics for MT Evaluation

Due to problems which manual/human intrinsic metrics deal with, automatic metrics have been widely used during MT evaluation campaigns. They compare MT output with human reference, which can comprise one single reference or multiple references for a single source sentence [22, 23].

Automatic evaluation of MT output can be conducted based on statistical principles (n-grams or edit distance), which means based on lexical similarities or on the use of deep linguistic structures (morphological, syntactic or semantic information), which means based on linguistic features.

In this paper we will focus only on automatic intrinsic metrics based on lexical similarity (n-grams which measure the overlap in word sequences and partial word order and also edit distance).

Precision, Recall and F-measure belong to standard and easy measures. Precision (P) and Recall (R) are based on the concordance of words in MT sentence (hypothesis) with the words in the reference, regardless of the position of the word in a sentence. They have a mutually inverse relationship, i.e. the higher precision, the lower recall and vice versa $Precision = \frac{correct\ words}{length\ hypothesis}$ and $Recall = \frac{correct\ words}{length\ reference}$.

F-measure is a combination of both, *Precision* and *Recall*. It originates in information retrieval and was adapted to machine translation. It is a weighted harmonic mean of precision and recall, $F-measure_\alpha = \frac{(1+\beta^2)*P*R}{R+\beta^2*P}$, where $\alpha, \beta \in R^+$ are parameters for the weights, whereby $\alpha = \frac{1}{1+\beta^2}$.

Bilingual Evaluation Understudy (BLEU) is a current standard and widely used metric for MT evaluation. It is a precision oriented metric, it is a geometric mean of *n*-gram i.e. it computes the number of n-grams in the MT output (hypothesis) which also occur in a reference (for n-gram of size 1-4 with the coefficient of brevity penalty).

$$BLEU(n) = BP \times exp \sum_{n=1}^{N} w_n * \log p_n,$$

where

$$BP = brevity\ penalty = \begin{cases} 1, & if\ hypothesis > reference \\ e^{1-\frac{reference}{hypothesis}}, & if\ hypothesis \leq reference \end{cases}$$

and

$$p_n = precision_n = \frac{\sum_{S\in C}\sum_{n-gram\in S} count_{matched}(n-gram)}{\sum_{S\in C}\sum_{n-gram\in S} count(n-gram)}.$$

Remark 1. S means hypothesis sentence in the complete corpus C.

The *BLEU* metric reflects two linguistic phenomena of manual evaluation metrics-adequacy and fluency, i.e. to semantically correct words and to word order. Lin and Och [22] or Papineni et al. [6] proved that shorter *n*-grams correlates better with adequacy with 1-gram being the best predictor, while longer n-grams has better fluency correlation.

Precision, Recall, F-measure and BLEU metrics are measures of accuracy based on lexical similarity. The other category of automatic intrinsic metrics of MT evaluation is a category based on edit distance. Metrics in this category are called metrics of error rates. They do not measure the concordance but they compute the minimum number of editing steps needed to transform MT output to reference.

Word Error Rate (WER) is based on the edit distance and takes into account the word order. Edit distance is the minimum number of edit operations like word insertions, substitutions and deletions necessary to transform the MT output into the reference. The number of edit operations is divided by the number of words in the reference. When multiple reference translations are given, the reported error for a translation hypothesis is the minimum error over all references

$WER(h, r) = \frac{min_{e \in E(h,r)} (insertion(e) + deletion(e) + substitution(e))}{|r|}$, where *insertion* (e) – number of adding words, *deletion* (e) – number of dropping words, *substitution* (e) – number of replacements (in sequence or path e), r is a reference of MT output h and $min_{e \in E(h,r)}$ is a minimal sequence of adding, dropping and replaced words necessary to transform the MT output (h) into the reference r.

Diversity of language expressions causes the existence of many correct translations even though they are marked as "wrong" order or "wrong" word choice by WER to the references.

Position-independent Error Rate (PER) is a solution for this problem. It does not take into account word order when matching MT output and reference [24]. It is similar to the recall measure. They use the same denominator. It computes the matches between words appearing in MT output (hypothesis) and in reference regardless of the word order in both sentences. It considers the reference and hypothesis as bags of words.

It takes into account excess words that are considered defective and should be removed for translations which are too long

$PER = 1 - \frac{correct - \max(0, length\ hypothesis - length\ reference)}{length\ reference}$.

Cover Disjoint Error Rate (CDER) is based on the Levenshtein distance. It uses the fact that the number of blocks in a sentence is the same as the number of gaps between them plus one. It does not add blocks movement to its calculation, it expresses that as a long jump operation (jump over the gaps between two blocks) and it does not penalize the transfer of entire blocks. *Long jump* is combined with the other steps of editing (*insertion*, *substitution* or *deletion*) and with the null operation in case of identity. In other words, it permits reordering of the whole blocks without penalization. Single words in the reference must be covered only once, while in the hypothesis they can be covered zero, one or more times

$CDER(h, r) = \frac{min_{e \in E(h,r)} (insertion(e) + deletion(e) + substitution(e) + longjump(e))}{|r|}$, where *insertion* (e) – number of adding words, *deletion* (e) – number of dropping words, *substitution* (e) – number of replacements (in sequence or path e) and *long jump* (e) – number of long jumps, r is reference translation of hypothesis h and $min_{e \in E(h,r)}$ is minimal sequence of adding, dropping and replaced words necessary to transform the MT output (h) into the reference r.

3 Method

In this experiment we examined the performance of the MT system – Google translation Api (GT), which is a free web translation service offering translation from/to Slovak. We used automatic intrinsic metrics of MT evaluation, namely- metrics of accuracy (precision, recall, F-measure and Bleu-n) and metrics of error rate (WER, PER and CDER), which are described in depth in the previous section. We assessed the translation quality from morphologically complex language into analytical. In other words, we evaluated the translation quality from European language- Slovak into English. Primarily Slovak language is very rich in inflectional and derivational forms contrary to English with its limited morphological system. Also Slovak has a loose word order compared to English which has a fixed word order. We chose this translation direction for better scores from the metrics WER and BLEU.

We developed a dataset consisting of 360 sentences derived from an original text written without using a control language. Sentences were translated using the above mentioned MT system. We chose only 360 sentences, because the experiment was limited by time (it was a part of another experiment focusing on human translation and post-editing of MT output). Human translators had only 90 min to translate the text into English and to provide a reference translation for MT evaluation. By the text representation we arose from the transaction-sequence model which is further described in [25–27]. We used our system of automatic MT evaluation, in which all algorithms representing the measures were implemented, to obtain scores of the examined metrics. As explained above, all metrics calculate the scores based on the comparison of MT output with one human reference. For sentence alignment the algorithm and software Hunalign was used [28]. Hunalign is an algorithm combining length-based [29, 30] and dictionary (translation) based [31–34] approaches for corpus alignment at the sentence level. For better performance, the algorithm does not take into account the possibility of more than two sentences matching into one sentence.

The first objective of the research was to determine the reliability of automatic intrinsic metrics for MT evaluation using the analysis of reliability and entropy. These metrics can be divided into metrics of error rate (the higher values of these metrics, the lower the translation quality) and metrics of accuracy. The second research objective targeted the identification of relevant and reductant automatic intrinsic metrics for MT evaluation. We tried to identify which metrics decrease the total score of reliability of automatic MT evaluation of machine translation from Slovak to English.

Entropy can be described as a measure of the expected content of the information or uncertainty probability distribution. It is also described as the degree of disorder or randomness in a system. Based on Shannon's definition [35, 36], given a class random variable C with a discrete probability distribution $\{p_i = Pr[C = c_i]\}_{i=1}^k, \sum_{i=1}^k p_i = 1$

where c_i is the ith class. Then the entropy $H(C)$ is defined as $H(C) = -\sum_{i=1}^{k} p_i \log p_i$, while the function decreases from infinity to zero and p_i takes values from interval 0–1 [35, 36].

4 Results

The analysis results showed that the examined automatic intrinsic metrics of error rate are considered highly reliable based on the direct estimation of reliability. As it is shown in Table 1, each metric correlates with the total score of the evaluation (*Avg inter-metrics correlation*: 0.885) and after their elimination the coefficient of reliability has not increased (*Cronbach's alpha*: 0.950; *Standardized alpha*: 0.953) except for the metric referring to word order (WER). After elimination of the metric *WER*, the coefficient of reliability- *Cronbach's alpha* increased from 0.947 to 0.964, which is insignificant. However, the metric WER is the most deviated from the others in the translation quality assessment.

Table 1. Statistics of automatic intrinsic metrics of error rate.

	Metrics-total correlation	Alpha if deleted	Metrics-total accuracy entropy
PER	0.878	0.934	0.934
WER	0.845	0.964	0.852
CDER	0.958	0.869	0.895

For the *entropy* calculation (Table 1), in the case of the analysis of automatic metrics characterizing the error rate of MT evaluation, individual metrics in comparison over accuracy metrics were used. *Entropy* was calculated for each sentence analysed using the specific metrics and for the comparison the average entropy of all sentences was used. From the definition [35] if the *entropy* is closer to 1, then the system is more irregular. The results of the *entropy* for each of the error rate metric correspond with the coefficient of reliability- *Cronbach's alpha*.

The same was shown by the metrics of accuracy. Based on the direct estimation of reliability, metrics precision, recall, F-measure and Bleu-n are considered highly reliable.

Each metric (Table 2) correlates (*Avg inter-metrics correlation*: 0.882) with the total score of evaluation and after their elimination, the *coefficient of reliability* has not increased (*Cronbach's alpha*: 0.975; *Standardized alpha*: 0.975) except for the metric *BLEU-4*. After the elimination of metric *BLEU-4*, the *coefficient of reliability- Cronbach's alpha* increased from 0.974 to 0.976, which is also insignificant (metric *BLEU-4* measures a score of sequence of four words including articles and prepositions).

After the first analysis concerning the reliability of metrics representing the error rate of MT output, we assumed, that the metrics Bleu-n would copy or behave like the

Table 2. Statistics of automatic intrinsic metrics of accuracy.

	Metrics-total correlation	Alpha if deleted	Metrics-total error rates entropy
Precision	0.854	0.973	0.832
Recall	0.939	0.967	0.859
F-measure	0.949	0.966	0.852
BLEU_1	0.933	0.967	0.855
BLEU_2	0.943	0.967	0.852
BLEU_3	0.900	0.970	0.763
BLEU_4	0.807	0.976	0.675

metric WER. This resulted from the fact (as we mentioned in the Sect. 2), that both measures refer to the syntactical structure of the sentence, namely to word order.

The estimations of the entropy of automatic metrics of accuracy (Table 2) were similarly calculated as in the case of the metrics of error rate. Also in this case, the average entropy of all sentences for each metric were used and the results relate with the coefficient of reliability- *Cronbach's alpha* with negligible variations. In case of *entropy*, it also showed that the metric *BLEU-4* deviates the most from the other metrics.

Based on the adjusted univariate test for repeated measures, the zero hypothesis reasoning that the score of automatic intrinsic measures of MT evaluation (PER, WER and CDER) does not depend on individual metrics of the error rate, is rejected at the 1 % significance level (*G-G Epsilon* = 0.6788, *G-G Adj. p* = 0.0000). The strictest metric of error rate was identified *WER* (approximately 63 %) and the loosest *PER* (approximately 47 %).

Based on the results of multiple comparisons- Tukey test (Table 3) three homogenous groups (*PER*), (*CDER*) and (*WER*) were identified in terms of the score of the automatic evaluation of MT. Statistically significant differences in the score between *PER/CDER/WER* and others were proved at the 5 % significance level.

Table 3. Homogeneous groups for automatic intrinsic metrics of error rate.

Metrics of error rate	Mean	1	2	3
PER	47.10	****		
CDER	56.98		****	
WER	63.06			****

Plot (Fig. 1) visualizes the differences between examined metrics. The means with error plot depicts the means and confidence intervals of metrics of error rate.

The metrics of error rate have a significant impact on the quality of MT evaluation, as well as that metrics PER, WER and CDER are relevant for automatic evaluation of MT output.

The second part of the analysis is similar to the previous, and differs only in the metrics. Based on the results of the adjusted univariate test for repeated measure

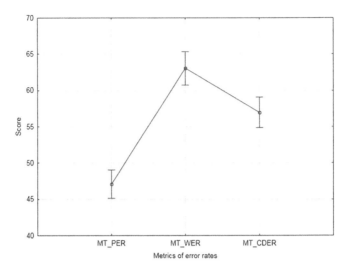

Fig. 1. The means with error plot for intrinsic metrics of error rates.

(*G-G Epsilon* = 0.3426, *G-G Adj. p* = 0.0000) the zero hypothesis reasoning that the score of automatic evaluation of MT does not depend on individual metrics of accuracy, is rejected at the 1 % significance level. The strictest metrics of accuracy (Table 4) were identified *BLEU-4*, *BLEU-3* and *BLEU-2* (approximately 14 %–32 %), and the loosest *Recall*, *BLEU-1*, *F-measure* and *Precision* (approximately 57 %–61 %).

Table 4. Homogeneous groups for automatic intrinsic metrics of accuracy.

Metrics of accuracy	Mean	1	2	3	4	5
BLEU-4	14.03		****			
BLEU-3	20.64			****		
BLEU-2	31.93				****	
Recall	56.86	****				
BLEU-1	57.50	****				
F-measure	58.25	****				
Precision	60.92					****

From multiple comparisons- Tukey test (Table 4) five homogenous groups (Recall, BLEU-1, F-measure), (*BLEU-4*), (*BLEU-3*), (*BLEU-2*) and (*Precision*) were identified in terms of the score of automatic evaluation of MT. Statistically significant differences were proved at the 5 % significance level in the score of automatic evaluation of MT between *BLEU-1/BLEU-2/BLEU-3* and others as well as between *Precision* and others.

The means that the error plot (Fig. 2) depicts means and confidence intervals of metrics of accuracy. The plot visualizes homogeneous groups as well as differences between examined metrics.

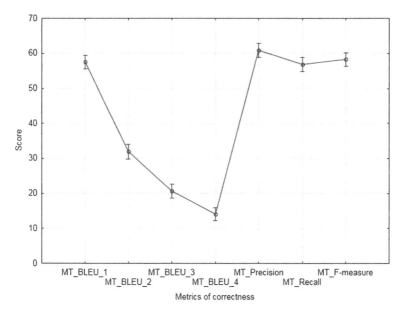

Fig. 2. The means with error plot for intrinsic metrics of accuracy.

Similarly to the metrics of error rate, metrics of accuracy have a significant impact on the quality of MT evaluation, except for metrics *BLEU-1* and *F-measure*. The metrics *BLEU-1* and *F-measure* were identified as redundant metrics of accuracy for MT evaluation.

5 Conclusion and Future Direction

Williams [37] claimed that the techniques and methods for translation quality assessment (TQA) must pass validity and reliability tests if we want TQA procedures to be as objective as possible.

For this reason, we carried out the evaluation of automatic intrinsic metrics for MT evaluation. Evaluation was realized over the textual data obtained from machine translation. Translation was done by MT system (free online machine translation service) and the translation direction was from Slovak (morphologically complex language and low resource) into English.

We showed a way how to identify relevant and redundant automatic metrics for MT evaluation. We presented two approaches to the identification, using the analysis of reliability and using entropy. We used three coefficients of reliability – *Cronbach's alpha*, *Standardized alpha* and *entropy* – to estimate reliability. All estimations were very similar, i.e. individual automatic metrics for MT evaluation have the same variability. The metrics of automatic evaluation have a significant impact on the quality evaluation of MT, except *BLEU-1* and *F-measure* (it was showed that both metrics are redundant in comparison to others).

In future work, we would apply automatic intrinsic metrics for the evaluation of MT output into our translation quality assessment model from Slovak to English and vice versa. In addition, for automatic errors identification and classification of MT output using these metrics which are interconnected to specific morphological and syntactical errors, as well as for the MT evaluation based on POS tagging and for the development of a tool for automatic error detection based on these metrics and morphological annotation of the reference, hypothesis and post-edited MT output.

Above that, MT systems and evaluation of their performance in the context of the inflectional Slovak language has not yet been investigated, which makes the research purposeful and innovative.

Acknowledgments. This work was supported by the Slovak Research and Development Agency under the contract No. APVV-14-0336 and Scientific Grant Agency of the Ministry of Education of the Slovak Republic (ME SR) and of Slovak Academy of Sciences (SAS) under the contracts No. VEGA-1/0559/14.

References

1. Liu, C., Dahlmeier, D., Ng, H.T.: Better evaluation metrics lead to better machine translation. In: Proceedings of Conference Empirical Methods in Natural Language Processing, pp. 375–384 (2011)
2. Babych, B., Hartley, A., Sharoff, S.: Translating from under-resourced languages: comparing direct transfer against pivot translation. In: Proceedings of the MT Summit XI. Citeseer, Copenhagen (2007)
3. Adly, N., Al Ansary, S.: Natural Language Processing and Information Systems. Springer, Heidelberg (2010)
4. Vandeghinste, V., Dirix, P., Schuurman, I., Markantonatou, S., Sofianopoulos, S., Vassiliou, M., Yannoutsou, O., Badia, T., Melero, M., Boleda, G., Carl, M., Schmidt, P.: Evaluation of a machine translation system for low resource languages: METIS-II (2008)
5. Vilar, D., Xu, J., D'Haro, L.F., Ney, H.: Error analysis of statistical machine translation output. In: Proceedings of the 5th International Conference on Language Resources and Evaluation (LREC-06), pp. 697–702, Genoa (2006)
6. Papineni, K., Roukos, S., Ward, T., Zhu, W.-J.: BLEU. In: Proceedings of the 40th Annual Meeting on Association for Computational Linguistics - ACL 2002, p. 311. Association for Computational Linguistics, Morristown, NJ, USA (2001)
7. Banerjee, S., Lavie, A.: METEOR: an automatic metric for MT evaluation with improved correlation with human judgments. In: Proceedings of the ACL Workshop on Intrinsic and Extrinsic Evaluation Measures for MT and/or Summarization (ACL 2005), pp. 65–72, Michigan (2005)
8. Doddington, G.: Automatic evaluation of machine translation quality using n-gram co-occurrence statistics, pp. 138–145 (2002)
9. Koehn, P.: Statistical Machine Translation. Cambridge University Press, Cambridge (2010)
10. Snover, M., Dorr, B., Schwartz, R., Micciulla, L., Makhoul, J.: A study of translation edit rate with targeted human annotation. In: Proceedings of Association for Machine Translation in the Americas, pp. 223–231 (2006)

11. Munkova, D., Munk, M.: An automatic evaluation of machine translation and Slavic languages. In: 2014 IEEE 8th International Conference on Application of Information and Communication Technologies (AICT), pp. 1–5. IEEE (2014)

12. Koehn, P., Monz, C.: Manual and automatic evaluation of machine translation between European languages (2006)

13. Callison-Burch, C., Fordyce, C., Koehn, P., Monz, C., Schroeder, J.: (Meta-) evaluation of machine translation (2007)

14. Callison-Burch, C., Fordyce, C., Koehn, P., Monz, C., Schroeder, J.: Further meta-evaluation of machine translation. In: Proceedings of the Third Workshop on Statistical Machine Translation, pp. 70–106. ACL (2008)

15. Callison-Burch, C., Koehn, P., Monz, C., Schroeder, J.: Findings of the 2009 workshop on statistical machine translation (2009)

16. Callison-Burch, C., Koehn, P., Monz, C., Peterson, K., Przybocki, M., Zaidan, O.F.: Findings of the 2010 Joint Workshop on Statistical Machine Translation and Metrics for Machine Translation. In: Proceedings of the Joint Fifth Workshop on Statistical Machine Translation and Metrics MATR, pp. 17–53. Association for Computational Linguistics (2010)

17. Callison-Burch, C., Koehn, P., Monz, C., Zaidan, O.F.: Findings of the 2011 workshop on statistical machine translation. In: Proceedings of Sixth Workshop Statistical Machine Translation, pp. 22–64 (2011)

18. Callison-Burch, C., Koehn, P., Monz, C., Post, M., Soricut, R., Specia, L.: Findings of the 2012 workshop on statistical machine translation (2012)

19. Bojar, O., Buck, C., Callison-Burch, C., Federmann, C., Haddow, B., Koehn, P., Monz, C., Post, M., Soricut, R., Specia, L.: Findings of the 2013 workshop on statistical machine translation. In: Proceedings of the Eighth Workshop on Statistical Machine Translation. pp. 1–44. Association for Computational Linguistics, Sofia, Bulgaria (2013)

20. Bojar, O., Buck, C., Federmann, C., Haddow, B., Koehn, P., Leveling, J., Monz, C., Pecina, P., Post, M., Saint-Amand, H., Soricut, R., Specia, L., Tamchyna, A.: Findings of the 2014 workshop on statistical machine translation. In: Proceedings of the Ninth Workshop on Statistical Machine Translation, pp. 12–58. Association for Computational Linguistics, Stroudsburg, PA, USA (2014)

21. Bojar, O., Chatterjee, R., Federmann, C., Haddow, B., Huck, M., Hokamp, C., Koehn, P., Logacheva, V., Monz, C., Negri, M., Post, M., Scarton, C., Specia, L., Turchi, M.: Findings of the 2015 workshop on statistical machine translation. In: Proceedings of the Tenth Workshop on Statistical Machine Translation, pp. 1–46. Association for Computational Linguistics, Stroudsburg, PA, USA (2015)

22. Lin, C.-Y., Och, F.J.: Automatic evaluation of machine translation quality using longest common subsequence and skip-bigram statistics. In: Proceedings of the 42nd Annual Meeting on Association for Computational Linguistics - ACL 2004, p. 605. Association for Computational Linguistics, Morristown, NJ, USA (2004)

23. Han, A.L.-F., Wong, D.F., Chao, L.S., He, L., Lu, Y.: Unsupervised quality estimation model for English to German translation and its application in extensive supervised evaluation. Sci. World J. **2014**, 760301 (2014)

24. Tillmann, C., Vogel, S., Ney, H., Zubiaga, A., Sawaf, H.: Accelerated DP based search for statistical translation (1997)

25. Munková, D., Munk, M., Adamová, L.: Modelling of language processing dependence on morphological features. In: Trajkovik, V., Anastas, M. (eds.) ICT Innovations 2013, pp. 77–86. Springer International Publishing, Heidelberg (2014)

26. Munková, D., Munk, M., Vozár, M.: Data pre-processing evaluation for text mining: transaction/sequence model. Procedia Comput. Sci. **18**, 1198–1207 (2013)

27. Munková, D., Munk, M., Adamová, L.: Influence of Stop-words removal on sequence patterns identification within comparable corpora. In: Trajkovik, V., Anastas, M. (eds.) ICT Innovations 2013: ICT Innovations and Education. Advances in Intelligent Systems and Computing, pp. 67–76. Springer International Publishing, Heidelberg (2014)

28. Varga, D., Németh, L., Halácsy, P., Kornai, A., Trón, V., Nagy, V.: Parallel corpora for medium density languages. Proc. RANLP **2005**, 590–596 (2005)

29. Brown, P.F., Lai, J.C., Mercer, R.L.: Aligning sentences in parallel corpora. In: Proceedings of the 29th Annual meeting on Association for Computational Linguistics, pp. 169–176. Association for Computational Linguistics, Morristown, NJ, USA (1991)

30. Tóth, K., Farkas, R., Kocsor, A.: Sentence alignment of Hungarian-English parallel corpora using a hybrid algorithm. Acta Cybern. **18**, 463–478 (2008)

31. Melamed, I.D.: Models of translational equivalence among words. Comput. Linguist. **26**, 221–249 (2000)

32. Melamed, I.D.: Statistical machine translation by parsing. In: Proceedings of the 42nd Annual Meeting on Association for Computational Linguistics - ACL 2004, p. 653–661. Association for Computational Linguistics, Morristown, NJ, USA (2004)

33. Moore, R.C.: A discriminative framework for bilingual word alignment. In: Proceedings of the conference on Human Language Technology and Empirical Methods in Natural Language Processing - HLT 2005, pp. 81–88. Association for Computational Linguistics, Morristown, NJ, USA (2005)

34. Yu, X., Wu, J., Zhao, W.: Dictionary-based Chinese-Tibetan sentence alignment. In: 2010 International Conference on Intelligent Computing and Integrated Systems, pp. 489–493. IEEE (2010)

35. Shannon, C.E.: A mathematical theory of communication. ACM SIGMOBILE Mob. Comput. Commun. Rev. **5**, 3 (2001)

36. Lima, C.F.L., Assis, F.M., Souza, C.P.: A Comparative study of use of Shannon, Rényi and Tsallis entropy for attribute selecting in network intrusion detection. In: Yin, H., Costa, J.A. F., Barreto, G. (eds.) IDEAL 2012. LNCS, vol. 7435, pp. 492–501. Springer, Heidelberg (2012). doi:10.1007/978-3-642-32639-4_60

37. Williams, M.: Translation quality assessment. Mutatis Mutandis **2**, 3–23 (2009)

Bankruptcy Prediction Using Memetic Algorithm

Nekuri Naveen[(⊠)] and Mamillapalli Chilaka Rao

Department of Computer Science and Engineering, Sasi Institute of Technology
and Engineering, Kadakatla, Tadepalligudem, Andhra Pradesh 534101, India
naveen.nekuri@gmail.com, chilakarao@gmail.com

Abstract. This paper proposes a new memetic algorithm using Cuckoo search algorithm and Particle Swarm Optimization algorithm. Training set is fed to the proposed method to get trained. The effectiveness of the proposed method is evaluated using three bankruptcy viz., Spanish banks, Turkish banks and US banks and three benchmark datasets namely, Iris, WBC and Wine datasets. We performed 10 Fold Cross Validation testing and observed that the results obtained by the proposed method in terms of the sensitivity, specificity and accuracy are encouraging when compared to that of the baseline decision tree.

Keywords: Memetic algorithm · Cuckoo search algorithm · Particle swarm optimization · Classification · Bankruptcy prediction · Data mining

1 Introduction

Data mining is the extraction of non-obvious, hidden and actionable knowledge from humongous datasets. The knowledge extracted is useful in providing decision support in many disciplines like banking, insurance, telecom etc. Bankruptcy prediction in banks and financial firms has become significant over the past two decades. Bankruptcy can affect each and every area where the sufferers like creditors, auditors, stockholders and senior management etc. The very particular method to monitor banks is by on-site examinations. The examinations are conducted by regulatory authority in banks for every 12–18 months, as it is mandated by the Federal Deposit Insurance Corporation Improvement Act of 1991. Regulators make use of a six part rating system to indicate the safety and soundness of the organization. Rating is referred to as the CAMELS rating, which evaluates the banks according to their basic functional areas: Capital adequacy, Asset quality, Management expertise, Earnings strength, Liquidity and Sensitivity to market risk. The motivation for this research work: 1. To find out which banks will succeed and which fail in the next few years. 2. To find out the specific characteristics that decides winners and losers.

Objective is to develop a memetic model using Cuckoo search algorithm and Particle Swarm optimization (PSO) algorithm. Cuckoos lay eggs and also search for the food for survival. Hence, two algorithms (Cuckoo search algorithm and PSO) are combined together to work in tandem.

The rest of the paper is structured as follows: Literature review on memetic algorithms using global optimization techniques is presented in Sect. 2. In Sect. 3, the

C. Sombattheera et al. (Eds.): MIWAI 2016, LNAI 10053, pp. 153–161, 2016.
DOI: 10.1007/978-3-319-49397-8_13

proposed memetic algorithm is presented. Results and discussion are presented in Sect. 4, followed by conclusions in Sect. 5.

2   Literature Review

Daniel-Stefan and Sorin-Iulian [1] used statistical models to assess the bankruptcy risk on the Romanian capital market. EI-Maleh et al. [2] proposed Cuckoo search optimization (CSO) algorithm is employed for solving the State Assignment (SA) problem. Yang and Deb in 2009, [3] used cuckoo search algorithm for solving optimization problems. Manoj and Rutupama [4] used a novel adaptive cuckoo search algorithm for intrinsic discriminate analysis based face recognition. Abd Elazim and Ali [5] used Cuckoo search algorithm for Optimal Power System Stabilizers design. Abdelaziz and Ali [6] proposed Cuckoo Search algorithm based load frequency controller (LFC) design for nonlinear interconnected power system. Liu and Fu [7] proposed Cuckoo search algorithm based on frog leaping local search and chaos theory. Huang et al. [8] used Chaos-enhanced Cuckoo search optimization algorithms for global optimization. Balasubbareddy et al. [9] used non-dominated sorting hybrid cuckoo search algorithm to solve single and multi-objective optimization problems. Kaveh and Ilchi [10] used cuckoo search algorithm to determine optimum design of structures for both discrete and continuous variables. Long et al. [11] used an effective hybrid cuckoo search algorithm for constrained global optimization. Xiangtao et al. [12] used Multi-objective cuckoo search algorithm for design optimization. Ehsan et al. [13] proposed improved cuckoo search algorithm for reliability optimization problems. Bing et al. [14] used PSO to develop novel feature selection approaches with the goals of maximizing the classification performance, minimizing the number of features and reducing the computational time. Ali [15] used cuckoo search algorithm for solving manufacturing optimization problems. Milan et al. [16] used Modified cuckoo search algorithm for unconstrained optimization problems. Walton et al. [17] used Modified cuckoo search: A new gradient free optimization algorithm.

Guido et al. [18] used Adaptive acceleration coefficients for a new search diversification strategy in PSO algorithms. Later, Tiago et al. [19] used PSO for extract the classification rules from data base. The potential if-then rules are encoded into real-valued particles that contain all types of attributes in data sets. Rule discovery task is formulated into an optimization problem with the objective to get the high accuracy, generalization performance, and comprehensibility, and then PSO algorithm is employed to resolve it. The advantage of this approach is that it can be applied on both categorical data and continuous data. Zhao et al. [20] uses binary version PSO based algorithm for fuzzy classification rule generation, also called fuzzy PSO. Classification rule generation problem is abstracted as a multiple objectives optimization problem. Namely the candidate rule should be accurate, general and interesting according to the sample data set. Then a fitness function, as a criterion to evaluate the strength of a certain rule, is designed. On this point, their algorithm has common background with the traditional evolutionary algorithm for knowledge discovery. The fitness function consists of three parts which are called accuracy, coverage and interestingness separately. It should be mentioned that in the interestingness part, we generalize the objective measure for the interestingness of a crisp classification rule to that for a fuzzy rule.

Rouhi and Jafari [21] used Classification of benign and malignant breast tumors based on hybrid level set segmentation. Voglis et al. [22] used a parallel memetic global optimization algorithm (p-MEMPSODE) for shared memory multicore system. Aliasghar and Alireza [23] used an adaptive gradient descent-based local search in memetic algorithm applied to optimal controller design. Psychas et al. [24] used Hybrid evolutionary algorithms for the Multi-objective Traveling Salesman Problem. Zahra et al. [25] used Memetic binary particle swarm optimization for discrete optimization problems. Michael et al. [26] used a new memetic pattern based algorithm to diagnose/exclude coronary artery disease. Li et al. [27] used a hybrid memetic algorithm for global optimization. Zhang et al. [28] proposed A Rule-Based Model for Bankruptcy Prediction Based on an Improved Genetic Ant Colony Algorithm to predict corporate Bankruptcy. Bao et al. [29] used A PSO and pattern search based memetic algorithm for SVMs parameters optimization. Wu and Hao [30] used Memetic search for the max-bisection problem. Caraffini et al. [31] used Parallel memetic structures for solving optimization problems. Qasem et al. [32] used Memetic multi-objective particle swarm optimization-based radial basis function network for classification problems. Cano et al. [33] used An interpretable classification rule mining algorithm for solving classification problems. Pedro et al. [34] used A two-stage evolutionary algorithm based on sensitivity and accuracy for multi-class problems. Voglis et al. [35] used MEMPSODE: A global optimization software based on hybridization of population-based algorithms and local searches. Senthamarai and Ramaraj [36] used a novel correlation based local memetic search algorithm for filter ranking. Ang et al. [37] used an evolutionary memetic algorithm for rule extraction. Lu and Jin-Kao [38] used a memetic algorithm for graph coloring. Chiam et al. [39] used a memetic model of evolutionary PSO for computational finance applications. Funda et al. [40] used a memetic random-key genetic algorithm for a symmetric multi-objective traveling salesman problem. Chen et al. [41] used Memetic algorithm for two real-world problems involving designing intrusion detection system and breast cancer classification. Wang et al. [45] has presented the bio-inspired algorithms like Ant Colony Optimization, Genetic Algorithm, Evolutionary Algorithm and PSO in web service composition.

3 Existing and Proposed Method

3.1 Cuckoo Search Algorithm

Cuckoo Search Algorithm was developed by Yang and Deb [3] in 2009. It is one of the population based optimization algorithm. Cuckoos are intriguing birds or extremely attention-grabbing birds. They make very beautiful sounds. Cuckoos lay their eggs in other birds' nests. This is by nature to get generating the family by some cuckoo species. Host birds can take on direct conflict with the cuckoos which lays eggs in their nests. Cuckoo Search for such breeding behavior.

If the host bird discovers the eggs which are not laid by their own, then they simply abandon its nest and re-build a new nest in a new location. Some female parasitic cuckoos are very specific in the mimicry in color and pattern of the eggs of a few chosen host species. Remove other bird eggs to increase the hatching probability of their own eggs. This reduces the probability of their eggs being abandoned and thus increases their re-productivity.

Cuckoo eggs hatches slightly earlier than the host bird eggs. First cuckoo chick is hatched, the first sense act it will take own eggs and throw out their eggs out of the nest blindly. This in turn increases the cuckoo chick's share of food provided by its host bird. A cuckoo chick can also mimic the call of host chick's in order to get more access to feeding opportunity.

Characteristics [3]
- Each egg in a nest represents a solution.
- Cuckoo egg represents a new solution.
- Each nest has one egg represents a single solution.
- Each nest has multiple eggs representing a set of solutions or group of solutions.

Cuckoo Search Rules [3]

It is based on three rules. They are,

1. Each cuckoo lays one egg at a time, and places its egg in a randomly chosen nest.
2. The best nest with high quality of eggs will carry over to the next generation.
3. The number of available host nests is fixed, and the egg laid by a cuckoo is discovered by the host bird with a probability of $Pa \in [0,1]$.

Based on the above three rules, the algorithm is presented below.

Algorithm (Yang and Deb, [3])

> Objective function f(x), $x=(x_1, x_2,......, x_d)^T$
> Generate initial population of n host nests xi
> **While**(t<MaxGenerations)
> > Get a cuckoo randomly /generate a solution by Levy light
> > > and then evaluated its quality/fitness F_i
> > Choose a nest among n (say,j) randomly
> > **if**($F_i > F_j$)
> > > Replace j by the new solution
> > **end**
> > A fraction (pa) of worse nests are abandoned
> > and new ones are built
> > Keep best solutions
> > Rank the solutions and find the current best
> **end** while
> Postprocess results and visualization

3.2 Particle Swarm Optimization

PSO [42] Algorithm was developed by Kennedy and Eberhart in 1995. It is also one of the population based optimization algorithm. PSO is initialized with a group of random particles and then searches for optimal solution for every iterations. In every iteration, each particle is updated by two "best" values. The first value is the best solution or best fitness it has achieved so far. This value is called "**Pbest**". Another "best" value that is obtained so far by any particle in the population. This best value is a global best or "**Gbest**".

Each particle is concerned towards the current global best g position and its own best position x_i. If a particle finds a location which is better than its previously best found locations, then it will be updated as the new current best location for particle "i". There is a current best for all "n" particles at any time "t" during iterations. The aim is to find the global best solution among all the current best solutions until the number of iterations completed. Velocity is updated as,

$$\mathbf{V}_{id}^{new} = \mathbf{w}\,\mathbf{V}_{id}^{old} + \mathbf{c}_1\mathbf{r}_1\left(\mathbf{P}_{id} - \mathbf{X}_{id}\right) + \mathbf{c}_2\mathbf{r}_2\left(\mathbf{P}_{gd} - \mathbf{X}_{id}\right) \tag{1}$$

w- Weighted acceleration, v-velocity, c_1, c_2 are acceleration coefficients,

r_1, r_2 random numbers between (0 1), p_{id} is the local best and p_{gd} is the global best position. Position is updated as,

$$\mathbf{X}_{id}^{new} = \mathbf{X}_{id}^{old} + \mathbf{V}_{id}^{new} \tag{2}$$

3.3 Proposed Memetic Algorithm

The term Memetic Algorithm is now widely used as a synergy of evolutionary or any population-based approach with separate individual learning or local improvement procedures for problem search. Cuckoo Search Algorithm and Particle Swarm Optimization Algorithm are combined to frame Memetic algorithm. Memetic algorithm is applied for bankruptcy prediction. This algorithm is used to find global optimum of a solution for a given optimization problems, which can predict bankruptcy.

Algorithm
    Begin
        Generate initial population of birds x_i
        While(t<MaxGenerations)
            Get a cuckoo randomly /generate a solution by Levy light
            and then evaluated its quality/fitness F_i
            Choose a nest among n (say,j) randomly
            if($F_i > F_j$)
                Replace j by the new solution
            End
            Update the velocity and positions of birds
            Calculate newFitness
            Find Best Global Solution
        end while

4   Results and Discussions

We worked on three different bankruptcy data sets viz., Spanish banks, Turkish banks, UK banks and three benchmark datasets viz., Iris, WBC, and Wine. Experimental setup is followed in a different fashion. We first divided the dataset into two parts in 80:20 ratios. 20 % data is then named validation set and stored aside for later use and 80 % of the data is used as training set. Then 10-fold cross validation was performed on the 80 % of the data i.e. training data for building the model. Later, the efficiency of the model is evaluated against validation set i.e. 20 % of the original data. Sensitivity, Specificity and Accuracy are the three metrics used to evaluate the performance of the proposed model.

Sensitivity is the measure of proportion of the true positives, which are correctly identified [43].

$$\text{Sensitivity} = \frac{\#\text{Total samples correctly classified Positive Samples}}{\#\text{Total number of Positive samples in the Validation set}} * 100$$

Specificity is the measure of proportion of the true negatives, which are correctly identified [43].

$$\text{Specificity} = \frac{\#\text{Total samples correctly classified Negative Samples}}{\#\text{Total number of Negative samples in the Validation set}} * 100$$

Accuracy is the measure of proportion of true positives and true negatives, which are correctly identified [43].

$$\text{Accuracy} = \frac{\#\text{Total samples correctly classified}}{\#\text{Total number samples in the Validation set}} * 100$$

Empirical results presented in this thesis, are the best Sensitivities, Specificities and Accuracies against validation set. User defined parameters in the memetic algorithm are taken as follows: The size of the population and the host bird nests are taken as 30 across all the data sets. C1 and C2 are taken with combination between [1 4]. Inertia (W) in PSO is taken in between (0 1). Random number is also taken between 0 and 1. The probability (Pa) to discover the eggs by the host bird nest is taken in between [0 1]. We got better results for Pa value 0.9. Proposed memetic algorithm is implemented in java. Results are tabulated in the Table 1 of all the data sets. The proposed method is compared with the decision tree results which are presented in Table 1. Decision Tree (DT) was introduced by the Quinlan in 1984 [44]. The public province implementation of decision tree is C4.5 algorithm. C4.5 is a supervised approach for classification problems. DT is well familiarized to the data mining researchers.

Our results indicate that the proposed method has yielded better results in the case of WBC and Turkish bank data sets with good sensitivity and accuracy. Where as in the case of US bank data set it is almost similar when compared with accuracy. Even though the proposed method didn't perform well in the case of Iris and Wine datasets, it has performed well for the two class classification problems in the study.

Table 1. Average results of 10-FCV

Data set	Decision tree			Proposed method		
	Sens*	Spec*	Acc*	Sens*	Spec*	Acc*
Iris	–	–	92.66	–	–	57.51
Wine	–	–	97.42	–	–	43.81
WBC	31.11	91.7	71.19	91.31	90.86	91.40
Spanish banks	71.66	100	86.91	67.66	81.67	72.56
Turkish banks	55.00	100	77.5	88.50	96.00	90.00
US Banks	93.07	85.37	89.22	81.25	94.36	89.04

*Sens = Sensitivity; Spec = Specificity; Acc = Accuracy.

5 Conclusions

Data mining applications are becoming very much common in every domain like banking and finance, biology, manufacturing, etc. The researchers have started using the optimization algorithms and also memetic algorithms to solve the data mining problems. In this new era, we proposed a new memetic algorithm by employing Cuckoo Search algorithm and PSO. The data sets taken were bankruptcy and bench-mark datasets. The proposed algorithm is compared with the Decision tree. By considering the sensitivity and accuracy in the bankruptcy data sets, memetic algorithm yielded better. It clearly shows that for the two class classification problems the proposed memetic algorithm works better. As the problems solved in banking and finance domain, management would be very much interested to have model to understand the behavior of the customer and the financial health of a bank and predict the future.

References

1. Daniel-Stefan, A., Sorin-Iulian, C.: An assessment of the bankruptcy risk on the Romanian capital market. Procedia-Soc. Behav. Sci. **182**, 535–542 (2015)
2. EI-Maleh, A.H., Sait, S.M., Bala, A.: State assignment for area minimization of sequential circuits based on cuckoo search optimization. Comput. Electr. Eng. **44**, 13–23 (2015)
3. Yang, X-S., Deb, S.: Cuckoo search via levy flights. Nature & Biologically Inspired Computing, pp. 210–214 (2009)
4. Manoj, K.N., Rutupama, P.: A novel adaptive cuckoo search algorithm for intrinsic discriminant analysis based face recognition. Appl. Soft Comput. **38**, 661–675 (2016)
5. Abd Elazim, S.M., Ali, E.S.: Optimal power system stabilizers design via Cuckoo Search algorithm. Int. J. Electr. Power Energy Syst. **75**, 99–107 (2016)
6. Abdelaziz, A.Y., Ali, E.S.: Cuckoo Search algorithm based load frequency controller (LFC) design for nonlinear interconnected power system. Int. J. Electr. Power Energy Syst. **73**, 632–643 (2015)
7. Lu, X., Fu, M.: Cuckoo search algorithm based on frog leaping local search and chaos theory. Appl. Math. Comput. **266**, 1083–1092 (2015)
8. Huang, L., Ding, S., Shouhao, Y., Wang, J., Ke, L.: Chaos-enhanced Cuckoo search optimization algorithms for global optimization. Appl. Math. Model. **40**(5–6), 3860–3875 (2016)

9. Balasubbareddy, M., Sivanagaraju, S., Chintalapudi, V.S.: Multi-objective optimization in the presence of practical constraints using non-dominated sorting hybrid cuckoo search algorithm. Int. J. Eng. Sci. Technol. **18**(4), 603–615 (2015)

10. Kaveh, A., Ilchi, M.G.: Cuckoo search optimization. In: Kaveh, A. (ed.) Advances in Metaheuristic Algorithms for Optimal Design of Structures, pp. 317–347. Springer International Publishing, Cham (2014)

11. Long, W., Ximing, L., Huang, Y., Chen, Y.: An effective hybrid cuckoo search algorithm for constrained global optimization. Neural Comput. Appl. **25**(3), 911–926 (2014)

12. Xiangtao, X.-S.Y., Suash, D.: Multiobjective cuckoo search algorithm for design optimization. Comput. Oper. Res. **40**(6), 1616–1624 (2013)

13. Ehsan, V., Saeed, T., Shahram, M., Atiyeh, H.: Improved Cuckoo search algorithm for reliability optimization problems. Comput. Ind. Eng. **64**(1), 459–468 (2013)

14. Bing, X., Zhang, M., Will, N.B.: Particle swarm optimization for feature selection in classification: novel initialization and updating mechanisms. Appl. Soft Comput. **18**, 261–276 (2014)

15. Ali, R.: Yildiz: Cuckoo search algorithm for the selection of optimal machining parameters in milling operations. Int. J. Adv. Manuf. Technol. **64**(1), 55–61 (2013)

16. Milan, T., Milos, S., Nadezda, S.: Modified Cuckoo search algorithm for unconstrained optimization problems. In: Proceedings of the European Computing Conference (2011)

17. Walton, S., Hassan, O., Morgan, K., Brown, M.R.: Modified cuckoo search: a new gradient free optimisation algorithm. Choas Solitons Fractals **44**(9), 710–718 (2011)

18. Guido, A., Giovanna, C., Giorgio, P.: Adaptive acceleration coefficients for a new search diversification strategy in particle swarm optimization algorithms. Inf. Sci. **299**, 337–378 (2015)

19. Tiago, S., Silva, A., Neves, A.: Particle swarm based data mining algorithms for classification tasks. Parallel Comput. **30**, 767–783 (2004)

20. Zhao, X., Zeng, J., Gao, Y., Yang, Y.: A particle swarm algorithm for classification rules generation. In: International Conference on Intelligent Systems Design and Applications (2006)

21. Rouhi, R., Jafari, M.: Classification of benign and malignant breast tumors based on hybrid level set segmentation. Expert Syst. Appl. **46**, 45–59 (2016)

22. Voglis, C., Hadjidoukas, P.E., Parsopoulos, K.E., Papageorgiou, D.G., Lagaris, I.E., Vrahatis, M.N.: p-MEMPSODE: Parallel and irregular memetic global optimization. Comput. Phys. Commun. **197**, 190–211 (2015)

23. Aliasghar, A., Alireza, A.: An adaptive gradient descent-based local search in memetic algorithm applied to optimal controller design. Inf. Sci. **299**, 117–142 (2015)

24. Psychas, I.-D., Eleni, D., Yannis, M.: Hybrid evolutionary algorithms for the multiobjective traveling salesman problem. Expert Syst. Appl. **42**(22), 8956–8970 (2015)

25. Zahra, B., Siti, M.S., Shafaatunnur, H.: Memetic binary particle swarm optimization for discrete optimization problems. Inf. Sci. **299**, 58–84 (2015)

26. Michael, J.Z., Miriam, B., Urs, B., Andrew, T., Peter, R., Matthias, E.P.: A new memetic pattern based algorithm to diagnose/exclude coronary artery disease. Int. J. Cardiol. **174**(1), 184–186 (2014)

27. Li, Y., Jiao, L., Li, P., Wu, B.: A hybrid memetic algorithm for global optimization. Neurocomputing **134**, 132–139 (2014)

28. Zhang, Y., Wang, S., Jiet, G.: A rule-based model for bankruptcy prediction based on an improved genetic ant colony algorithm to predict corporate bankruptcy. Math. Probl. Eng., 10 (2013)

29. Bao, Y., Hu, Z., Tao, X.: A PSO and pattern search based memetic algorithm for SVMs parameters optimization. Neurocomputing **117**, 98–106 (2013)

30. Wu, Q., Hao, J.-K.: Memetic search for the max-bisection problem. Comput. Oper. Res. **40** (1), 166–179 (2013)
31. Caraffini, F., Neri, F., Lacca, G., Mol, A.: Parallel memetic structures. Inf. Sci. **227**, 60–82 (2013)
32. Qasem, S.N., Shamsuddin, S.M., Hashim, S.Z.M., Darus, M., Al-Shammari, E.: Memetic multiobjective particle swarm optimization-based radial basis function network for classification problems. Inf. Sci. **239**, 165–190 (2013)
33. Cano, A., Zafra, A., Ventura, S.: An interpretable classification rule mining algorithm. Inf. Sci. **240**, 1–20 (2013)
34. Pedro, A.G., Cesar, H.M., Jose, F., Mariano, C.: A two-stage evolutionary algorithm based on sensitivity and accuracy for multi-class problems. Inf. Sci.: Int. J. **197**, 20–37 (2012)
35. Voglis, C., Parsopoulos, K.E., Papageorgiou, D.G., Lagaris, I.E., Vrahatis, M.N.: MEMPSODE: a global optimization software based on hybridization of population-based algorithms and local searches. Comput. Phys. Commun. **183**(5), 1139–1154 (2012)
36. Senthamarai, S.K., Ramaraj, N.: A novel hybrid feature selection via Symmetrical Uncertainty ranking based local memetic search algorithm. Knowl.-Based Syst. **23**(6), 580–585 (2010)
37. Ang, J.H., Tan, K.C., Mamun, A.A.: An evolutionary memetic algorithm for rule extraction. Expert Syst. Appl. **37**(3), 1302–1315 (2010)
38. Lu, Z., Jin-Kao, H.: A memetic algorithm for graph coloring. Eur. J. Oper. Res. **203**(1), 241–250 (2010)
39. Chiam, S.C., Tan, K.C., Mamun, A.A.: A memetic model of evolutionary PSO for computational finance applications. Expert Syst. Appl. **36**(2), 3695–3711 (2009)
40. Funda, S., William, G.F., Mary, E.K.: A memetic random-key genetic algorithm for a symmetric multi-objective traveling salesman problem. Comput. Ind. Eng. **55**(2), 439–449 (2008)
41. Chen, Y., Ajith, A., Yang, B.: Feature selection and classification using flexible neural tree. Neurocomputing **70**(1–3), 305–313 (2006)
42. Eberhart, R.C., Kennedy, J.: A new optimizer using particle swarm theory. In: Proceedings of the ISMMHS, pp. 39–43 (1995)
43. Fawcett, T.: An introduction to ROC analysis. Pattern Recogn. Lett. **27**, 861–874 (2006)
44. Quinlan, J.R.: C4.5: Programs for Machine Learning. Morgan Kaufmann, San Mateo (1992)
45. Wang, L., Shen, J., Yong, J.: A survey on bio-inspired algorithms for web service composition. In: Proceeding of the 2012 IEEE 16th International Conference on Computer Supported Cooperative Work in Design, pp. 569–574, USA (2012)

Crowd Simulation in 3D Virtual Environments

Somnuk Phon-Amnuaisuk[1,3](✉), Ahmad Rafi[2], Thien-Wan Au[3], Saiful Omar[3], and Nyuk-Hiong Voon[3]

[1] Media Informatics Special Interest Group, Centre for Innovative Engineering,
Universiti Teknologi Brunei, Gadong, Brunei
`somnuk.phonamnuaisuk@utb.edu.bn`
[2] Faculty of Creative Multimedia, Multimedia University, Cyberjaya, Malaysia
`rafi@mmu.edu.my`
[3] School of Computing and Informatics,
Universiti Teknologi Brunei, Gadong, Brunei
{`twan.au,saiful.omar,jennifer.voon`}`@utb.edu.bn`

Abstract. Realistic animation of agents' activities in a 3D virtual environment has many useful applications, for examples, creative industries, urban planning, military simulation and disaster management. It is tedious to manually pre-program each agent's actions, its interactions with other agents and with the environment. Simulation is a good approach in this kind of domain since complex global behaviors emerge from the local interactions. We simulate a crowd movement using a multi-agent approach where each agent is situated in the virtual environment. An agent can perceive and interact with other agents and with the environment. Complex behaviors emerging from these interactions are from local rules and without any central control. These behaviors reveal the complexity of the domain without explicitly programming the system. In this work, we investigate (i) the navigation of the agents and (ii) the corresponding animations of each agent's behaviors. Simulation results under different parameters are presented and discussed.

Keywords: Crowd simulation · Boids framework · Behavior-based animation · 3D virtual environments

1 Introduction

Given a complex system, where there are interactions among components in the system, it is a great challenge to define the underlining functions governing the complex interactions among them. Contemporary works in this area favor the explanation of complex interactions based on the self-organising paradigm. Complex behaviors observed in a self-organising model emerges from simple interactions among the components in the system without any central control. For example, fish schooling and bird flocking can be successfully modelled based on simple local rules such as the one proposed in the Boid framework [1].

Simulation is an effective approach to model a complex system since emerging complex behaviors in the system are not required to be pre-programmed; instead,

© Springer International Publishing AG 2016
C. Sombattheera et al. (Eds.): MIWAI 2016, LNAI 10053, pp. 162–172, 2016.
DOI: 10.1007/978-3-319-49397-8_14

they emerge from the simulation process. Computational approaches such as cellular automata [2], social cognitive simulation [3] and agent/multi-agent systems [4], have been developed in recent decades to deal with simulations of agents in static and dynamic environments. In this work, crowd behaviors are simulated by adapting ideas from the Boids framework. The crowd movement in a 3D virtual environment is navigated using steering forces. Detailed animations of agents' behaviors are animated using a behavior-based Finite State Machine (FSM). The term behavior-based FSM used here implies that, in each FSM state, the agent's action is abstracted at a particular behavioral unit e.g., walk, run, etc. This abstraction allows us to animate an agent according to its behavioral state. We simulate crowd behaviors in two main scenarios (i) crowd movement in an open space and (ii) evacuation movement of crowd.

The rest of the paper is organized into the following sections: Sect. 2 discusses the background of crowd behaviors simulation in a 3D virtual environment; Sect. 3 discusses our approach and gives the details of the techniques behind it; Sect. 4 provides the experimental design and provides a critical discussion of the output from the proposed approach; and finally, the conclusion and further research are presented in Sect. 5.

2   Crowd Behaviors Simulation

The simulation of crowd behavior has been investigated by researchers from different disciplines. Creative programmers simulate crowd behaviors for their art-work or other creative visual effects [5]; game and virtual reality developers implement crowd simulation to enhance visual experiences [6,7]; cognitive scientists study crowd psychology and cognitive models of herd behaviors [8]; and real estate companies simulate a walk-through of their properties for marketing purposes and simulate an evacuation plan in case of fire or other disasters to obtain a 'fit for dwelling' license from their government.

Depending on the domain of interest, agents' behaviors and their control languages should be abstracted at an appropriate granularity according to their tasks and their design frameworks. For example, the subsumption architecture [9] requires a reactive control on each individual layer and the mechanisms to pass and to suppress control signals between layers. In such a system, a steering behavior may be abstracted at the control granularity of signal strengths that feed to the left/right wheels of a robot.

There are many existing design frameworks which can be classified in the spectrum, at one end is the deliberative framework such as the STRIPS system [10] and at the other end of the spectrum is the reactive framework. A pure deliberative system will not react to stimuli but deliberate for the best action according to perceptions and its internal world model. The knowledge used to infer deliberative actions may be learned and represented as policies [11], finite state machines, behavior trees [6,12], or emerging swarm behaviors [13], etc. On the other hand, a pure reactive system does not require any world model nor any planing but just react to any stimulus. Most implementations are a hybrid between these two components of the deliberative and the reactive. From the literature, research activities in this area span over various topics: deliberative planning process [10], reactive subsumption architecture [9], emerging behaviour [1], behavior-based agent, [12,14], motion synthesis, motion planning, character animation [15–17] and behaviors of autonomous agent [18]. This is a rich research area and there is a large volume of work.

How could crowd behaviors be best simulated? If software agents and their environment could be implemented such that their interactions are completely governed under the laws of a physical world, then all behaviors can be simulated by a computer simulation. This is not feasible due to high computational expenses and its complexity. However, we can choose to describe the physics instead of implementing the physics. In other words, the behaviors of an agent under different stimuli can be described as a program. This alternative is feasible but the level of fidelity will be compromised since it is still not feasible to describe a complete world model, e.g., a rag doll can be under the physics but it is still not feasible now to let the avatar walk and run under the laws of physics.

This work focuses on a 3D virtual environment. The vocabulary in our universe of discourse is abstracted at the *names of symbolic properties* and *action words* granularity levels. For examples, the agents perceive their environment as wall, floor, stair, door, etc., and interact with the environment according to their policies. The agents' actions, e.g., walk, run, jump, etc., can be either reactive or deliberative. The reactive actions tend to be local in nature while their deliberative actions are determined using more information about the world model. A navigation path obtained from the shortest path algorithm such as A* search [19] is an example of a deliberative action plan commonly used to navigate the agents.

3   Our Approach

Crowd simulations have been investigated using various techniques: flow-based modelling [20], Boids-based modelling [1] and the agent-based modelling [21]. The agent-based model has gained popularity in recent years due to advances in game engine platforms. This work explores techniques drawn from both Boids-based and agent-based approaches which will be described in this section.

Given a crowd with the goal of navigating around an area, each agent in the crowd must navigate itself from its current location to a new location. This is a very rich domain and there are many plausible simulations. Self-interested agents will compete to gain the best utility (i.e., for an evacuation example, this could be to get to safety as soon as possible) while collaborative agents may choose to optimize overall the survival rate of the crowd, etc. We can pose various agents' personalities that will better reflect real life situation. However, it is beyond the scope of this work to examine different agents' personalities and their simulation results. Here, we assume that agents are rational and they act according to their policies. In the big picture, each agent's behaviors in the crowd at each time step can be summarized as follows: (i) sensing the environment, (ii) reacting and deliberating appropriate actions and (iii) executing an action.

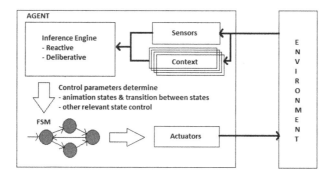

Fig. 1. Agent architecture: an agent perceives and acts in the environment. Actions can be either deliberative or reactive. An agent's behaviors are animated according to their corresponding states in the Finite State Machine (FSM).

Figure 1 shows a simplified block diagrams of the behavior-based agent employed for crowd simulation in this experiment. It should be clear that in the multi-agent system, each agent cannot plan their movements in isolation with other agents' actions. Besides, if the environment is dynamic, it will not be fruitful to plan many moves ahead into the future since the environment will change and the plan must be revised anyway. Hence, simulation is commonly carried out at each time step and behavior-based agents are reactive by nature while keeping sight of a long term goal.

3.1   Controlling Behavior-Based Animations

To animate the agents' behaviors, the agents' perceptions are translated into control parameters which are used to condition the agents' reactive actions as well as used to infer the agents' deliberative actions. The agents' actions are grounded to animation clips where various techniques such as blending and masking actions can be employed to increase the realism of the animations.

Let $\mathbf{S}_t = <s_1, ..., s_i>_t$ be an input vector from a sensor at the time t, let $C = \{panic, calm, fight, flight, ...\}$ be a set of context and let $\mathbf{A}_B$ denote an animation clip of behavior B such that $\mathbf{A}_{walk} = [walk_1, ..., walk_f]$ portrays a *walking* behavior in f animation frames. Since it is not desired to store all possible animations in the database (it is not a feasible option), variations in the animation can be obtained from a finite set of animation clips using techniques such as layers blending, and masking, etc.

In this simulation, the agents' perceptions are obtained through *ray casting*[1] where the agents continuously sense changes in the environment. The agents' perceptions and context are mapped to control parameters $\mathcal{K}$ which are employed by a FSM. This mapping is handled by the *inference engine* $\mathcal{I}$–see Fig. 1.

---

[1] Unity game engine provides the *Raycaster* event system which allows an agent to query about game objects in the scene.

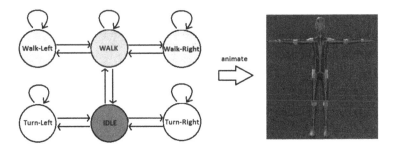

Fig. 2. Behavior Control through Control Parameters: FSM starts at the *Idle* state, and transits to various states according to its perceptions and its policies. Each state animates the avatar using a corresponding animation clip of the state.

$$\mathcal{K}_{FSM} \leftarrow \mathcal{I}(\mathbf{S}_t, C)$$

Figure 2 shows how control parameters control the agent's appearance behaviors by controlling the state transition and state animation properties of the FSM. For example, if an agent is at the WALK state and is walking toward the wall, it should sense the presence of the wall ahead and the appropriate control parameters $\mathcal{K}_{FSM}$ should be inferred. The agent will then be transited to a new appropriate state with an appropriate animation.

3.2   Boids-Based Navigation

The Boids framework [1] successfully simulates flocking behaviors by controlling the following three essential concepts: alignment, separation and cohesion. Inspired by the Boids framework and the recent implementation by [5] where the swarming concept has been applied to create spectacle visual effects, we explore the usage of the Boids framework in the agent navigation task.

Navigating a crowd between two locations requires a steering force that drives the crowd to the destination. Hence, the alignment concept is implemented in this work as a force driving each agent n toward its goal destination.

$$\mathbf{f}_a = k_a(p_n.goal() - p_n.loc())$$

where p_n denotes the agent n, $p_n \in \mathbf{P}$ and $p_n.goal()$ and $p_n.loc()$ denotes the goal position and the current position respectively. Each member in the crowd is attracted (coherent) and repelled (separation) from the group under the following separation force

$$\mathbf{f}_s = \sum_{i=1}^{|\mathbf{P}|} \mathcal{N}_i k_s(p_d.loc() - p_i.loc())$$

and the coherent force

$$\mathbf{f}_c = \sum_{i=1}^{|\mathbf{P}|} \mathcal{N}_i k_c(p_i.loc() - m_{\mathcal{N}}.loc())$$

where $\mathcal{N}_i = 1$ if p_i is within the desired neighborhood of p_d, else $\mathcal{N}_i = 0$, $m_{\mathcal{N}}$ is the center of all particles in the neighborhood of p_d which is a sphere of radius r (arbitrarily set by the users), k_a, k_s and k_c is a normalisation factor that moderates the effect of the distances $\|p_n.goal() - p_n.loc()\|$, $\|p_i - m_{\mathcal{N}}\|$ and $\|p_d - p_i\|$ respectively. The forces $\mathbf{f}_s$ and $\mathbf{f}_c$ control the direction and the speed of the agents.

Table 1. Control parameters

Information	Experiment 1	Experiment 2
Objective	Pursuing 4 locations	Evacuate the area
Number of agents	30, 90, 150	5, 25, 50, 100, 200
Agents' percept	Ray cast	Ray cast
FSM transition controls		
$S(f_a + f_c + f_s) \rightarrow$ speed	Speed $\in [0,100]$	Speed $\in [0,100]$
$W(f_a + f_c + f_s) \rightarrow$ walk	Walk $\in [0,100]$	Walk $\in [0,100]$
$T(f_a + f_c + f_s) \rightarrow$ turn	Turn $\in [0,100]$	Speed $\in [0,100]$

*where $S(\cdot)$, $W(\cdot)$ and $T(\cdot)$ are functions that return corresponding *Speed, walk and turn* FSM transition control signals.

The essential concepts of the Boids framework can be summarized as follows: alignment enforces the agents to steer along the group direction, separation enforces the agents to steer away from each other if the agents are too close to each other while cohesion enforces the agents to steer toward a group center. The steering parameters in our behavior-based approach include *speed, walk, turn left* and *turn right* and all the Boids concepts are manifested through these control parameters. This is quite different from the original Boids framework since the steering forces are not directly derived from positions and velocities. Table 1 summarizes the information about the experiments and their relevant control parameters.

4   Experimental Design and Results

In a static environment with a single agent, an agent's optimal navigation route may be easily computed. For example, the most effective route between any two points in the environment can be computed using A* algorithm. Although the path-finding facility has been provided in commercial packages, it is not of our interest here to examine it since its application would be suitable for a static environment. Here, we explore the multi-agent systems which are dynamic by nature. In order for the agent to obtain optimal adaptive behavior in a dynamic environment, it must continuously observe its environment and recalculate its actions. This is a very expensive task and it is our interest to investigate the navigation adapted from the Boids framework.

4.1   Experimental Setup

In this work, various crowd movements have been simulated using the Boid-based navigation together with the behavior-based animations. Two scenarios have been investigated (i) crowd movement in an open space and (ii) evacuation movement of a crowd. In the first scenario, the environment is an open space of size 50×50 unit^2; each member of the crowd is randomly spawned at the circular area with 20 unit diameter at the center of the environment (positions and initial directions are randomly initialized). Four goals are set at the four corners at $(-20, 0, -20), (-20, 0, 20), (20, 0, -20)$ and $(20, 0, 20)$ respectively. Each agent knows that it must achieve these four goals in sequence. Once the agent has completed all the goals, its goals are refreshed and it repeats the task. In the second scenario, we simulate the evacuation of different crowd sizes 5, 10, 50, 100 and 200 agents evacuating an area through a door. Figure 3 shows the environment employed in this experiment. Both scenarios are simulated using different crowd sizes and the time steps taken by the different crowds are recorded.

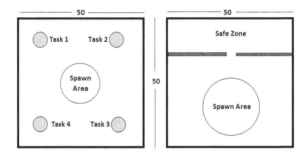

Fig. 3. Environment: left pane – for crowd movement in an open space scenario; right pane – evacuation movement of a crowd.

4.2   Results and Discussion

Figure 4 shows the crowd traversing the four corners in the environment. Top row: agents form a clear line (effects from a strong alignment and a weak separation forces). This results in a more organized patrolling behavior. The crowd takes less time to make a complete round. Bottom row: agents form a loose crowd (effects from a strong separation force). This results in a less organized patrolling behavior and the crowd takes a longer time to make a complete round. Figure 5 shows the crowd escaping to a safety zone. Four snap shots (from left to right and top to bottom) illustrate the crowd at the start of the simulation and when all agents have completed their goals.

Since the Boids-based navigation has three main control parameters: alignment, separation and cohesion, varying these parameters yields different crowd behaviors. Figure 6 summarizes the results from the two experiments. The left pane shows the time the crowd takes to complete each of the 4 tasks (traverse the four waypoints). The result is logical - a bigger crowd size takes a longer time.

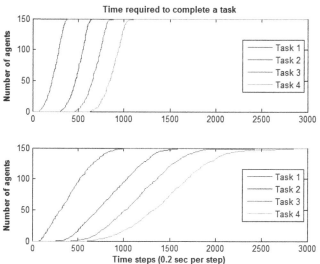

Fig. 4. Agents behave differently under different control parameters.

The right pane (of Fig. 6) shows the time the crowd takes to evacuate the area. The results show a similar pattern. As the size of the crowd grows, more time is needed. However, we observe that the average step that the crowd takes to escape to the safety zone appears to converge to 250 steps. This issue must be further investigated. We suggest that this converging point depends on many factors: the environment e.g., the width of the exit door, the distance to the exit door; the number of agents; the initial position spawned and the directions of each agent. With a small number of agents, each agent walks to the exit door

Fig. 5. From left to right and top to bottom: (i) agents are randomly spawned, (ii, iii, iv) each agent moves toward the exit door while interacting with each other and the environment.

at a constant speed. As the number of agents grows, all the interactions among the agents affect the outcome. The agents' speed may be increased (walk faster) if they are free to walk; their speed may be decreased (walk slowly) if they are clustered together, or they turn away from each other if they are about to crash into each other, and so on. Hence, it takes longer to enter the safe zone.

However, when the number of agents grows larger than a particular point in a certain environmental configuration, the overall time taken is not linearly increased since each agent may speed up or slow down and since the size of the area is constant. The time taken by the crowd should be convergent and this is observed here.

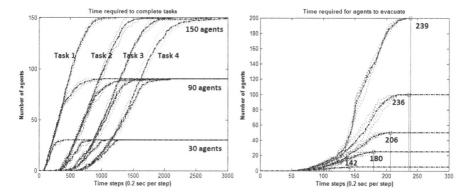

Fig. 6. Summary of results from the two experiments: (i) left-pane, agents pursuing four waypoints in an open space and (ii) right pane, agents evacuate the area. The black dash lines are the average value from 5 runs.

5 Conclusion and Future Direction

In this paper, we simulate crowd behaviors under two scenarios: traversing waypoints and evacuate an area. The approach taken here animates each agent's actions according to its navigational behaviors i.e., walk at different speeds, turns, etc. This offers a realistic animation of simulated agents. The experimental results highlight the potential of the approach for applications in creative industries, urban planning, and disaster management. In future work, we wish to explore further into increasing the number of behavioral states and their corresponding animations.

Acknowledgments. We wish to thank anonymous reviewers for their comments that have helped improve this paper. We would like to thank the GSR office for their partial financial support given to this research.

References

1. Reynolds, C.W.: Flocks, heards, and schools: a distributed behavioral model. Comput. Graph. **21**(4), 25–34 (1987)

2. Wolfram, S.: A New Kind of Science. Wolfram Media, Champaign (2002)
3. Ron, S.: Cognition and Multi-agent Interaction: From Cognitive Modeling to Social Simulation. Cambridge University Press, Cambridge (2006)
4. Yoav, S., Kevin, L.B.: Multiagent Systems: Algorithmic, Game-Theoretic, and Logical Foundations. Cambridge University Press, Cambridge (2008)
5. Phon-Amnuaisuk, S., Palaniappan, R.: Exploring swarm-based visual effects. In: Lavangnananda, K., Phon-Amnuaisuk, S., Engchuan, W., Chan, J.H. (eds.) IES 2015. PALO, vol. 5, pp. 333–341. Springer, Heidelberg (2016). doi:10.1007/978-3-319-27000-5_27
6. Isla, D.: Handling complexity in the Halo 2 AI. In: Proceedings of the Game Developers Conference (2005)
7. Pettré, J., Ciechomski, P.H., Maïm, J., Yersin, B., Laumond, J.P., Thalmann, D.: Real-time navigating crowds: scalable simulation and rendering. Comput. Animation Virtual Worlds **17**(3–4), 445–455 (2006)
8. Raafat, R.M., Chater, N., Frith, C.: Herding in humans. Trends Cogn. Sci. **13**(10), 420–428 (2009)
9. Brooks, R.A.: Intelligence without representation. Artif. Intell. **47**(1–3), 139–159 (1991)
10. Fikes, R.E., Nilsson, N.J.: STRIPS: a new approach to the application of theorem proving to problem solving. Artif. Intell. **2**, 189–209 (1971)
11. Phon-Amnuaisuk, S.: Learning chasing behaviours of non-player characters in games using SARSA. In: Squillero, G., Burelli, P. (eds.) EvoApplications 2016. LNCS, vol. 9598, pp. 133–142. Springer, Heidelberg (2011). doi:10.1007/978-3-642-20525-5_14
12. Pereira, R.P., Engel, P.M.: A framework for constrained and adaptive behavior-based agents. The Computing Research Repository (CoRR) abs/1506.02312 (2015)
13. Adham, A., Phon-Amnuaisuk, S., Ho, C.K.: Navigating a robotic swarm in an uncharted 2D landscape. Appl. Soft Comput. **10**(1), 149–169 (2010)
14. Balch, T., Arkin, R.C.: Behavior-based formation control for multi-robot teams. IEEE Trans. Robot. Autom. **14**(6), 926–939 (1998)
15. Pejsa, T., Pandzic, I.S.: State of the art in example-based motion synthesis for virtual characters in interactive applications. Comput. Graph. Forum **29**(1), 202–226 (2010)
16. Shapiro, A.: Building a character animation system. In: Kallmann, M., Bekris, K. (eds.) MIG 2012. LNCS, vol. 7660, pp. 98–109. Springer, Heidelberg (2011). doi:10.1007/978-3-642-25090-3_9
17. Mizuguchi, M., Buchanan, J., Calvert, T.: Data driven motion transitions for interactive games. In: Proceedings of the European Association for Computer Graphics (Eurographics 2011) (2001). https://diglib.eg.org/handle/10.2312/egs20011039
18. Roberson, D., Hodson, D., Peterson, G., Woolley, G.: The unified behavior framework for the simulation of autonomous agents. In: Proceedings of the International Conference on Scientific Computing (CSC 2015), pp. 49–58 (2015)
19. Dechter, R., Pearl, J.: Generalized best-first search strategies and the optimality of A*. J. ACM **32**(3), 505–536 (1985)
20. Kisko, T.M., Francis, R.L., Nobel, C.R.: Evacnet4 users guide (1998). http://www.ise.ufl.edu/kisko/files/evacnet/EVAC4UG.HTM
21. Russell, S., Norvig, P.: Artificial Intelligence: A Modern Approach, 3rd edn. Pearson, Upper Saddle River (2009)

Filter-Based Feature Selection Using Two Criterion Functions and Evolutionary Fuzzification

Ohm Sornil[(✉)]

Graduate School of Applied Statistics,
National Institute of Development Administration, Bangkok, Thailand
osornil@as.nida.ac.th

Abstract. Real world problems often contain noise features which can decrease effectiveness of classification models. This article proposes a filter-based technique to select a minimal set of features for classification problems. The proposed method employs fuzzification of original features based on irregular-shaped membership functions created by genetic algorithm and particle swarm optimization, and a feature selection process using two criterion functions to evaluate feature subsets. The first function is applied to eliminate features with redundant effects, and the second function is applied to select a feature subset that maximizes inter-class distances and minimize intra-class distances. Standard machine learning data sets in various sizes and complexities are used in experiments. The results show that the proposed technique is effective and performs well in comparisons with other research.

1 Introduction

A feature selection method selects a small subset of highly predictive features from the original set of features. It most of the time yields better results due to reduction of noises and distractions, and takes less training time for a classifier than using the entire set of features. Feature selection approaches can be classified into three categories which are wrapper, filter, and hybrid approaches.

Given a classification problem, a wrapper method incorporates the classification itself in the feature evaluation process. To evaluate a candidate feature subset, a classification model is built and used to evaluate the set. Maroño et al. [11] propose a wrapper based feature selection using ANOVA decomposition and functional networks to calculate global sensitivity indices. Features with high index values are selected. Zhuo et al. [19] use a genetic algorithm (GA) to optimize a support vector machine (SVM) kernel parameters for selecting a feature subset. The fitness function is accuracy which also is used as the criterion function for selecting features. The wrapper approach is expected to return a subset of features that yields high accuracy since every candidate feature set is evaluated by the classifier that is used in the problem. Since classification models are trained and tested many times, and data becomes larger in dimensionality and number of instances, this approach takes a long time for learning such data and in many cases is inapplicable.

© Springer International Publishing AG 2016
C. Sombattheera et al. (Eds.): MIWAI 2016, LNAI 10053, pp. 173–183, 2016.
DOI: 10.1007/978-3-319-49397-8_15

In a filter method, instead of performing classification as part of the feature selection process, a quality measure is used to evaluate each feature set. The filter-based approach composes two important components which are a selection algorithm and a criterion function. The selection algorithm creates candidate features while the criterion function selects features and evaluates feature subsets. The criterion function can be independent from the classification model, but it should be suitable for the problem. The filter-based approach generally takes less time than does the wrapper approach since no classifier is trained and tested as in the wrapper approach. It is more preferable for real-world problems, especially those with large data sets. Many researchers find that it yields subsets with lower accuracy than do the other two approaches. However, it is not true to state that the filter approach always gives lower accuracy. Some criterion functions may return subsets with equivalent or better performance than other approaches.

Yu and Liu [15] use symmetrical uncertainty as the measure to select features relevant to classes which are not redundant with other selected features. Zhou et al. [18] propose a forward algorithm to select features using conditional maximum entropy modeling to approximate the gain for features. Fleuret [4] uses conditional mutual information (CMI) as the criterion function to fasten the forward search process. Haindl et al. [6] propose a backward filter-based feature selection method based on mutual correlation, a similarity measure between two variables, to select features which are uncorrelated.

The hybrid approach takes advantage of both the wrapper and the filter approaches. It applies a filter-based technique to select highly significant features and applies a wrapper-based technique to add candidate features and evaluate candidate sets. Zhang et al. [17] apply the RELIEFF algorithm to estimate the quality of attributes according to how well their values distinguish between instances that are close to each other, and apply GA with classifier accuracy as the fitness function to search for an optimal feature subset. Somol et al. [13] present a hybrid floating search, named hSFFS, by applying a filter criterion function first to filter some features and applying a wrapper criterion to generate a candidate set. After that a wrapper criterion function is applied to select the best feature from the candidate set. This is a wrapper-dominating hybrid method. Gan et al. [5] propose an alternative to hSFFS, which is a filter-dominating hybrid method. A filter criterion is used to select the best feature from an unselected set, and a wrapper criterion is then used to evaluate a feature subset.

Problems usually found in real-world applications are mixtures of ambiguous and noisy data. This results in an inaccurate classification model. Fuzzy Logic, which is a multi-value logic that allows intermediate values to be defined between conventional crisp evaluations, e.g., true/false, yes/no, etc., provides a simple way to define conclusions based upon vague, ambiguous, imprecise, noisy, or missing input information [3]. Membership functions for fuzzy sets can be of any shape or type, such as triangular, trapezoidal, and Gaussian-shaped, as determined by experts in the domain over which the sets are defined.

This paper proposes a feature selection technique for classification using two criterion functions and feature fuzzification using irregular-shaped membership functions, evolved by genetic algorithm and particle swarm optimization. The technique is evaluated using standard machine learning data sets in various sizes and complexities.

2 Proposed Feature Fuzzification

Irregular-shaped membership functions for every continuous attribute are evolved. Values of those attributes are fuzzified to create a suitable set of value ranges. All attributes are then fed into the filter-based feature selection algorithm which employs two criterion functions to generate the best set of predictive features.

The membership function (MF) shape determined in advance by experts may not be suitable for a specific problem at hand, especially those with large and complex search spaces. We convert the wrapper-based hierarchical co-evolutionary by Huang et al. [7] for generating irregular-shaped membership functions (ISMFs) into a filter-based algorithm using two optimization techniques: genetic algorithm and particle swarm optimization, where a criterion function is used in order to improve efficiency. An MF shape is represented as one pivot point, left shoulder points, and right shoulder points, depicted in Fig. 1.

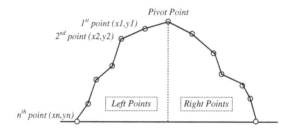

Fig. 1. An irregular-shaped membership function

2.1 Membership Function Evolution by Genetic Algorithm

A genetic algorithm can be employed to create membership functions for continuous variables. An irregular-shaped MF is represented as one pivot point, left shoulder points and right shoulder points, as shown in Fig. 2(a). Fuzzy partitions on each input variable are encoded in genetic segmentations and concatenated into one chromosome in the first level (L1-level) for the corresponding variable. A chromosome in the second level (L2-level) composes of genes pointing to chromosomes for all variables in L1-level. An L2-level gene contains the integer value of an index in the L1-level chromosome. With GA operations (crossover, mutation and selection), coordinates of points will be changed, and it results in changing shapes. Constraints and repairing schemes are applied before decoding the genetic representation.

The algorithm partitions and encodes possible solutions as populations in different levels, allowing for different kinds of chromosomes and genetic operations. A higher level chromosome selects a set of lower-level chromosomes to form a solution. In this case, a highly complicated search task can be partitioned into several subtasks which are simultaneously and effectively handled. The structure of the chromosome is shown in Fig. 2(b).

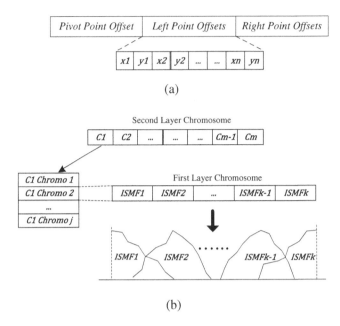

(a)

(b)

Fig. 2. Chromosome structure

2.2 Membership Function Evolution by Particle Swarm Optimization

Particle swarm optimization (PSO) [2] can be used to evolve optimal locations of points on ISMFs for a feature. PSO is a heuristic global optimization method based on swarm intelligence. Potential solutions, called particles, fly through the problem space by following the current optimum particles. Each particle keeps track of its coordinates in the problem space which are associated with the best solution (fitness) it has achieved so far. This value is called *pbest*. Another best value that is tracked by the particle swarm optimizer is the best value obtained so far by any particle in the neighbors of the particle. In this research, a particle takes the entire population as its topological neighbors, the best value is a global best and is called *gbest*. At each time step, we change the velocity of (accelerating) each particle toward its pbest and the gbest locations. Acceleration is weighted randomly toward the pbest and the gbest locations. Content of a PSO particle for generating ISMFs is shown in Fig. 3.

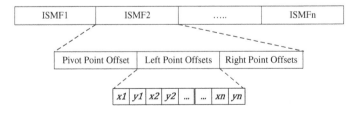

Fig. 3. Particle content

Evolution in PSO is the process to update particles' positions. A particle position is updated as follows:

$$x_{ij}(t+1) = x_{ij}(t) + v_{ij}(t+1)$$

and

$$v_{ij}(t+1) = wv_{ij}(t) + c_1r_{1j}(t)\left[b_{ij}(t) - x_{ij}(t)\right] + c_2r_{2j}(t)\left[\hat{b}(t) - x_{ij}(t)\right]$$

where x_{ij} is the vector of i-th particle with j dimensions, t denotes a discrete time step or iteration, w is the inertia weight, r is a random number in the range [0..1] sampled from a uniform distribution, c is an acceleration constant, $\hat{b}(t)$ is the best position among all particles, and b_{ij} (t) is the best position of the i-th particle.

To determine the values of three important parameters needed by PSO which are w, c_1 and c_2, Zhang et al. [16] constructs a relationship between the dynamic process of particle swarm optimization and the transition process of a control system. It reduces the three parameters to the percentage overshoot, and from their experiments the value should fall between 0.6 and 0.8. The percentage overshoot allows us to determine the values of w and c, and further $c = c_1 = c_2$. Comparing to other parameter setting strategies, this method leads to similar optimization results but faster convergence.

Opposition-based PSO technique [8] is used to initialize particles to preserve the coverage. The process can be described as:

1. Randomly initialize n particles.
2. Calculate opposite particles of first n particles.
3. Evaluate $2n$ particles from steps 1 and 2, and select the best n particles to be in the swarm of optimization process.

3 Proposed Feature Selection Process

We improve the filter-based sequential forward floating search algorithm [12] by employing two criterion functions with different characteristics to complement each other and allowing more thorough search for features by introducing candidate sets. Conditional mutual information (CMI) is employed as the first criterion function. It measures dependency between two variables with respect to a class, conditional to the response of features already picked [4]. CMI selects features which maximize MI to the target class where such information must not have been caught by features already selected to reduce redundant features. It generates a candidate set of features which are suitable to be added to or removed from a selected subset instead of examining one feature at a time. Using the candidate sets makes the search more thorough. The second criterion function selects a feature to be added or removed from this set.

Input to the algorithm consists of the original feature set S, the first criterion function J_1, and the second criterion function J_2. Let D be the total number of original features. d_{sel} is the number of selected features. d_{cand} is the number of features in a candidate set where $d_{cand} \geq 1$. S_{sel} is the selected feature subset. S_{cand}^- is the candidate set in the backward step, and S_{cand}^+ is the candidate set in the forward step.

In the forward step, unselected features are evaluated by the a criterion function J_1 and sorted in descending order. A candidate feature set is created as follows:

$$S^+_{cand} = \{x_n | x_n \in S \backslash S_{sel} \text{ and } n = [1\ldots d_{cand}] \text{ and } J_1(x_1) \geq J_1(x_2) \cdots \geq J_1(x_n)\}$$

where $J_1 = min\, I(Y; X_n | S_{sel})$ where $X_n \in S \backslash S_{sel}$

As mentioned earlier, CMI is used as J_1, and it can be calculated as follows:

$$I(Y; X_n | X_m) = H(Y, X_m) - H(X_m) - H(Y, X_n, X_m) + H(X_n, X_m)$$

where $I(Y; Xn|Xm)$ is the conditional mutual information between Y and X_n given X_m, and H is an entropy function. For more information on how to compute CMI, see [4].

The feature selected is the one when combined with the previously selected subset of size k gives the best subset when evaluated with J_2, forming the selected subset of size $k + 1$. Then the algorithm compares the new subset with the previously selected subset of size $k + 1$ and retains the better one.

In the backward step, a feature to be removed must be the one providing the least information to target classes, and its information has been caught by features already picked. Therefore, J_1 in the backward step is calculated as follows:

$$J_1 = max\, I(Y; X_n | S_{sel} \backslash X_n) \text{ where } X_n \in S_{sel}$$

Selected features are evaluated by J_1 and sorted in ascending order. A candidate set is generated as follows:

$$S^-_{cand} = \{x_n | x_n \in S \backslash S_{sel} \text{ and } n = [1\ldots d_{cand}] \text{ and } J_1(x_1) \leq J_1(x_2) \leq \cdots \leq J_1(x_n)\}$$

The feature to be removed is the one when removed from the selected subset yields the best subset with k features according to J_2. The algorithm compares the new subset and the previously selected subset of size k and retains the better one. The exclusion step continues to smaller subsets if the new subset is better, or else the algorithm goes back to the inclusion step. The algorithm terminates when the selected subset size is $d_{sel} + \Delta$.

3.1   The Second Criterion Function

As part of the feature selection process, the second criterion function (J_2)'s role is to select a feature subset that maximizes inter-class distances and minimizes intra-class distances. Three effective measures are studied as candidates for J_2.

Mutual Information (MI). MI can be calculated as follows:

$$I(Y; X_n) = H(Y) + H(X_n) - H(Y, X_n)$$

where H is an entropy function, Y is a class attribute, and X_n is the feature to be selected.

Jeffreys-Matusita Distance Bound to the Bayes Error (JMBH). JMBH can be calculated as follows:

$$J_{bh} = \sum_{i=1}^{c} \sum_{j=1}^{c} \sqrt{P(\omega_i)P(\omega_j)} J_{ij}^2$$

$$J_{ij} = \left[2(1 - e^{-B_{ij}})\right]^{1/2}$$

$$B_{ij} = \frac{1}{8}(m_i - m_j)^t \left(\frac{\Sigma_i + \Sigma_j}{2}\right)^{-1} (m_i - m_j) + \frac{1}{2}\log\left[\frac{\left(\frac{\Sigma_i + \Sigma_j}{2}\right)}{\sqrt{|\Sigma_i||\Sigma_j|}}\right]$$

Where m_i, m_j and Σ_i, Σ_j are mean vectors and covariance matrices for the classes ω_i and ω_j, respectively.

Mahalanobis Distance (MAHA). MAHA can be calculated as follows:

$$D_M(x) = \sqrt{(x - \mu)^T S^{-1} (x - \mu)}$$

where μ is the mean vector, and S is the covariance matrix for a group.

3.2   Classification Model

Classification and Regression Trees (CART) was introduced by Breiman et al. [1]. CART is based on a fundamental idea that each split should be selected so that the data in each descendant subset is purer than the data in the parent node. The node impurity is largest when all classes are equally mixed together and smallest when the node contains only one class. CART produces binary splits. Hence, it produces binary trees. CART uses Gini impurity index as an attribute selection measure to build a decision tree. Consider a parent node m, which contains the data that belongs to the jth class. The impurity function for node t is given by $i(t) = 1 - \sum_i p^2(j|m)$. The decrease of split impurity is given by $\Delta i(\delta, t) = i(t) - p_L i(m_L) - p_R i(m_R)$, where t is a parent node using a splitting coefficient δ to split into two nodes m_L and m_R. The split with the largest decrease in impurity is chosen for that particular node.

4   Experimental Evaluation

The data sets used in the experiments using standard data sets from the UCI machine learning repository. For any data set without a separate test set provided, a 10-fold cross validation is employed to measure performance. The stopping criterion for genetic algorithm and PSO is set at 100 iterations.

Table 1. Classification accuracy of applying and not applying feature fuzzification by genetic algorithm and particle swarm optimization, with 3 different J_2 criterion functions

Data Set	Feature fuzzification by Genetic algorithm					
	MI		JMBH		MAHA	
	Fuzzified Features	Non-Fuzzified Features	Fuzzified Features	Non-Fuzzified Features	Fuzzified Features	Non-Fuzzified Features
Wine	94.44	88.89	100	100	100	88.89
Pima	75.33	75.33	79.22	76.63	75.33	75.33
Image segmentation	92.57	90.62	92.57	90.19	92.57	90.62
Breast cancer	96.49	92.98	98.24	96.49	98.25	96.49
Sonar	95.23	80.95	95.23	85.71	95.23	80.95
Hill with noise	59.4	56.6	59.9	56.6	59.08	56.6
Arrhythmia	80	62.22	82.22	60	66.67	62.22
Madelon	78.33	69.83	84.83	74	81.5	76
Data set	Feature fuzzification by Particle swarm optimization					
	MI		JMBH		MAHA	
	Fuzzified features	Non-fuzzified features	Fuzzified features	Non-fuzzified features	Fuzzified features	Non-fuzzified features
Wine	94.44	94.44	100	94.44	100	94.44
Pima	75.33	76.32	76.62	73.68	75.33	76.32
Image segmentation	91.48	91.48	90.81	90.81	90.95	90.95
Breast cancer	94.74	91.23	96.49	94.74	96.49	94.74
Sonar	80.95	100	88.28	85	85.71	100
Hill with noise	58.58	58.58	61.67	57.43	57.43	57.43
Arrhythmia	74.19	65.22	80.64	60.87	67.74	69.56
Madelon	80.17	80.17	82.67	82.67	85.67	85.67

4.1 Effectiveness of Feature Fuzzification

In this experiment, we study the effectiveness of the fuzzification process using both genetic algorithm and particle swarm optimization against not using the fuzzification at all in different classification problems. An initial study shows that the swarm size of 30 particles gives the highest accuracy and will be used in all experiments. The results are shown in Table 1. We can see that in almost all configurations using fuzzification yields higher accuracy then not using it, thus the fuzzification is useful. In addition, the configuration that gives the best results is fuzzification using GA and JMBH as J_2 function (referred to as Fuzzified GA+JMBH).

4.2 Performance of the Proposed Technique

Since fuzzification and JMBH are beneficial to the performance of the proposed feature selection technique, in this section we focus more on the feature reduction abilities of fuzzification by genetic algorithm and particle swarm optimization. The results in

Table 2 show that although GA yields higher accuracies in general, however, PSO tends to give better feature reduction rates.

Table 2. Accuracies and feature reduction abilities of fuzzification by GA and PSO

Data Set		Original features	Fuzzified GA +JMBH	Fuzzified PSO +JMBH
Wine	Accuracy	83.33	100	100
	Features	14	3	2
Pima	Accuracy	70.13	79.22	76.62
	Features	8	4	2
Image segmentation	Accuracy	90.29	92.57	90.81
	Features	20	9	9
Breast cancer	Accuracy	89.47	98.24	96.49
	Features	31	4	3
Sonar	Accuracy	71.43	95.23	88.28
	Features	61	5	5
Hill valley with noise	Accuracy	60.07	59.9	61.67
	Features	101	12	1
Arrhythmia	Accuracy	66.67	82.22	80.64
	Features	280	20	6
Madelon	Accuracy	75.67	84.83	82.67
	Features	501	12	9

4.3   Comparisons with Other Research

Lastly, the proposed method (Fuzzified GA+JMBH) is compared against three recent research on fuzzy-based feature selection which are: Jalali et al. [9], Vieira et al. [14], Li and Wu [10], using the performance numbers reported in each paper. The results (in Table 3) show that the proposed method outperforms [9] and [10] in all common data sets. Comparing with [14], we find that the proposed method gives higher accuracy in

Table 3. Results of (Fuzzified GA + JMBH) compared to other previous fuzzy-based research. Feature reduction percentages relative to the original feature sets are shown in parentheses.

Data Set	Original number of features	Jalali et al. [9]	Vieira et al. [14]	Li and Wu [10]	Proposed Method
Pima	8	–	–	71.51 (89.17)	80.52 (50)
Wine	14	95.4 (65.38)	96 (69.23)	90.99 (83.59)	100 (76.92)
Breast cancer	31	63.6 (91.66)	98 (86.67)	95.97 (96.67)	98.24 (87.09)
Sonar	61	–	86 (86.66)	68.06 (96.78)	95.24 (93.33)
Arrhythmia	280	–	87 (95.69)	–	82.22 (92.83)

3 out of 4 data sets. Thus, the proposed technique is shown to perform very well across different data sets and in comparison with other techniques.

5   Conclusion

As data sets grow in size and complexity, a feature selection technique is needed to select a small subset of highly predictive features from the entire set of features. The technique is expected to reduce noises and distractions, thus improve both effectiveness and efficiency of machine learning. This paper presents a new filter-based technique to select a minimal set of features for classification problems. The proposed technique employs fuzzification of original features using irregular-shaped membership functions evolved by genetic algorithm and particle swarm optimization, and a filter-based feature selection using two criterion functions where the first function is applied to eliminate features with redundant effects, and the second function is used to select a feature subset that maximizes inter-class distances and minimize intra-class distances. The technique is evaluated using standard UCI data sets and compared to recent fuzzy-based feature selection research papers. The results show that feature selection improves classification accuracy; that the use of evolutionary feature fuzzification and two criterion functions enhances the performance of feature selection; and that the best configuration is using Jeffreys-Matusita Distance Bound to the Bayes Error as the second criterion function and genetic algorithm to evolve irregular-shaped fuzzy membership functions. In addition, the proposed technique performs well in comparison to previous research on common data sets.

References

1. Breiman, L., Friedman, J.H., Olshen, R.A., Stone, C.J.: Classification and Regression Trees. Wadsworht, Pacific Grove (1984)
2. Kennedy, J., Eberhart, B.: Particle swarm optimization. In: Proceedings of IEEE International Conference on Neural Network, Pert, Australia, pp. 1942–1948 (1995)
3. Engelbrecht, A.P.: Computational Intelligence: An Introduction, 2nd edn. Wiley, New York (2007)
4. Fleuret, F.: Fast binary feature selection with conditional mutual information. J. Mach. Learn. Res. **5**(11), 1531–1555 (2004)
5. Gan, J.Q., Awwad Shiekh Hasan, B., Tsui, C.S.L.: A hybrid approach to feature subset selection for brain-computer interface design. In: Yin, H., Wang, W., Rayward-Smith, V. (eds.) IDEAL 2011. LNCS, vol. 6936, pp. 279–286. Springer, Heidelberg (2011). doi:10.1007/978-3-642-23878-9_34
6. Haindl, M., Somol, P., Ververidis, D., Kotropoulos, C.: Feature selection based on mutual correlation. In: Martínez-Trinidad, J.F., Carrasco Ochoa, J.A., Kittler, J. (eds.) CIARP 2006. LNCS, vol. 4225, pp. 569–577. Springer, Heidelberg (2006). doi:10.1007/11892755_59
7. Huang, H., Pasquier, M., Quek, C.: HiCEFS – A Hierarchical Coevolutionary Approach for the Dynamic Generation of Fuzzy System, pp. 3426–3443. IEEE Congress on Evolutionary Computation, CEC (2007)

8. Jabeen, H., Jalil, Z., Baig, A.: Opposition based initialization in particle swarm optimization (O-PSO). In: Proceedings of Genetic and Evolutionary Computation Conference, Montreal, Canada, pp. 2047–2052 (2009)
9. Jalali, L., Nasiri, M., Minaei, B.: A hybrid feature selection method based on fuzzy feature selection and consistency measures. In: Intelligent Computing and Intelligent System (ICIS), pp. 718–722 (2009)
10. Li, Y., Wu, Z.F.: Fuzzy feature selection based on min-max learning rule and extension matrix. Pattern Recogn. **41**, 217–226 (2008)
11. Maroño, N.S., Betanzos, A.A., Castillo, E.: A new wrapper method for feature subset selection. In: Proceedings-European Symposium on Artificial Neural Networks, pp. 515–520 (2005)
12. Pudil, P., Novovičová, J., Kittler, J.: Floating search methods in feature selection. Pattern Recogn. Lett. **15**, 1119–1125 (1994)
13. Somol, P., Novovičová, J., Pudil, P.: Flexible-hybrid sequential floating search in statistical feature selection. In: Yeung, D.-Y., Kwok, J.T., Fred, A., Roli, F., Ridder, D. (eds.) SSPR/SPR 2006. LNCS, vol. 4109, pp. 632–639. Springer, Heidelberg (2006). doi:10.1007/11815921_69
14. Vieira, S.M., Sousa, J.M.C., Kaymak, U.: Fuzzy criteria for feature selection. Fuzzy Sets Syst. **189**, 1–18 (2012)
15. Yu, L., Liu, H.: Feature selection for high-dimensional data: a fast correlation-based filter solution. In: Proceedings of the Twentieth International Conference on Machine Learning (ICML-2003) (2003)
16. Zhang, W., Ma, D., Wei, J., Liang, H.: A parameter selection strategy for particle swarm optimization based on particle positions. Expert Syst. Appl. **41**, 3576–3584 (2014)
17. Zhang, L.X., Wang, J.X., Zhao, Y.N., Yang, Z.H.: A novel hybrid feature selection algorithm: using relief estimation for GA-wrapper search. In: Proceedings of the Second International Conference on Machine Learning and Cybernetics, pp. 380–384 (2003)
18. Zhou, Y., Weng, F., Wu, L., Schmidt, H.: A fast algorithm for feature selection in conditional maximum entropy modeling. In: Proceedings of the 2003 Conference on Empirical Methods in Natural Language Processing, pp. 153–159 (2003)
19. Zhuo, L., Zheng, J., Wang, F., Li, X., Ai, B., Qian, J.: A genetic algorithm based wrapper feature selection method for classification of hyperspectral images using support vector machine. Int. Arch. Photogramm. Remote Sens. Spat. Inf. Sci. **XXXVII Par B7**, 397–402 (2008)

Resolving the Manufacturing Cell Design Problem Using the Flower Pollination Algorithm

Ricardo Soto[1], Broderick Crawford[1], Rodrigo Olivares[1,2(✉)],
Michele De Conti[1], Ronald Rubio[1], Boris Almonacid[1],
and Stefanie Niklander[3]

[1] Pontificia Universidad Católica de Valparaíso, Valparaíso, Chile
{ricardo.soto,broderick.crawford}@pucv.cl,
{michele.de,ronald.rubio.h,boris.almonacid}@mail.pucv.cl
[2] Universidad de Valparaíso, Valparaíso, Chile
rodrigo.olivares@uv.cl
[3] Universidad Adolfo Ibañez, Viña Del Mar, Chile
stefanie.niklander@uai.cl

Abstract. The Manufacturing cell design problem focuses on the creation of an optimal distribution of the machinery on a productive plant, through the creation of highly independent cells where the parts of certain products are processed. The main objective is to reduce the movements between this cells, decreasing production times, costs and getting other advantages. To find solutions to this problem, in this paper, the usage of the Flower Pollination Algorithm is proposed, which is one of the many nature-based algorithms, which in this case is inspired in the Pollination of the flowers, and has shown great capacities in the resolution of complex problems. Experimental results are shown, with 90 instances taken from Boctor's experiments, where the optimum is achieved in all them.

Keywords: Manufacturing cell design problem · Flower pollination algorithm · Metaheuristics · Optimization

1 Introduction

In the manufacturing industry, one of the main concerns is determinate efficient ways to group the productive machinery, and a proved way to do this is the usage of the Cellular Manufacturing (CM) [1], which consists in the decomposition of the manufacturer system in subsystems called *Cells* which are easier to administrate than the system as a whole. The objective is to find the optimal cell configuration, where manipulation times, waste and the amount of part exchange between cell are minimized and the administrative word is improved. This problem is known as the Manufacturing Cell Design Problem (MCDP).

To solve extensive problems as the one described early, humanity has observed the behavior of many phenomena which nature has perfected through millions of years of evolution and based on those natural processes a bast amount

© Springer International Publishing AG 2016
C. Sombattheera et al. (Eds.): MIWAI 2016, LNAI 10053, pp. 184–195, 2016.
DOI: 10.1007/978-3-319-49397-8_16

of algorithms were developed, like genetic ones, based on the Darwinian evolution of species [2], the ones based on swarms of birds or fishes [3,4], or even based in the flashing patterns of the tropical fireflies [5] among many other examples. From all these nature-inspired algorithms, this paper will focus en the Flower Pollination Algorithm (FPA), which is based in the way the flowers are pollinated.

This paper is organized as follows: In the Sects. 2 and 3, we introduce the related works and a deeper explanation of the MCDP, including the mathematical formulation. Section 4 describes the FPA and how does it work. Finally, in the Sects. 5 and 6 we present the experimental results and conclusions, respectively.

2   Related Work

The cell formation has been researched for many years. One of the first investigations focused on resolving it is Burbidge's work in 1963 [6]. He proposes the usage of the incidence matrix, which is reorganized in a Block Diagonal Form [6] (BDF). In recent years, many methods were used to resolve the MCDP. For example, the Genetic Algorithm [7] (GA) inspired in the biological evolution and its genetic-molecular base. The Neural Network [8] (NN) that takes the behavior of the neurons and the connections of the human brain. Constraint Programming [9] (CP) where the relationships between variables are expressed as restrictions. Artificial Fish Swarm [10] (AFSA) which studies the intelligent behavior of a fish swarm. Shuffled Frog Leaping Algorithm [11] (SFLA) inspired in the memetic of frogs, where they try to jump into the search domain in search of a better solution until the stop condition is met. Finally, Migrating Birds Optimization [12,13] (MBO) inspired by migrating birds to fly.

3   Manufacturing Cell Design Problem

The Cellular Manufacturing, the convenient division of the machinery into cells, is based in the Group Technology (GT) principles, proposed by Flanders in 1925 [14], the goal of this manufacturing method is the division of a set of parts into families which share similitude in it's geometry, fabrication process and/or functions, and are produced in the same place. From the GT, it follows the creation of highly independent cells that groups families of machinery, reducing the setting time, work, re-work, waste material, delivery time, production cost, increasing the flexibility, improving the product's quality and the production control. There is abundant literature about this specific topic, for example [15–17].

Find the best way to group the machines that processes the part's families and minimize the movement and interchange of parts between cells is the main objective of the Manufacturing Cell Design Problem, to do this, the first thing is schematize the problem into a matrix, called part-machine, where from the reorganization of the points inside the matrix, the number of movements between cells can be minimized, from this two new matrix are created, called the machine-cell and the part-cell. To explain this goal this optimization model [18, 19] is used. Let:

M, is the number of machines.
P, is the number of parts.
C, is the number of cells.
i, is the index of machines ($i = 1, 2, ..., M$).
j, is the index of parts ($i = 1, 2, ..., P$).
k, is the index of cells ($i = 1, 2, ..., C$).
M_{max}, is the maximum number of machines per cell.
$A = [a_{ij}]$, is the binary machine-part incidence matrix, where:

$$a_{ij} = \begin{cases} 1 & \text{if machine } i \text{ process the part } j \\ 0 & \text{otherwise} \end{cases} \tag{1}$$

$B = [b_{ik}]$, is the binary machine-cell incidence matrix, where:

$$b_{ik} = \begin{cases} 1 & \text{if machine } i \text{ belongs to cell } k \\ 0 & \text{otherwise} \end{cases} \tag{2}$$

$C = [c_{jk}]$, is the binary part-cell incidence matrix, where:

$$c_{jk} = \begin{cases} 1 & \text{if part } j \text{ belongs to cell } k \\ 0 & \text{otherwise} \end{cases} \tag{3}$$

The objective function models the minimization of the part movements between cells as depicted in Eq. 4.

$$min \sum_{k=1}^{C} \sum_{i=1}^{M} \sum_{j=1}^{P} a_{ij} c_{jk} (1 - b_{ik}) \tag{4}$$

This objective function is under to three constraints where: the Eq. 5 states that each machine belongs to only one cell. The Eq. 6 states that each part is assigned to only one cell, and Eq. 7 determines the maximum number of machines that a cell could has.

$$\sum_{k=1}^{C} b_{ik} = 1, \forall i \tag{5}$$

$$\sum_{k=1}^{C} c_{jk} = 1, \forall j \tag{6}$$

$$\sum_{i=1}^{M} b_{ik} \leq M_{max}, \forall k \tag{7}$$

4   Flower Pollination Algorithm

The Flower pollination Algorithm [20] (FPA) is based in the pollination of flowers, process focused on the survival of the fittest and the optimal reproduction of plants in number and aptitude terms. The flower's main purpose is the reproduction via pollination. This is made by the association with different kinds of insects, birds, bats, and other animals, namely pollinators. This process can be made in two different ways, biotic or abiotic. 90 % of flowering plants uses the biotic pollination, in other words, with the pollinator's help. The other 10 % uses the abiotic way, that consist in the diffusion principally through air and water [21].

Some pollinator tends to visit only some species of flowers, avoiding in other species on their route, this behavior is denominated flower constancy [22]. This maximizes the pollen transfer between flowers of the same family, increasing the reproduction chances, the pollinators benefit themselves increasing the nectar harvest and minimizing the learning and exploration costs. The pollination process can be conceived by cross-pollination or self-pollination. The cross-pollination is when the pollination occurs between the pollen of different plants of the same species, while the self-pollination is the fertilization of the same or different flower, but of the same plant.

The pollinators in the biotic pollination can travel great distances, and are considered global pollinators. The movement patterns of birds and bees can be explained with the Lévy Flight [23] and their jumps or flight distance obey a Lévy distribution. Finally, the flower constancy is used as an incremental step using the similitude of difference between two flowers.

The previously described characteristics are used to construct an algorithm that emulates said process, and follow the next rules:

i The biotic and cross-pollination are considered as a global pollination process, in which the pollen is carried by pollinator performing a Lévy Flight.
ii The abiotic and self-pollination are considered local pollination.
iii The flower Constancy can be considered as the reproduction probability, which is proportional to the similarity of the two flowers involved.
iv The global and local pollination is controlled by a probability switch p between 0 and 1. $p \in [0, 1]$. Due physical proximity and other factors like wind, the local pollination may have a great fraction p in the general pollination activities.

In reality, one plant has many flowers, and every flower can free millions and even billions of pollen gametes. To simplify this is assumed that every plant has only one flower and each flower produce only one pollen gamete. In other words, pollen, gamete, flower and plant or solution (X_i) are considered the same thing.

There are two main steps in FPA, the global and local pollination, in the first one the pollen is carried by pollinators, traveling long distances. This assures reproduction and pollination of the fittest, and it will be denominated as g_*. Mathematically, the fist rule, together with the flower constancy can be represented as:

$$X_i^{t+1} = X_i^t + L(g_* - X_i^t) \tag{8}$$

where X_i^{t+1} represents the pollen or solution vector, X_i the iteration t, and g_* is the best solution found inside the solutions of the current iteration. The parameter L is the pollination strength, which essentially is the size of the step. As pollinators can move using different step sizes, the Lévy Flight is used to imitate efficiently this characteristic. Thus, an $L > 0$ is created from a Lévy distribution.

$$L \sim \frac{\lambda \Gamma(\lambda) sin(\pi\lambda/2)}{\pi} \frac{1}{s^1 + \lambda}, (S >> S_0 > 0). \tag{9}$$

where $\Gamma(\lambda)$ is the standard gamma function, and this distribution is valid for long steps $s > 0$.

The local pollination and the flower constancy can be represented as:

$$X_i^{t+1} = X_i^t + \epsilon(X_j^t - X_k^t) \tag{10}$$

Mathematically, if X_j^t and X_k^t are of the same species, or are selected from the same neighborhood, this would be a random local travel if an ϵ is created from a distribution in $[0, 1]$. Many pollination activities can happen locally or globally. In practice, near flowers or from the nearby neighborhoods will have the higher chance of being pollinated than the ones that are further away. To emulate this a probability switch is used to determinate the chance to get any of the two pollination processes.

Based in the rules set in the upper section, the following pseudo-code is presented:

Algorithm 1. Flower Pollination Algorithm

1: *Objective* min or max $f(x), x = (x_1, x_2, ..., x_d)$
2: *Initialize a population of n flowers/pollen gametes with random solutions*
3: *Define a switch probability $p \in [0, 1]$*
4: **while** $(t < MaxGeneration)$ **do**
5:     **for** $i = 1 : n$ *(all n flowers in the population)* **do**
6:         **if** *rand $< p$* **then**
7:             *Draw a (d-dimensional) step vector L which obeys a Lévy distribution*
8:             Global *pollination via $X_i^{t+1} = X_i^t + L(g_* - X_i^t)$*
9:         **else**
10:            *Draw ϵ from a uniform distribution in [0,1]*
11:            *Randomly choose j and k among all the solutions*
12:            *Do local pollination via $X_i^{t+1} = X_i^t + \epsilon(X_j^t - X_k^t)$*
13:        **end if**
14:        Evaluate new solutions, If new solutions are better, update the population
15:    **end for**
16:    Find the current best solution g_*
17: **end while**

4.1   Integration Between FPA and MCDP

As the Flower Pollination Algorithm uses real numbers, is necessary make some changes to solve the Manufacturing Cell Design which is a problem with a binary domain. To overcome this issue, a vector is used to represent the input of the problem (the matrix machine-cell). This vector will be as long as the number of machines, every position in the vector will represent one machine, and the value stored will represent in which cell the machine is currently assigned. From this point, this vector will be denoted as V and every value stored will be v_i, where $v_i \in [1, C]$ and C is the number of cells (Fig. 1). This will be explained in the following example:

- Quantity of Cells (C): 5
- Quantity of Machines (M): 10

Machines

Cells $\boxed{2\,5\,1\,2\,3\,4\,5\,1\,3\,3}$

Fig. 1. Vector representation with $C = 5$ and $M = 10$

As the C matrix (part-cell) is originated from the machine-cell matrix, is not necessary change it to do the movement. Once the previous vector is formed, the movements given by the metaheuristic can be made. This movements will follow the exactly same principles as explained early, but in this case vectors will be used. Every movement applies the modification to every position of the vector, where the Eq. 11 explains the Lévy step, and the 12 the local step:

$$V_j^{t+1}[k] = Approximate(V_j^t[k] + L * (bestV[k] - V_j^t[k])) \tag{11}$$

$$V_j^{t+1}[k] = Approximate(V_j^t[k] + E * (Vt_m[k] - V_n^t[k])) \tag{12}$$

where:

t, is the current iteration.
j, is the solution to modify at it's k-th.
$bestV$, is the best solution in the current iteration, which is also a vector.
L, is the size of the Lévy step.
E, is the size of the local step.
m and n are chosen randomly among the population.

The approximate function fixes the range of the solution given by the movement. Thus if the new value exceeds the maximum number of cells given by the problem. This value is changed to the number of cells, and in the opposite case is changed to zero.

5 Experimental Results

The FPA was codded in Java 1.8 and the program was executed in a notebook with an Intel Core i5 3230M (2600 MHz–3200 MHz) processor, 8 GB RAM DDR3 1600 MHz and with Windows 10 as operative system. 90 problems where tested (10 instances considering 5 maximum values for the machines with 2 cells, and 10 instances considering 4 maximum values for the machines for 3 cells). These problems where obtained from the Boctor's experiments [19]. To obtain the results the following values where used: $n = 160$, $iterations = 100$, $p = 0.1$, $delta = 1.5$. Every experiment was executed 30 times.

The results are evaluated using the relative percentage deviation (RPD). RPD value quantifies the deviation of the objective value Z from Z_{opt} that in our case is the best known value for each instance and it is calculated as follows:

$$RPD = \left(\frac{Z - Z_{opt}}{Z_{opt}} \right) \times 100 \qquad (13)$$

Tables 1 and 2 contrast the values obtained from the metaheuristic versus the known global optimum for each problem. Also, a comparison between two classical techniques such as SA and PSO, and a comparison with two recent techniques like MBO and SFLA. Each column represent respectively the instance of

Table 1. Experiments using $C = 2$: Optimum values for Flower Pollination (FPA), Simulated Annealing (SA), Particle Swarm Optimization (PSO), Migrating Birds Optimization (MBO), and Shuffled Frog Leaping Algorithm (SFLA).

Boctor Problem	M_{max}	Optimum Value	FPA Optimum	Average	RPD%	SA	PSO	MBO	SFLA
1	8	11	11	11.40	0	11	11	11	11
1	9	11	11	11.03	0	11	11	11	11
1	10	11	11	11.07	0	11	11	11	11
1	11	11	11	11.17	0	11	11	11	11
1	12	11	11	11.30	0	11	11	11	11
2	8	7	7	7.07	0	7	7	7	7
2	9	6	6	6.33	0	6	6	6	6
2	10	4	4	4.77	0	10	5	4	4
2	11	3	3	4.07	0	4	4	3	3
2	12	3	3	3.33	0	3	4	3	3
3	8	4	4	4.43	0	5	5	4	4
3	9	4	4	4.37	0	4	4	4	4
3	10	4	4	4.23	0	4	5	4	4
3	11	3	3	3.80	0	4	4	3	3
3	12	1	1	2.17	0	4	3	1	1
4	8	14	14	14.00	0	14	15	14	14

Table 1. (*continued*)

Boctor Problem	M_{max}	Optimum Value	FPA Optimum	Average	RPD%	SA	PSO	MBO	SFLA
4	9	13	13	13.10	0	13	13	13	13
4	10	13	13	13.23	0	13	13	13	13
4	11	13	13	13.47	0	13	13	13	13
4	12	13	13	13.00	0	13	13	13	13
5	8	9	9	9.73	0	9	10	9	9
5	9	6	6	6.57	0	6	8	6	6
5	10	6	6	6.90	0	6	6	6	6
5	11	5	5	5.90	0	7	5	5	5
5	12	4	4	5.10	0	4	5	4	4
6	8	5	5	5.17	0	5	5	5	5
6	9	3	3	3.70	0	3	3	3	3
6	10	3	3	3.80	0	5	3	3	3
6	11	3	3	3.43	0	3	4	3	3
6	12	2	2	3.10	0	3	4	2	2
7	8	7	7	7.13	0	7	7	7	7
7	9	4	4	4.17	0	4	5	4	4
7	10	4	4	4.30	0	4	5	4	4
7	11	4	4	4.57	0	4	5	4	4
7	12	4	4	4.83	0	4	5	4	4
8	8	13	13	13.10	0	13	14	13	13
8	9	10	10	10.93	0	20	11	10	10
8	10	8	8	8.63	0	15	10	8	8
8	11	5	5	6.60	0	11	6	5	5
8	12	5	5	5.70	0	7	6	5	5
9	8	8	8	9.77	0	13	9	8	8
9	9	8	8	8.40	0	8	8	8	8
9	10	8	8	8.63	0	8	8	8	8
9	11	5	5	6.50	0	8	5	5	5
9	12	5	5	6.87	0	8	8	5	5
10	8	8	8	8.37	0	8	9	8	8
10	9	5	5	6.30	0	5	8	5	5
10	10	5	5	5.63	0	5	7	5	5
10	11	5	5	5.80	0	5	7	5	5
10	12	5	5	5.83	0	5	6	5	5

Table 2. Experiments using $C = 3$: Optimum values for Flower Pollination (FPA), Simulated Annealing (SA), Particle Swarm Optimization (PSO), Migrating Birds Optimization (MBO), and Shuffled Frog Leaping Algorithm (SFLA).

Boctor Problem	Mmax	Optimum Value	FPA Optimum	Average	RPD%	SA	PSO	MBO	SFLA
1	6	27	27	27.57	0	28	-	27	-
1	7	18	18	18.07	0	18	-	18	-
1	8	11	11	13.23	0	11	-	11	-
1	9	11	11	12.13	0	11	-	11	-
2	6	7	7	7.27	0	7	-	7	-
2	7	6	6	6.07	0	6	-	6	-
2	8	6	6	7.77	0	7	-	6	-
2	9	6	6	6.93	0	12	-	6	-
3	6	9	9	9.23	0	8	-	9	-
3	7	4	4	4.07	0	8	-	4	-
3	8	4	4	4.57	0	4	-	4	-
3	9	4	4	4.77	0	27	-	4	-
4	6	27	27	27.03	0	18	-	27	-
4	7	18	18	18.00	0	14	-	18	-
4	8	14	14	14.73	0	13	-	14	-
4	9	13	13	14.13	0	11	-	13	-
5	6	11	11	11.70	0	9	-	11	-
5	7	8	8	8.60	0	9	-	8	-
5	8	8	8	10.03	0	8	-	8	-
5	9	6	6	7.47	0	8	-	6	-
6	6	6	6	6.83	0	5	-	6	-
6	7	4	4	4.10	0	5	-	4	-
6	8	4	4	4.50	0	4	-	4	-
6	9	3	3	4.17	0	11	-	3	-
7	6	11	11	11.60	0	5	-	11	-
7	7	5	5	5.27	0	5	-	5	-
7	8	5	5	6.07	0	5	-	5	-
7	9	4	4	4.20	0	14	-	4	-
8	6	14	14	14.47	0	11	-	14	-
8	7	11	11	11.63	0	11	-	11	-
8	8	11	11	12.70	0	10	-	11	-
8	9	10	10	11.67	0	12	-	10	-
9	6	12	12	12.60	0	12	-	12	-

Table 2. (*continued*)

Boctor Problem	Mmax	Optimum Value	FPA Optimum	Average	RPD%	SA	PSO	MBO	SFLA
9	7	12	12	12.53	0	13	-	12	-
9	8	8	8	8.40	0	8	-	8	-
9	9	8	8	8.60	0	8	-	8	-
10	6	10	10	10.13	0	10	-	10	-
10	7	8	8	8.57	0	8	-	8	-
10	8	8	8	8.93	0	8	-	8	-
10	9	5	5	7.30	0	5	-	5	-

the Boctor problem, the maximum number of machines per cell, the known optimum, the next columns present the results of other techniques (Migrating Birds Optimization (MBO), Simulated Annealing (SA), and Particle Swarm Optimization (PSO)). As it can be appreciated, the FPA achieve the optimum in every instance tested. These experimental results show that the algorithm is able to give good results with an Relative Percentage Deviation of 0 %.

On further inspection, it can be observed that the FPA has a quick convergence to the optimum value. As an example in the 56 instance of the problem, the algorithm converges in the eleventh iteration (see Fig. 2) and the 71 instance achieve the optimum value in the 12 iterations (see Fig. 3). In this regard, in the problems with two cells the best solution is achieved in average in 6.4 iterations and in the configuration used, takes 318.4 ms of execution time. In the case of the problems with 3 cells, the algorithm uses in average 18.3 iterations and 521.3 ms of execution time.

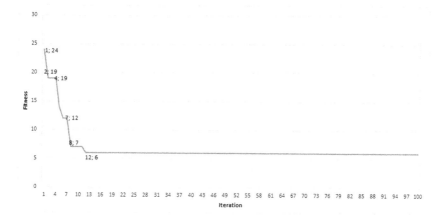

Fig. 2. Performing graph of the FPA with C = 3 and M = 6.

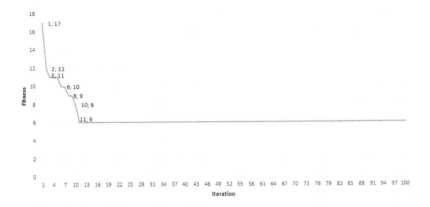

Fig. 3. Performing graph of the FPA with C = 3 and M = 7.

6 Conclusion and Future Work

During the development of this work, it has been proven once more that the Flower Pollination Algorithm is a highly promising metaheuristic, able to deliver results to complex problems in an efficient and precise way. The MCDP although highly investigated through the years, is complex enough to be a good benchmark to the algorithm's properties. To prove this it was used the vector representation of the problem because it fit the movements described by the algorithm in a more direct way. Finally, the integration between the problem and metaheuristic was performed successfully. Also, FPA showed great speed convergence, reaching optimal solutions in 90 test instances.

Acknowledgments. Ricardo Soto is supported by Grant CONICYT/FONDECYT/REGULAR/1160455, Broderick Crawford is supported by Grant CONICYT/FONDECYT/REGULAR/1140897, Rodrigo Olivares is supported by Postgraduate Grant Pontificia Universidad Católica de Valparaíso, Chile (INF-PUCV 2015), and Boris Almonacid is supported by Postgraduate Grant Pontificia Universidad Católica de Valparaíso, Chile (VRIEA-PUCV 2016 and INF-PUCV 2015), by Animal Behavior Society, USA (Developing Nations Research Awards 2016) and by Ph.D (h.c) Sonia Alvarez, Chile.

References

1. Storch, R.L.: Group technology. College of Engineering, University of Washington (2010)
2. Atmani, A., Lashkari, R., Caron, R.: A mathematical programming approach to joint cell formation and operation allocation in cellular manufacturing. Int. J. Prod. Res. **33**(1), 1–15 (1995)
3. Adil, G., Rajamani, D., Strong, D.: A mathematical model for cell formation considering investment and operational costs. Eur. J. Oper. Res. **69**(3), 330–341 (1993)
4. Kusiak, A., Chow, W.: Efficient solving of the group technology problem. J. Manuf. Syst. **6**(2), 117124 (1987)

5. Purcheck, G.F.K.: A linear-programming method for the combinatorial grouping of an incomplete set. J. Cybern. **5**(4), 51–78 (1975)
6. Medina, P.D., Cruz, E.A., Pinzon, M.: Generacin de celdas de manufactura usando el algoritmo de ordenamiento binario (aob). Scientia et Technica Ao **XVI**(44), 106–110 (2010)
7. Kusiak, A.: The part families problem in flexible manufacturing systems. Ann. Oper. Res. **3**(6), 277–300 (1985)
8. Shargal, M., Shekhar, S., Irani, S.: Evaluation of search algorithms and clustering efficiency measures for machine-part matrix clustering. IIE Trans. **27**(1), 43–59 (1995)
9. Seifoddini, H., Hsu, C.-P.: Comparative study of similarity coefficients and clustering algorithms in cellular manufacturing. J. Manuf.Syst. **13**(2), 119–127 (1994)
10. Srinivasan, G.: A clustering algorithm for machine cell formation in group technology using minimum spanning tree. Int. J. Prod. Res. **32**(9), 2149–2158 (1994)
11. Deutsch, S., Freeman, S., Helander, M.: Manufacturing cell formation using an improved p-median model. Comput. Ind. Eng. **34**(1), 135–146 (1998)
12. Soto, R., Crawford, B., Almonacid, B., Paredes, F.: A migrating birds optimization algorithm for machine-part cell formation problems. In: Sidorov, G., Galicia-Haro, S.N. (eds.) MICAI 2015. LNCS (LNAI), vol. 9413, pp. 270–281. Springer, Heidelberg (2015). doi:10.1007/978-3-319-27060-9_22
13. Soto, R., Crawford, B., Almonacid, B., Paredes, F.: Efficient parallel sorting for migrating birds optimization when solving machine-part cell formation problems. Sci. Program. **2016** (2016). https://www.hindawi.com/journals/sp/2016/9402503/
14. Xambre, A.R., Vilarinho, P.M.: A simulated annealing approach for manufacturing cell formation with multiple identical machines. Eur. J. Oper. Res. **151**(2), 434–446 (2003)
15. Yang, X.-S.: Nature-Inspired Metaheuristic Algorithms. Luniver Press, Bristol (2008)
16. Beni, G., Wang, J.: Swarm intelligence in cellular robotic systems. Adv. Sci. Lett. **102**, 703–712 (1993)
17. Li, L., Shao, Z., Quian, J.: An optimizing method based on autonomous animals: fish-swarm algorithm. Syst. Eng. Theory Pract. **22**(11), 32–38 (2002)
18. Oliveira, S., Ribeiro, J., Seok, S.: A spectral clustering algorithm for manufacturing cell formation. Comput. Ind. Eng. **57**(3), 1008–1014 (2009)
19. Durn, O., Rodriguez, N., Consalter, L.: Collaborative particle swarm optimization with a data mining technique for manufacturing cell design. Expert Syst. Appl. **37**(2), 1563–1567 (2010)
20. Shafer, S.M., Rogers, D.F.: A goal programming approach to the cell formation problem. J. Oper. Manag. **10**(1), 28–43 (1991)
21. Aljaber, N., Baek, W., Chen, C.-L.: A tabu search approach to the cell formation problem. Comput. Ind. Eng. **32**(1), 169–185 (1997)
22. Lozano, S., Daz, A., Egua, I., Onieva, L.: A one-step tabu search algorithm for manufacturing cell design. J. Oper. Res. Soc. **50**(5), 209–516 (1999)
23. Wu, T., Chang, C., Chung, S.: A simulated annealing algorithm for manufacturing cell formation problems. Expert Syst. Appl: Int. J. **34**(3), 1609–1617 (2008)

A New Multiple Objective Cuckoo Search for University Course Timetabling Problem

Thatchai Thepphakorn[1,2], Pupong Pongcharoen[1(✉)],
and Srisatja Vitayasak[1]

[1] Department of Industrial Engineering, Centre of Operations Research
and Industrial Applications (CORIA), Faculty of Engineering,
Naresuan University, Phitsanulok 65000, Thailand
pupongp@nu.ac.th
[2] Faculty of Industrial Technology, Pibulsongkram Rajabhat University,
Phitsanulok 65000, Thailand

Abstract. University course timetabling problem (UCTP) is classified into combinatorial optimisation problems involving many criteria to be considered. Due to many conflict objectives or difference objective units, combining conflicting criteria into a single objective (weight sum approach) may not be the best way of optimisation. The UCTP is well known to be Non-deterministic Polynomial (NP)-hard problem, in which the amount of computational time required to find the solution increases exponentially with problem size. Solving the UCTP manually with/without course timetabling tool is extremely difficult and time consuming. A new multiple objective cuckoo search based timetabling (MOCST) tool has been developed in order to solve the multiple objective UCTP. The cuckoo search via Lévy flight (CSLF) and cuckoo search via Gaussian random walk (CSGRW) using the Pareto dominance approach were embedded in the MOCST program for determining the set of non-dominated solutions. Eleven datasets obtained from Naresuan University in Thailand were conducted in computational experiment. It was found that the CSLF outperformed the CSGRW for almost all datasets whilst the computational times required by the proposed methods were slightly difference.

Keywords: Course timetabling · Cuckoo search · Random walk · Multiple objectives

1 Introduction

University course timetabling problem (UCTP) is one of the most challenging scheduling problems due to its complexity and constraints [1]. This problem arises every semester and is solved either manually by academic staff or using automatic course timetabling tool [2, 3]. The UCTP is known to be a non-deterministic polynomial (NP) hard problem, which means that the computational time required to find the solution increases exponentially with problem size [4]. Solving large course timetabling problems without efficient timetabling program is extremely difficult and may require a group of people to work for several days [5].

© Springer International Publishing AG 2016
C. Sombattheera et al. (Eds.): MIWAI 2016, LNAI 10053, pp. 196–207, 2016.
DOI: 10.1007/978-3-319-49397-8_17

The UCTP is also classified into combinatorial optimisation problems [1] involving many criteria to be considered. There are two general approaches to solve multiple objective optimisation including: (i) to combine the individual objective functions into a single composite function (such as weight sum method); and (ii) to determine the set of Pareto optimal solutions [6]. Due to some considered objectives conflicting with each other, optimising an individual solution with respect to a single objective may produces unacceptable results for other objectives [6]. Moreover, the precise and accurate selection of the weights responding to the preferences of decision-maker are very difficult [6]. A set of Pareto optimal solutions is often preferred to a single solution because: (i) all solutions contained in a Pareto optimal set are non-dominated with respect to each other; and (ii) the final solution of the decision-maker is always a trade-off [6]. However, the size of the Pareto set usually increases with the increase in the number of objectives [6].

There has been very few research works reported on the multiple objective UCTP. For example, non-dominated sorting genetic algorithm-II (NSGA-II) for resource allocation problem (NSGA-II RAP) was developed to solve two real-world instances [7]. NSGA-II with a variable population size (NSGA-II VPS) was proposed to solve post enrolment course timetabling problem [8]. Guided search NSGA (GSNSGA) was introduced for solving the benchmark datasets [1]. Local search algorithm was also applied to improve the utilisation of university teaching space in Australia [9].

Computational intelligence has been extended to many applications directly or indirectly [10]. Machine learning, classifications and cluster, data mining, neural networks are included in the classical methods [10]. Nowadays, nature-inspired meta-heuristic algorithms have become powerful and popular in computational intelligence [10] such as genetic algorithms (GA), ant colony optimisation (ACO), particle swarm optimisation (PSO), bat algorithm (BA), artificial bee colony (ABC) algorithm, firefly algorithm (FA), harmony search (HS), cuckoo search (CS), and so on. These algorithms have been widely used to solve NP hard problems within acceptable computational time, but they do not guarantee optimum solutions [11].

The CS is a one of nature-inspired metaheuristic algorithms and proposed by Yang and Deb in 2009 [12]. During the last few years, CS has received high attention in science and engineering because of several advantages including: (i) it can solve both continuous and combinatorial problem domains [10]; (ii) it is a population-based algorithm [13] that performs multiple directional searches using a set of population (host nests); (iii) it uses some sort of elitism and/or selection similar to that used in HS [12]; (iv) the randomisation in CS is more efficient as the step length is heavy-tailed, any large step is possible [13]; (v) CS has the number of parameters fewer than GA and PSO to be tuned [12]; and (vi) it is potentially more generic to apply to more categories of optimisation problems [13].

Unfortunately, the applications of CS for solving the UCTP have been rarely reported. Teoh et al. [14] have developed an adapted cuckoo optimisation algorithm (ACOA) based on Lévy flight and elitist mechanisms to solve five datasets of the UCTP in Malaysia. The numbers of hard and soft constraint violations were simultaneously considered into a single function with different the penalty weights. The ACOA outperformed the GA for almost all datasets, especially in the large problem size.

Therehas been no report on the application of multiple objective cuckoo search using Pareto dominance for solving real-world university course timetabling problem.

The objectives of this paper were to: (i) develop a new multiple objective cuckoo search based timetabling (MOCST) tool using the Pareto dominance approach for solving real-world course timetabling datasets obtained from the Faculty of Engineering, Naresuan University; and (ii) compare the movement performances of the cuckoo search using either Lévy flight (CSLF) or Gaussian random walk (CSGRW) in terms of the solution quality and computational time required. The next section of this paper briefly explains the CS algorithm. Section 3 describes the UCTP followed by the procedures of the MOCST tool. Section 5 presents the experimental results and analysis followed by conclusions.

2   Cuckoo Search (CS) Algorithm

Cuckoo search (CS) is one of the nature-inspired metaheuristic algorithms [13], inspired by the obligate brood parasitism of some cuckoo species by laying their eggs in the nests of other host birds [15]. Each egg in a host nest represents a solution, and a cuckoo egg represents a new solution [13]. The aim is to create the new and potentially better solutions (cuckoos) to replace not-so-good solutions in the host nests [16]. For simple CS algorithm, there are mainly three principle rules: (i) each cuckoo lays one egg at a time, and dumps its egg in a randomly chosen nest [17]; (ii) the best nests with high-quality eggs will be maintained to the next generation [17]; and (iii) the number of available host nests is fixed, and the egg laid by a cuckoo is discovered by the host bird with a probability, in this case, the host bird can either get rid of the laid egg or abandon the nest and build a new nest [17]. The pseudo code for simple CS procedure based on these principle rules is shown in Fig. 1.

```
Begin Objective function F(x), x = (x₁, . . . ,xₐ)
   Generate initial population xᵢ(i = 1,2,…,P)
   While (t < Max_Iteration: I) or (stop criterion)
       Get a cuckoo (say, xᵢ) randomly
       Generate a solution xᵢ by Lévy flights
       Evaluate new solution xᵢ or F(xᵢ)
       Choose a nest among n (say, xⱼ) randomly
       If F(xᵢ) > F(xⱼ)
              Replace xⱼ by the new solution xᵢ
       End if
       A fraction (Pₐ) of worse nests are abandoned,
              and new ones/solutions are built/generated
       Keep the best solution (or nest)
       Rank the solutions,
              and find the current best solution
   End while
   Postprocess results and visualisation
End
```

Fig. 1. Pseudo code of the simple CS algorithm [12]

In the initial process, objective function $F(x)$ is specified. Then, each solution x_i ($i = 1, 2,..., P$) is generated randomly and evaluated its fitness before identifying the current best solution. When generating new solutions $x_i^{(t+1)}$ for a cuckoo i, the CS via Lévy flight (CSLF) is performed by using Eq. (1) [17].

$$x_i^{(t+1)} = x_i^{(t)} + \alpha \oplus L\acute{e}vy\,(\beta) \tag{1}$$

The product $\oplus$ means entrywise multiplications that is similar to those used in PSO [13]. Where $\alpha > 0$ is the step size which is related to the scales of the problem of interest [17]. In order to accommodate the difference between solution quality, the α can be defined as Eq. (2) [17].

$$\alpha = \alpha_0(x_j^{(t)} - x_i^{(t)}) \tag{2}$$

Where α_0 is a constant, while the term in the bracket corresponds to the randomly solution difference between x_j and x_i [17]. The $L\acute{e}vy\,(\beta)$ is the random walk that random step lengths is drawn from a Lévy distribution in Eq. (3) [17].

$$L\acute{e}vy \sim u = t^{-1-\beta}, (0 < \beta \le 2) \tag{3}$$

Which has an infinite variance with an infinite mean, in which the consecutive steps of a cuckoo essentially form a random walk process with obeys a power-law step-length distribution with a heavy tail [17]. Therefore, the generation of step size s samples can be summarised by Eq. (4) [17].

$$s = \alpha_0(x_j^{(t)} - x_i^{(t)}) \oplus L\acute{e}vy\,(\beta) \sim 0.01 \frac{u}{|v|^{1/\beta}}(x_j^{(t)} - x_i^{(t)}) \tag{4}$$

Where u and v are drawn from normal distributions following in Eqs. (5) and (6) [17]:

$$u \sim N(0, \sigma_u^2), v \sim N(0, \sigma_v^2) \tag{5}$$

$$\sigma_u = \left\{ \frac{\Gamma(1+\beta)\sin(\pi\beta/2)}{\Gamma[(1+\beta)/2]\beta 2^{(\beta-1)/2}} \right\}^{1/\beta}, \ \sigma_v = 1 \tag{6}$$

Where Γ is the standard Gamma function [17]. In case of CS via Gaussian random walks (CSGRW), a new solution $x_i^{(t+1)}$ generated from a cuckoo i is performed by using Eq. (7) [12].

$$x_i^{(t+1)} = x_i^{(t)} + \alpha \oplus \varepsilon_t \tag{7}$$

Where, ε_t obeys a Gaussian distribution, this becomes a standard random walk [12]. The α is defined as Eq. (2) [17]. After preforming the Lévy flight or Gaussian random walks, the fitness value of new solution x_i or $F(x_i)$ is evaluated. A nest or solution among current population is randomly selected (called solution x_j) in order to compare the solution quality with a new solution x_i. The new x_i will be accepted and replaced to the x_j if the $F(x_i)$ is better than the $F(x_j)$.

Next step, the worse nests or solutions will be abandoned according to a probability $P_a \in [0,1]$ [17]. The new nests will be built at new locations by using random walks or random permutation according to the similarity/difference to the host eggs [17]. After that, the population will be sorted basing on the solution quality before identifying the best so far solution. These processes are repeated until getting to the maximum iteration (I) or stop criterion.

3   University Course Timetabling Problem (UCTP)

In educational institutions, timetabling courses and examinations is a crucial activity, which assigns appropriate timeslots for students, lecturers, and classrooms [18]. In this research, the real-world course timetabling problem from Naresuan University (NU) in Thailand was considered because of many reasons: (i) comprising very large course timetabling data; (ii) lacking an efficient and automatically timetabling tool due to the current semi-automatic program not only unable to construct the best course time-tabling but also requires many staffs and long consuming time to schedule in every semester; and (iii) including constraint variants to schedule.

For examples: (i) there are multiple lecturers per a course, especially found in laboratory periods; (ii) a course having multiple sections and teaching by the same lecturers must be assigned only one section at the same time; and (iii) a course taught by lecturers who being administrative position of university or external special lecturers, their requirements (such as available days and periods, building locations, and classroom facilities) must be obeyed. Therefore, the large number and variety of timetabling constraints related with course structures and special requirements found in real-world problems will increases the difficulty and complexity to find a practicable timetable [19].

The general constraints in course timetabling can be classified into two types: hard constraints (HC) and soft constraints (SC) [11]. Hard constraints are the most important and must be satisfied to have a feasible timetable whereas soft constraints are more relaxed as some violations are acceptable, however the number of violations should be minimised by algorithms [18]. The course timetabling constraints considered in this research can be described into the HC and SC as following:

HCs considered were:

- all lectures within a course must be scheduled and assigned to distinct periods (HC_1);
- students and lecturers can only attend one lecture at a time (HC_2);
- only one lecture can take place in a room at a given time (HC_3);
- lecturers and students must be available for a lecture to be scheduled (HC_4);
- all courses must be assigned into the classrooms according to their given requirements including building location, room facilities, and room types (HC_5); and
- all lectures within a course required consecutive periods must be obeyed (HC_6).

In additions, SCs considered were:

- all courses should be scheduled in the appropriate classroom in order to avoid unnecessary operating or renting costs (per hour) (SC_1);
- the courses taught by the given lecturer(s) should be assigned into their available or preferred day and periods in order to save the hiring costs (per hour) (SC_2);
- the classrooms should be scheduled in consecutive working periods of a day in order to reduce the number of times to clean or setup after using the rooms (per time) (SC_3); and
- the classrooms must have sufficient seats for the students on the module (SC_4).

HC_1–HC_6 determine whether potential solutions are feasible. HC_1–HC_3 are the fundamental timetabling constraints (called "event-clash" or binary constraint) that can be found in almost all university timetabling problems [11] whilst HC_4–HC_6 are individual requirements and timetabling policy found in the NU. SC_1–SC_4 are represented by the objective functions in this work. SC_1–SC_3 under consideration aimed to minimise the total university operating costs determined from the constructed course timetables whereas SC_4 is to minimise the total inadequate seats. Due to difference in the measurement units between costs (currency) and seat numbers, these SCs should be determined in the multiple objective functions following (8);

$$f_1(x) = W_1 SC_1 + W_2 SC_2 + W_3 SC_3$$

$$f_2(x) = W_4 SC_4$$

$$Minimise \ \ F(x) = \ (f_1(x), f_2(x)) \tag{8}$$

Subject to:

$$HC_k = 0, \quad \forall k, \tag{9}$$

Equation (8) is the multiple objective functions that evaluated the total university operating costs of the SC_1–SC_3, called $f_1(x)$, and the total number of inadequate chairs of the SC_4, called $f_2(x)$. The weightings $(W_1$–$W_4)$ for each SC are not restricted and depend upon the user preferences for each institution. In this work, W_1–W_4 were specified at 50 (per hour), 300 (per hour), 2.5 (per time), and 1 (per time), respectively. Equation (9) checks a timetable to be a feasible timetable, in which all hard constraints must be satisfied. Where k is an index relating to the kth hard constraint ($k = 1, 2, 3, ..., H$), where H is the number of hard constraints.

4   Multiple Objective Cuckoo Search Based Timetabling Tool

The multiple objective cuckoo search based timetabling (MOCST) program has been coded in modular style using a general purpose programming language called TCL/TK with C extension [20]. It was developed in order to solve the multiple objective UCTP by using a cuckoo search (CS) algorithm. The main procedures within the MOCST program are included in six steps and shown in Fig. 2.

```
Begin                                                    /*Step 1*/
    Input university course timetabling data
    Set CS's parameters
    Sort a list of courses using heuristic orderings
    Create initial population, x_i (i = 1,2,…,P)
    Generate random keys for each x_i
    While t < Max_Iteration(I) do                        /*Step 2*/
        For (i=1, i<= Max_Pop(P), i++) do
            Get a cuckoo (say, x_i) randomly
            Generate a new solution x_i' using Eq.(1) or Eq.(7)
            If (x_i' = an infeasible timetable) do
                Repair x_i' to be a feasible timetable
            End if                                       /*Step 3*/
            Evaluate objective functions F(x_i')         /*Step 4*/
            Choose a nest among P (say, x_j) randomly
            If F(x_i') dominate F(x_j) do
                Replace the x_j by the new solution x_i'
            End if
        End for
        Rank the population using                        /*Step 5*/
            Pareto dominate approach
        If (rand < P_a) do                               /*Step 6*/
            Build/generate new solutions x_new
            Replace the x_worse by new solution x_new
        End if
        Update the set of global non-dominated solutions
    End while
    Postprocess results and visualisation
End
```

Fig. 2. Pseudo code of the MOCST tool

Step 1: after uploading course timetable data and assigning CS's parameters, the total number of events (n) is determined from the number of teaching periods required for all modules (courses). Then, an event list containing a set of n events was initialised. The event sequence in the list was sorted by using the Largest unpermitted period degree (LUPD) first heuristic [21]. This rule reduces the probability of getting infeasible timetables that generally occur in the process of solution initialisation.

Next step is to create an empty timetable (solution), in which the length of that is calculated taking into account the numbers of timeslots per day, working day per week, and given classrooms. Then, all events according to the sorted list were inserted into an empty timetable in order to produce an initial population x_i ($i = 1, 2, 3,…, P$) that represents a set of possible timetables. Next step is to create a new list having the same length of solution, in which each timeslot of a solution is assigned random numbers uniformly distributed between 0 and 1, called random key technique [22].

Step 2: this step is the evolution process of the CS algorithm. A cuckoo x_i is randomly selected from the current population. Then, the evolution of the x_i is produced by using Lévy flight from Eq. (1) or Gaussian random walks from Eq. (7).

Step 3: after evolution process, a new solution (x_i') may be either feasible or infeasible timetable. The repair process was therefore design and embedded in the MOCST program in order to rectify infeasible solutions.

Step 4: the solution quality of the x_i' can be measured by using Eq. (8). After that, a cuckoo x_j is randomly selected from the current population. Due to multiple objective UCTP, Pareto dominance approach was adopted to compare the dominance between x_i' and x_j. If $F(x_i')$ dominates $F(x_j)$, a cuckoo x_j is replaced by the x_i', otherwise x_i' is discarded. This processes will be repeated until all cuckoos in the population are improved.

Step 5: the current population are increasingly ranked by using Pareto dominance approach. A set of solutions at the first ranking is usually called the set of non-dominated solutions.

Step 6: a solutions at the last ranking (x_{worse}) are randomly selected to produce the host nest abandon process. If random value $(rand)$ obtained from the $U[0,1]$ is lower or equal to the P_a parameter, the x_{worse} will be replaced by a new solution (x_{new}). Otherwise, the x_{worse} will be kept. Then, the set of global non-dominated solutions is updated. These processes (Step 2 to Step 6) will be repeated until reach the maximum iterations before showing the results.

5   Experimental Results and Analysis

The objective of the MOCST program is to construct the set of non-dominated timetables that minimise both the university operating costs, $f_1(x)$, and the total number of inadequate chairs, $f_2(x)$. Eleven course timetabling datasets obtained from the Faculty of Engineering, Naresuan University (NU) in Thailand (as shown in Table 1) were used in the computational experiment.

Table 1. Characteristics of the proposed UCTP

Datasets	Characteristics of university course timetabling problems						
	Courses	Events	Classrooms	Days/week	Periods/day	Lecturers	Curricula
1	56	173	53	5	10	30	19
2	103	323	77	7	10	62	36
3	123	353	86	7	10	49	27
4	124	380	74	7	11	56	35
5	144	452	91	7	10	78	43
6	162	486	99	7	10	71	34
7	163	499	88	7	11	72	38
8	204	639	114	7	10	96	52
9	208	647	99	7	11	102	56
10	221	687	108	7	12	94	44
11	323	1,009	142	7	13	143	66

The experiment was aimed to investigate and compare the moving performances of cuckoo search using either Lévy flight (CSLF) or Gaussian random walks (CSGRW) to solve the proposed eleven datasets using the multiple objective functions. Personal computer with Core 2 Duo 2.67 GHz CPU and 4 GB RAM was used to determine the computational time required to execute experimental runs.

The factors of the CSLF and CSGRW included (i) the combination of population sizes (host nests) and the number of iterations (PI), which determines the total number of solutions generated (or amount of search) and the execution time, this computational experiment the value was fixed at 24,000 to limit the time taken for computational search; and (ii) the probability of alien egg discovery (P_a). The appropriate parameter settings of both methods for each problem were adopted from previous work [23].

Each dataset, the CSLF and CSGRW were repeated thirty replications using different random seed numbers due to stochastic optimisation method. The sets of non-dominated solutions obtained from thirty replications were combined together before determining a set of Pareto optimal solutions (called X) again. After that, the global non-dominated solutions (called Z) considered from the X of both CSLF and CSGRW were also identified. The numbers of X contained in the Z (called Y) for each algorithm were counted before calculating the Ratio Y/X [24]. The Ratio value is distributed between 0 and 1. An algorithm having the higher values of Ratio indicated that the performance of that algorithm to find the global non-dominated solutions is better. The average execution times (*Time*: minute unit) required to solve the proposed datasets were also determined and shown in Table 2.

Table 2. Comparative results to find the set of global non-dominated solutions

Datasets	No. Global non-dominated solutions (Z)	CSLF				CSGRW			
		No. Non-dominated solutions (X)	No. X found in Z (Y)	Ratio Y/X	*Time* (*min*)	No. Non-dominated solutions (X)	No. X found in Z (Y)	Ratio Y/X	*Time* (*min*)
1	10	9	6	**0.67**	3.54	14	4	0.29	3.24
2	16	10	8	0.80	10.95	9	8	**0.89**	10.59
3	4	3	3	**1.00**	14.81	13	1	0.08	16.63
4	5	6	4	**0.67**	12.34	5	1	0.20	11.14
5	14	14	14	**1.00**	17.16	6	0	0.00	15.14
6	4	3	3	**1.00**	28.04	10	1	0.10	26.37
7	11	10	10	**1.00**	17.81	5	1	0.20	15.94
8	5	4	0	0.00	36.94	5	5	**1.00**	35.12
9	13	11	8	**0.73**	24.29	13	5	0.38	24.88
10	3	3	3	**1.00**	36.79	13	0	0.00	35.54
11	7	6	6	**1.00**	107.22	5	1	0.20	92.60

From Table 2, it can be seen that the CSLF outperformed the CSGRW to find the set of global non-dominated solutions for most datasets because of the higher values of Ratio Y/X, whereas the CSGRW outperformed the CSLF for problem numbers 2 and 8. Moreover, six of eleven datasets solved by the CSLF had the highest value of Ratio Y/X whilst the CSGRW had only one dataset. The computational times required by CSGRW and CSLF were slightly difference for all datasets.

The sets of non-dominated solutions obtained from the CSGRW and the CSLF can be represented by Pareto frontiers. For examples, dataset numbers 5, 7, 8, and 9 related with medium and large size problems were selected to demonstrate both dominated and non-dominated solutions generated from the CSLF and the CSGRW (shown in Figs. 3 and 4). It can be seen that the frontier curves of non-dominated solutions obtained from CSLF and CSGRW were obvious difference for each dataset. Pareto frontier curves belonging to the CSLF and CSGRW were dominated with each other for some datasets (for example, see the 9th dataset in Fig. 4), whereas Pareto frontier curves obtained from both methods were obviously split into two curves for some datasets (for example, see the 8th dataset in Fig. 4). Moreover, the number of dominated solutions obtained from both methods was also diversity for all datasets.

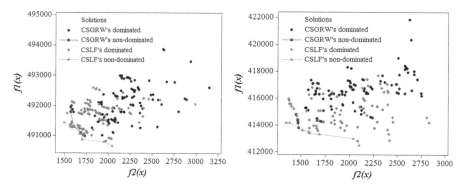

Fig. 3. Pareto frontiers for the 5th dataset (left) and the 7th dataset (right)

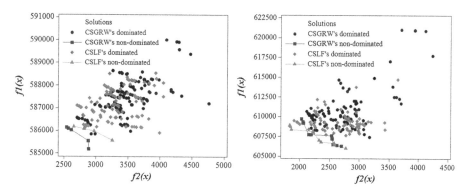

Fig. 4. Pareto frontiers for the 8th dataset (left) and the 9th dataset (right)

6   Conclusions

A new multiple objective cuckoo search based timetabling (MOCST) tool has been developed in order to solve the multiple objective UCTP. The solutions or timetables generated by the MOCST program were measured by minimising both the total university operating costs, $f_1(x)$, and the total number of inadequate chairs, $f_2(x)$. Cuckoo search via Lévy flight (CSLF) and cuckoo search via Gaussian random walk (CSGRW) using the Pareto dominance approach were embedded in the MOCST program for determining the set of non-dominated solutions. Eleven datasets obtained from real-world university course timetabling data in Thailand were computationally conducted using the proposed program. Experimental results indicated that the CSLF outperformed the CSGRW for almost all datasets. The computational times required by both methods were slightly difference.

Acknowledgement. This work was partly supported by the Naresuan University Research Fund; grant number R2558C156.

References

1. Jat, S.N., Yang, S.: A guided search non-dominated sorting genetic algorithm for the multi-objective university course timetabling problem. In: Merz, P., Hao, J.-K. (eds.) EvoCOP 2011. LNCS, vol. 6622, pp. 1–13. Springer, Heidelberg (2011). doi:10.1007/978-3-642-20364-0_1

2. Thepphakorn, T., Pongcharoen, P., Hicks, C.: Modifying regeneration mutation and hybridising clonal selection for evolutionary algorithms based timetabling tool. Math. Probl. Eng. **2015**, 16 (2015)

3. Lutuksin, T., Pongcharoen, P.: Best-worst ant colony system parameter investigation by using experimental design and analysis for course timetabling problem. In: 2nd International Conference on Computer and Network Technology, ICCNT 2010, pp. 467–471 (2010)

4. Pongcharoen, P., Promtet, W., Yenradee, P., Hicks, C.: Stochastic optimisation timetabling tool for university course scheduling. Int. J. Prod. Econ. **112**, 903–918 (2008)

5. MirHassani, S.A.: A computational approach to enhancing course timetabling with integer programming. Appl. Math. Comput. **175**, 814–822 (2006)

6. Konak, A., Coit, D.W., Smith, A.E.: Multi-objective optimization using genetic algorithms: a tutorial. Reliab. Eng. Syst. Saf. **91**, 992–1007 (2006)

7. Datta, D., Fonseca, C.M., Deb, K.: A multi-objective evolutionary algorithm to exploit the similarities of resource allocation problems. J. Sched. **11**, 405–419 (2008)

8. Abdullah, S., Turabieh, H., McCollum, B., McMullan, P.: A multi-objective post enrolment course timetabling problems: a new case study. In: IEEE Congress on Evolutionary Computation (CEC 2010), pp. 1–7 (2010)

9. Beyrouthy, C., Burke, E.K., Landa-Silva, D., McCollum, B., McMullan, P., Parkes, A.J.: Towards improving the utilization of university teaching space. J. Oper. Res. Soc. **60**, 130–143 (2009)

10. Yang, X.-S., Chien, S.F., Ting, T.O.: Computational intelligence and metaheuristic algorithms with applications. Sci. World J. **2014**, 4 (2014)

11. Lewis, R.: A survey of metaheuristic-based techniques for university timetabling problems. OR Spectr. **30**, 167–190 (2008)

12. Yang, X.-S., Deb, S.: Engineering optimisation by cuckoo search. Int. J. Math. Model. Numer. Optim. **1**, 330–343 (2010)

13. Yang, X.-S.: Nature-Inspired Metaheuristic Algorithms, 2nd edn. Luniver Press, University of Cambridge, Cambridge (2010)

14. Teoh, C.K., Wibowo, A., Ngadiman, M.S.: An adapted cuckoo optimization algorithm and genetic algorithm approach to the university course timetabling problem. Int. J. Comput. Intell. Appl. **13**, 1450002 (2014)
15. Li, X., Yin, M.: Modified cuckoo search algorithm with self adaptive parameter method. Inf. Sci. **298**, 80–97 (2015)
16. Valian, E., Tavakoli, S., Mohanna, S., Haghi, A.: Improved cuckoo search for reliability optimization problems. Comput. Ind. Eng. **64**, 459–468 (2013)
17. Yang, X.-S., Deb, S.: Multiobjective cuckoo search for design optimization. Comput. Oper. Res. **40**, 1616–1624 (2013)
18. Thepphakorn, T., Pongcharoen, P., Hicks, C.: An ant colony based timetabling tool. Int. J. Prod. Econ. **149**, 131–144 (2014)
19. Murray, K., Müller, T., Rudová, H.: Modeling and solution of a complex university course timetabling problem. In: Burke, E.K., Rudová, H. (eds.) PATAT 2006. LNCS, vol. 3867, pp. 189–209. Springer, Heidelberg (2007). doi:10.1007/978-3-540-77345-0_13
20. Ousterhout, J.K., Jones, K.: TCL and the TK Toolkit, 2nd edn. Addison-Wesley, New York (2009)
21. Thepphakorn, T., Pongcharoen, P.: Heuristic ordering for ant colony based timetabling tool. J. Appl. Oper. Res. **5**, 113–123 (2013)
22. Khadwilard, A., Chansombat, S., Thepphakorn, T., Thapatsuwan, P., Chainate, W., Pongcharoen, P.: Application of firefly algorithm and its parameter setting for job shop scheduling. J. Ind. Technol. **8**, 49–58 (2012)
23. Thepphakorn, T.: Solving complex university course timetabling using metaheuristics. Doctor of Philosophy, Department of Industrial Engineering, Faculty of Engineering, Naresuan University, Phitsanulok, Thailand (2016)
24. Khadwilard, A.: Multiple objective genetic algorithms for production scheduling in capital goods industries. Master of engineering, Department of Industrial Engineering, Faculty of Engineering, Naresuan University, Phitsanulok, Thailand (2007)

Application of Genetic Algorithm for Quantifying the Affect of Breakdown Maintenance on Machine Layout

Srisatja Vitayasak and Pupong Pongcharoen[(✉)]

Faculty of Engineering, Department of Industrial Engineering,
Centre of Operations Research and Industrial Applications,
Naresuan University, Phitsanulok 65000, Thailand
{srisatjav,pupongp}@nu.ac.th

Abstract. Computational intelligence (CI) has been successfully applied to solve a variety of optimisation problems under uncertain conditions in manufacturing domains. Uncertainties have an impact on product mix and material flow in manufacturing shop floor. The effective layout can reduce material handling costs, and lead to reduce up to 50 % of the total operating expenses. Machine layout design is classified into non-deterministic polynomial-time hard problem. This paper presents the application of Genetic Algorithm for generating robust layouts under uncertainties on customers' demand and breakdown maintenance. This was aimed to minimise both cost and handling distance of material flow. The experimental programme was conducted using eight different-size datasets. Minimising trade-off between material travelling distance and costs of relocating machines/equipment has been discussed. This provides a decision framework for evaluating the investment on facility re-layout design.

Keywords: Genetic Algorithm · Robust layout · Machine re-layout · Demand uncertainty · Maintenance

1 Introduction

Computational intelligence (CI), which is one important sub-branch of artificial intelligence, is the study of the design of intelligent agents [1]. An intelligent agent is an adaptive mechanism that can enable or facilitate intelligent behavior in complex and changing environments [2]. Applications of CI can be found in prediction of stock market [3], intrusion detection [4], protein classification [5], gene identification [6], traffic signal [7], facility layout [8, 9], scheduling [10], timetabling [11], vehicle routing [12], travelling salesman [13], and container packing [14]. CI has been also applied to solve the problems in uncertain conditions [15].

Uncertainties may be caused by internal and external forces [16, 17]. Internal disturbances include variable task times, rejects, rework, and machine breakdowns. Reduced number of available machines due to breakdown machines can cause longer flow time, lower productivity and higher production costs. Once materials flow is interrupted, successive resources will be idle, which can reduce utilisation. In this case,

C. Sombattheera et al. (Eds.): MIWAI 2016, LNAI 10053, pp. 208–218, 2016.
DOI: 10.1007/978-3-319-49397-8_18

the alternative machine is introduced to overcome the difficulties but material handling time and distances are likely to change [17].

External forces consist of competitive environment, availability of resources, product prices, product mix, and demand variability, which may be unknown. Uncertainty may arise out of actions of competitors, changing consumer preferences, technological innovations, and new regulations [17]. Variations in the level of customer demand and product mix can disrupt the efficient flow of materials within a facility. These changes affect production performance and the layout [18]. However, none of the papers took into account both demand uncertainty and machine maintenance in optimisation problems such as facility layout problem.

The objective of this paper is to demonstrate the application of Genetic Algorithm (GA) for quantifying the affect of breakdown maintenance on machine layout design. The remaining sections are organised as follows: Sect. 2 is comprehensive literature reviews including computational intelligence, production condition in facility layout problem, and machine breakdown. Application of GA on machine layout design problems is shown in Sect. 3. The experimental results are presented in Sect. 4 and finally, conclusions are drawn in Sect. 5.

2 Comprehensive Literature Review

The related literature reviews on application of computational intelligence, production condition in facility layout problem, and machine breakdown were presented in subsections as follows:

2.1 Computational Intelligence

In IEEE Word Congress on Computational Intelligence (1994), computational intelligence was defined as science-based approaches and technologies for analysis, design, and development of intelligent systems. CI tools based on natural inspirations can be classified as the human-mind model based (e.g. fuzzy set theory), the artificial immune system based (e.g. evolutionary computing), the swarm intelligence based (e.g. Ant Colony Optimisation), and the emerging geo-sciences based (e.g. Earthquakes) [19]. These natural-inspiration techniques (so called metaheuristics) are an important part of CI [20].

Genetic Algorithm (GA) [21], one of metaheuristics in evolutionary computing, is a population-based, nature-inspired algorithm. A set of candidate solutions is generated as an initial set of solutions, which then undergoes an evolutionary search process. Exploitation and exploration processes are carried out simultaneously via crossover and mutation operations, respectively [22], in which GA uses probabilistic (non-deterministic) transition rules to guide exploitative search and also performs a multiple directional search by maintaining a population of potential solutions. This mechanism can be adjusted for helping to escape from local optimal.

There are a number of literatures related to the application of GA on production and operation management in the last few decades [23]. Lenin et al. [24] demonstrated the effectiveness of GA to solve single-row layout design problem. The results obtained were more favorable than other approaches. Kia et al. [25] proposed that GA can find near-optimal solutions much less computational time than CPLEX for most datasets. Rose and Coenen [26] investigated the performance comparison of GA, and other algorithms (e.g. Particle Swarm Optimisation) for ship outfitting scheduling. The results indicated that GA was more efficient in terms of computational time.

2.2   Production Condition in Facility Layout Problem

Facility layout design involves arrangement of facilities into a limited manufacturing shop floor. The effective layout can reduce material handling costs, and lead to quicker transfer times between facilities, better productivity, and reduce up to 50 % of the total operating expenses [27]. When the flow of materials between the facilities does not change during the planning horizon, this problem is known as the static facility layout problem (SFLP). The dynamic facility layout problems (DFLP) take into account possible changes in the material handling over multiple periods. Variation in material flow has been resulted from customer demand, product mix, number of machines, and routing flexibility. Changing the processing route due to machine maintenance can affect the flow intensity over time period but the consideration of maintenance in FLP has not been reported in literature.

The facility layout problems (FLPs) are classified as Non-deterministic Polynomial time hard problem [28]. The approximation algorithms have been applied to solve these problems. A systematic literature review was undertaken using the ISI Web of Science database for the period 2001 to February 2016. It can be found that GA has been extensively applied to solve FLPs both in static [24] and dynamic [25] scenarios.

2.3   Machine Breakdown

Machine breakdown is a stochastic event that is a major concern in industry. If operations are interrupted, it may be necessary to revise the schedule to re-optimise the remaining operations taking into account the machine downtime. The easiest solution is often to apply some dispatching rule to sequence operations immediately after the breakdown occurs [29]. A number of parameters have been used to model machine maintenance problems, for example machine failure rate has often been represented by the Poisson distribution or generated randomly. Machine lifetime is commonly modelled using the Weibull distribution. Mean time to failure has been represented by the normal distribution or the exponential distribution. Breakdown maintenance has also been considered in the context of robust scheduling [30].

3   Application of GA on Machine Layout Design Problems

3.1   Problem Statement

Machine layout design (MLD) under dynamic production condition can be robust, and re-layout. Changing production condition due to external forces is product demand. Demand profiles can be obtained from empirical data (the demand value is known in advance and changes over time periods) or by using different types of distributions (exponential, normal distribution, or uniform) [31]. Breakdown maintenance (BM) can be considered as one of internal production conditions. Once a machine is broken down, an alternative machine will be assigned for production. This will effect to material handling distance on the changing route of machine sequence. For example, machine sequence is M1-M2-M3, if the M2 is unavailable because of BM, the alternate machine (M9) is therefore selected. The machine sequence is adapted as M1-M9-M3 by which, material handling distance is changed. The evaluation function (Z) for the efficiency of robust layout design can be used to minimise total material handling distance (MHD) as defined by Eq. (1)

$$Minimise\ Z\ =\ \sum_{i=1}^{M}\sum_{j=1}^{M}\sum_{g=1}^{N}\sum_{k=1}^{P} d_{ij}f_{ijgk}D_{gk} \qquad (1)$$

M is the number of machines, i and j are machine indexes (i and j = 1, 2, 3, ..., M). N is the number of product types, g is a product index (g = 1, 2, 3, ..., N) and P is the number of time periods, k is a time period index (k = 1, 2, 3, ..., P). d_{ij} is the distance from machines i to j (i ≠ j), f_{ijgk} is the frequency of material flow of product g from machines i to j on period k, and D_{gk} is the customer demand of product g on period k.

Rectangular machine layout design is concerned with the placement of machines into a limited shop floor area having gap between machines. Machines are sequentially arranged row by row, from left to right, starting at the first row and respecting floor length and the gap [32]. When there is not enough area for placing the next machine at the end of the row, it is placed in the next row. Vehicles moving between rows move to the left or right side of the row and then up or down to the destination row.

3.2   Genetic Algorithm on MLD Problems

The GA pseudo-code for the proposed MLD is shown in Fig. 1 [33]. It comprises the following steps: (i) encode the problem, which produces a list of genes using a numeric string. Each chromosome contains a number of genes, each representing a machine number, so that the length of each chromosome is equal to the total number of machines needed to be arranged; (ii) prepare input data: the number of machines (M), the dimensions of machines (width: M_W x length: M_L), maintenance plan, alternative machines, the number of products (N) and the machine sequences (M_S); (iii) specify parameters: the population size (Pop), the number of generations (Gen), the probability of crossover (P_c), the probability of mutation (P_m), floor length (F_L), floor width (F_W), the gap between machines (G), and the number of periods (P); (iv) create the demand levels of each product in each period (D_{gk}); (v) randomly generate an initial population based on the defined Pop; (vi) apply crossover and mutation operators to generate new offspring considering P_c and P_m respectively; (vii) arrange machines row by row based on F_L and F_W; (viii) evaluate the fitness function value; (ix) select the best chromosome having the shortest material handling distance using the elitist selection mechanism; (x) choose chromosomes for next generation by using the roulette wheel selection; and (xi) stop the GA process according to the number of generations. When the GA process is terminated, the best-so-far solution is reported.

```
Input problem dataset and Parameter setting (Pop, Gen, Pc, Pm, FL, Fw, G, P)
Create demand level (Dgk) for each product associated with demand distribution
Randomly create initial population (Pop)
Set a = 1 (first generation)
While a ≤ Gen do
        For b = 1 to cross do (cross = round ((Pc x Pop)/2)))
                Crossover operation
        End loop for b
        For c = 1 to mute do (mute = round(Pm x Pop))
                Mutation operation
        End loop for c
        Arrange machines row by row based on FL , Fw and G
        Calculate material handling distance based on either re-layout or robust layout
        Selection of the best solution using elitist selection
        Chromosome selection using roulette wheel method
        a = a + 1
End loop while
Output the best solution
```

Fig. 1. Pseudo-code of Genetic Algorithm

4   Experimental Design and Analysis

The computational experiments were conducted using eight datasets, all of which had different numbers of non-identical machines with various product types [34]. Each type of product had different demand profiles and machine sequences. The program was developed and coded in modular style using the Tool Command Language and Tool Kit programming language [35]. The experiments were conducted on a personal computer with an Intel Core i5 2.8 GHz CPU and 4 GB DDR3 RAM. The combination of population size and the number of generations (Pop*Gen) determines the amount of search and the computational time required. In this work Pop*Gen was set to 2,500 solutions, probabilities of crossover and mutation were set at 0.9 and 0.5 [31]. The appropriate genetic operators were presented in the following subsection.

4.1   Genetic Operators on GA Performance

A dataset (9 products to be processed on 15 machines) was used to compare the solution quality obtained from crossover operators (Enhanced edge recombination crossover: EERX [36] and two-point centre crossover: 2PCX [37]) and mutation operators (Two operations adjacent swap: 2OAS and two operations random swap: 2ORS [37]). The EERX and 2OAS were the best operators for scheduling problem [38]. The FLP was solved by GA with 2PCX [39] and 2ORS [40].

The computational experiment was repeated five times. The results in term of mean, standard deviation (SD), maximum (Max) and minimum (Min) of total material handling distances are summarised in Table 1. This suggested that the appropriate crossover and mutation operators for MLD problem were 2PCX and 2ORS. This also confirmed that the GA performance depends on the selection of the genetic operators.

Table 1. Relative performance of genetic operators (unit: metre)

Crossover operator	Mutation operator	Mean	SD	Max	Min
EERX	2OAS	1,464.25	35.56	1,411.15	1,498.25
	2ORS	1,477.35	20.43	1,442.95	1,497.05
2PCX	2OAS	1,437.33	37.68	1,373.55	1,469.45
	2ORS	**1,421.35**	15.17	1,397.35	1,436.55

For the next experiment, two scenarios under ten time-periods were considered: robust design, with no relocation when demand changed; and re-layout after demand changes. In order to denote the difference on the affect of breakdown maintenance (BM), the symbol (*) was introduced as the last letter of the variable only in BM scenario. For example, the material handling distance calculated without BM is denoted as MHD while in BM scenario, it is termed MHD*. In each period, the percentage of breakdown maintenance machines (%BMM) was determined at 10 %, 20 % and 30 %. During periods of maintenance, alternative machines were used. Each experiment was replicated thirty times using different random seeds. With eight datasets, thirty replications, three values of %BMM and two types of layout, a total of 1,440 (8 × 30 3 × 2) computational runs have been carried out. Each solution was evaluated on the ratio of the distance travelled with/without maintenance (MHD/MHD*) as shown in the subsection as follows.

4.2   Material Handling Distance (MHD) Based on Re-layout and Robust Layout

The second experiment aimed to minimise the MHD based on re-layout and robust layout design. The Mean, SD, Min, and Max of the distances travelled are shown in Table 2, where the lowest mean value of MHD for each dataset is indicated in bold. They were analysed using an analysis of variance (ANOVA) to calculate P values. The number of machines moved (NMM) and the machine movement distances (MMD) by re-layout between the periods are included in Table 2.

The average total distance for re-layout was shorter than with the robust layout in almost all datasets because the layout was redesigned according to the production flow over time period. The Student's t-test was applied to compare the differences in means of MHD. There were statistically significant differences between re-layout and robust layout with a 95 % confidence interval except for 40M20N and 40M40N. The robust layout produced a lower distance than re-layout with some datasets. The process of re-layout generated movement of machines between the periods, which effects the MMD and NMM. These costs of movements are considered in the next section.

In most datasets, the distance (MHD*) for re-layout was shorter than the robust layout. The ANOVA showed that the %BMM ratios significantly affected the material handling distance with a 95 % confidence interval (since the P values were less than 0.05 for all datasets). An increase in the number of BM machines caused more changes in machine sequences, so MHD* increased. However, the machine sequences depended upon the alternative machines defined.

Table 2. Values of total material handling distance (MHD) (unit: metre)

Dataset	Value	Robust layout				Re-layout						P-value
		MHD	MHD* based on %BMM			MHD	MMD	NMM	MHD* based on %BMM			
			10%	20%	30%			(machines)	10%	20%	30%	
10 machines	Mean	531,623.4	595,992.6	648,008.0	718,537.9	523,545.7	589.3	62.6	587,690.4	642,451.2	701,071.5	0.001
5 products	SD	11,023.7	18,962.6	22,179.8	19,694.1	6,716.8	202.0	10.4	12,336.1	12,828.6	16,117.7	
(10M5N)	Min	523,969.5	578,595.5	629,746.0	703,029.4	515,607.4	188.2	35.0	569,744.6	624,781.7	679,311.9	
	Max	565,028.6	649,594.9	701,533.4	774,072.3	542,662.9	960.7	81.0	623,345.7	675,020.9	737,729.3	
20M10N	Mean	3,375,077.5	3,542,104.0	3,628,745.5	3,941,721.4	3,291,791.1	2,128.8	178.4	3,480,769.7	3,590,226.8	3,943,083.0	0.000
	SD	66,940.2	70,083.5	90,727.4	93,797.2	31,598.7	141.0	1.4	44,159.4	62,496.9	88,685.8	
	Min	3,246,870.4	3,405,568.5	3,491,732.5	3,751,832.7	3,222,192.6	1,863.7	175.0	3,393,370.5	3,484,739.5	3,804,493.3	
	Max	3,525,702.9	3,697,350.4	3,825,789.1	4,136,080.9	3,354,304.0	2,490.0	180.0	3,593,512.0	3,738,508.9	4,135,547.0	
20M20N	Mean	10,258,008.5	10,886,407.5	11,491,341.5	10,040,630.5	10,040,630.52	1,987.2	177.3	10,235,922.3	10,665,973.00	11,374,110.73	0.000
	SD	174,047.6	377,961.7	379,507.2	64,521.2	64,521.22	178.5	2.3	147,267.5	244,816.87	206,137.12	
	Min	9,849,434.2	10,135,672.5	10,873,095.3	9,911,473.3	9,911,473.34	1,623.0	172.0	9,987,599.7	10,141,010.20	10,940,283.42	
	Max	10,614,291.9	11,917,048.8	12,629,320.2	10,145,030.9	10,145,030.94	2,266.9	180.0	10,548,10856	11,224,764.25	11,741,392.68	
20M40N	Mean	19,594,343.6	20,347,121.1	21,261,068.2	20,815,688.1	19,344,232.0	1,922.1	177.4	20,139,726.0	21,155,152.9	20,714,559.3	0.000
	SD	231,398.2	318,814.4	355,670.7	361,285.2	86,578.0	192.8	2.7	162,503.9	285,552.8	212,577.9	
	Min	19,246,235.6	19,887,820.5	20,614,947.8	20,049,924.7	19,196,354.8	1,441.9	166.0	19,776,615.4	20,599,823.7	20,343,304.6	
	Max	20,141,262.1	21,181,869.2	22,179,471.9	21,493,810.5	19,559,491.6	2,321.7	180.0	20,451,603.2	21,706,222.8	21,159,455.8	
30M15N	Mean	7,895,278.6	8,276,170.0	8,489,326.7	9,056,617.6	7,751,286.3	3,445.4	268.2	8,171,360.7	8,413,654.6	8,989,992.9	0.001
	SD	190,541.8	213,714.2	209,076.6	197,351.2	84,920.4	304.2	3.1	101,229.6	116,774.8	149,437.6	
	Min	7,477,879.1	7,915,148.6	8,041,397.4	8,663,135.8	7,602,228.3	2,608.1	253.0	7,997,155.8	8,170,185.6	8,740,363.1	
	Max	8,205,071.8	8,642,410.7	8,884,425.3	9,448,623.8	7,908,281.0	3,985.0	270.0	8,361,727.9	8,675,979.0	9,276,844.5	
40M20N	Mean	15,209,235.0	17,166,328.0	17,680,469.6	19,807,215.4	15,009,095.03	5,386.4	358.2	16,976,417.6	17,412,457.8	16,976,417.5	0.093
	SD	604,071.8	596,320.1	711,828.6	628,823.9	197,095.41	230.8	1.4	231,094.9	249,105.6	231,094.9	
	Min	14,168,156.5	16,275,826.0	16,382,959.9	18,444,970.6	14,526,384.17	4,899.7	355.0	16,516,726.2	16,862,780.6	16,516,726.2	
	Max	16,423,379.4	18,793,575.0	19,249,222.1	20,954,182.8	15,415,321.48	5,791.1	360.0	17,584,220.7	17,824,532.5	17,584,220.7	
40M40N	Mean	27,952,468.5	30,354,735.4	32,107,292.1	34,014,129.5	28,081,365.5	5,389.3	358.5	30,461,892.5	32,183,090.3	33,912,727.2	0.393
	SD	665,215.8	710,209.0	732,868.0	913,114.6	478,026.0	225.2	1.3	540,129.6	476,920.7	536,309.0	
	Min	26,735,184.2	28,861,378.3	30,625,120.7	32,160,002.9	27,262,325.6	4,911.6	355.0	29,295,863.1	30,959,845.3	32,852,483.2	
	Max	29,016,247.4	31,531,345.4	33,591,573.4	36,340,903.2	29,379,487.2	5,871.5	360.0	31,606,931.3	33,303,980.1	35,149,160.4	
50M25N	Mean	25,216,694.2	27,178,003.2	30,270,244.4	30,870,860.1	24,834,671.4	7,446.8	448.5	26,918,721.8	29,659,717.6	30,636,330.6	0.017
	SD	789,782.7	668,805.8	1,470,163.5	1,062,502.1	279,895.3	312.5	2.0	456,708.4	507,280.9	598,307.5	
	Min	23,928,776.0	25,890,014.3	28,608,583.1	29,248,548.6	23,883,084.6	6,891.8	443.0	25,888,161.1	28,725,879.4	29,174,212.3	
	Max	26,950,652.2	28,643,707.9	34,650,791.3	33,655,833.1	25,311,940.1	8,248.3	450.0	27,726,425.7	30,672,417.1	32,139,498.5	

4.3 Total Cost for Robust Layout and Re-layout

The results obtained from the above section will be used to determine the total costs of both robust and re-layout approaches. For robust layout, the total cost was determined from the MHD multiplied by the material handling cost (C_{MH}), in which the C_{MH} was assumedly equals to one currency unit per metre. The total cost (TC) of re-layout included material handling cost and machine shifting costs as shown in Eqs. (2) and (3). The shifting costs were related to the NMM and the MMD. The average shifting costs based on the NMM (C_{MV}) and the MMD (C_{MD}) were set at 1,000 currency units per machine moved [18], and 50 currency units per metre, respectively. Both total costs without maintenance (TC) and total costs with maintenance (TC*), which were calculated using the MHD and MHD* obtained from the previous experiment, are shown in Table 3.

$$Total\ cost\ of\ re-layout = C_{MH} \sum_{i=1}^{M} \sum_{j=1}^{M} \sum_{g=1}^{N} \sum_{k=1}^{P} d_{ij} f_{ijgk} D_{gk} + \sum_{k=1}^{P} C_{MV}(NMM_{k+1})$$

(2)

$$Total\ cost\ of\ re-layout = C_{MH} \sum_{i=1}^{M} \sum_{j=1}^{M} \sum_{g=1}^{N} \sum_{k=1}^{P} d_{ij} f_{ijgk} D_{gk} + \sum_{k=1}^{P} C_{MD}(MMD_{k+1})$$

(3)

Total cost (TC) of robust layout was lower than TC of re-layout both in terms of the NMM and the MMD except for 20M20N, 20M40N and 50M25N. A Student's t-test showed that there were statistically significant differences in the mean of total cost between robust layout and re-layout in some types of shifting cost. The mean total cost had no statistically significant difference between the layout designs for the 40M20N and 50M25N cases, so the layout was robustly designed or redesigned. The re-layout

Table 3. Values of total cost (unit: currency unit)

Dataset	Value	Robust layout TC	TC* based on %BMM 10	20	30	Re-layout TC (MMD)	TC* (MMD) based on %BMM 10	20	30	TC (NMM)	TC* (NMM) based on %BMM 10	20	30
10M5N	Mean	531,623.4	595,992.6	648,008.0	718,537.9	553,009.4	617,154.1	671,914.9	730,535.2	586,145.7	650,290.4	705,051.2	763,671.5
	SD	11,023.7	18,962.6	22,179.8	19,694.1	14,336.7	19,091.1	19,243.6	21,080.1	15,286.6	19,789.0	20,922.9	22,120.9
	Min	523,969.5	578,595.5	629,746.0	703,029.4	526,644.5	584,482.9	634,191.7	695,406.9	552,234.5	610,072.9	659,781.7	727,311.9
	Max	565,028.6	649,594.9	701,533.4	774,072.3	588,964.2	662,230.7	720,037.3	775,130.1	622,662.9	703,345.7	755,020.9	816,245.1
20M10N	Mean	3,375,077.5	3,542,104.0	3,628,745.5	3,941,721.4	3,397,598.8	3,586,577.4	3,696,034.5	4,048,890.7	3,470,191.1	3,659,169.7	3,768,626.8	4,121,483.0
	SD	66,940.2	70,083.5	90,727.4	93,797.2	31,005.2	44,299.4	61,865.7	88,405.8	31,228.0	44,107.5	62,585.4	88,702.6
	Min	3,246,870.4	3,405,568.5	3,491,732.5	3,751,832.7	3,331,036.6	3,496,907.5	3,588,321.5	3,908,075.3	3,401,192.6	3,569,370.5	3,660,739.5	3,980,493.3
	Max	3,525,702.9	3,697,350.4	3,825,789.1	4,136,080.9	3,463,198.0	3,703,134.0	3,848,130.9	4,244,894.5	3,533,304.0	3,772,512.0	3,917,508.9	4,314,547.0
20M20N	Mean	10,258,008.5	10,886,407.5	11,491,341.5	10,040,630.5	10,139,991.7	10,335,283.5	10,765,334.2	11,473,471.9	10,217,897.2	10,413,189.0	10,843,239.7	11,551,377.4
	SD	174,047.6	377,961.7	379,507.2	64,521.2	68,532.2	149,103.1	245,520.8	206,918.0	64,320.5	147,020.8	244,641.4	205,679.7
	Min	9,849,434.2	10,135,672.5	10,873,095.3	9,911,473.3	10,012,763.3	10,091,920.2	10,231,535.7	11,023,521.9	10,090,473.3	10,164,599.7	10,318,010.2	11,120,283.4
	Max	10,614,291.9	11,917,048.8	12,629,320.2	10,145,030.9	10,254,087.9	10,647,773.1	11,305,913.8	11,852,185.2	10,324,030.9	10,722,108.6	11,398,764.3	11,919,392.7
20M40N	Mean	19,594,343.6	20,347,121.1	21,261,068.2	20,815,688.1	19,440,338.1	20,235,832.1	21,251,259.0	20,810,665.4	19,521,665.4	20,317,159.4	21,332,586.3	20,891,992.7
	SD	231,398.2	318,814.4	355,670.7	361,285.2	88,273.5	165,232.7	286,826.0	212,576.5	86,839.1	162,608.5	286,568.2	212,810.9
	Min	19,246,235.6	19,887,820.5	20,614,947.8	20,049,924.7	19,284,193.3	19,864,453.9	20,696,530.7	20,447,300.6	19,373,354.8	19,953,615.4	20,773,823.7	20,522,304.6
	Max	20,141,262.1	21,181,869.2	22,179,471.9	21,493,810.5	19,655,657.1	20,537,783.2	21,799,237.3	21,255,621.3	19,738,491.6	20,627,603.2	21,886,222.8	21,338,455.8
30M15N	Mean	7,895,278.6	8,276,170.0	8,489,326.7	9,056,617.6	7,923,556.1	8,343,630.4	8,585,924.3	9,162,262.6	8,019,453.0	8,439,527.3	8,681,821.3	9,258,159.5
	SD	190,541.8	213,714.2	209,076.6	197,351.2	89,296.9	101,249.5	120,358.7	144,596.4	84,689.9	101,180.5	115,640.4	148,779.8
	Min	7,477,879.1	7,915,148.6	8,041,397.4	8,663,135.8	7,772,775.7	8,177,523.4	8,343,097.6	8,912,985.1	7,870,228.3	8,265,155.8	8,439,185.6	9,008,363.1
	Max	8,205,071.8	8,642,410.7	8,884,425.3	9,448,623.8	8,079,591.0	8,545,580.9	8,866,431.5	9,468,480.5	8,176,281.0	8,631,727.9	8,942,979.0	9,545,844.5
40M20N	Mean	15,209,235.0	17,166,328.0	17,680,469.6	19,807,215.4	15,278,413.5	17,245,736.0	17,681,776.2	17,245,735.9	15,367,295.0	17,334,617.6	17,770,657.8	17,334,617.5
	SD	604,071.8	596,320.1	711,828.6	628,823.9	203,181.5	237,005.0	254,929.8	237,005.0	197,200.6	231,222.1	249,132.9	231,222.1
	Min	14,168,156.5	16,275,826.0	16,382,959.9	18,444,970.6	14,777,255.2	16,767,597.2	17,127,857.6	16,767,597.2	14,885,384.2	16,875,726.2	17,222,780.6	16,875,726.2
	Max	16,423,379.4	18,793,575.0	19,249,222.1	20,954,182.8	15,693,211.5	17,862,110.7	18,107,170.5	17,862,110.7	15,774,321.5	17,943,220.7	18,181,532.5	17,943,220.7
40M40N	Mean	27,952,468.5	30,354,735.4	32,107,292.1	34,014,129.5	28,350,828.1	30,731,355.1	32,452,552.9	34,182,189.7	28,439,832.2	30,820,359.2	32,541,557.0	34,271,193.9
	SD	665,215.8	710,209.0	732,868.0	913,114.6	482,626.3	544,456.3	479,500.9	540,295.4	477,948.1	540,169.0	476,529.2	536,438.5
	Min	26,735,184.2	28,861,378.3	30,625,120.7	32,160,002.9	27,532,346.1	29,541,444.1	31,231,225.3	33,114,762.2	27,621,325.6	29,652,863.1	31,319,845.3	33,210,483.2
	Max	29,016,247.4	31,531,345.4	33,591,573.4	36,340,903.2	29,651,844.7	31,895,652.8	33,576,337.6	35,442,737.4	29,738,487.2	31,964,931.3	33,662,980.1	35,508,160.4
50M25N	Mean	25,216,694.2	27,178,005.2	30,270,244.4	30,870,860.1	25,207,009.7	27,291,060.2	30,032,056.0	31,008,669.0	25,283,171.4	27,367,221.8	30,108,217.6	31,084,830.6
	SD	789,782.7	668,805.8	1,470,163.5	1,062,502.1	281,574.1	459,718.3	504,468.1	603,017.6	280,087.1	456,842.0	507,530.8	598,949.3
	Min	23,928,776.0	25,890,014.3	28,608,583.1	29,248,548.6	24,237,851.6	26,242,928.1	29,081,217.4	29,529,550.3	24,330,084.6	26,335,161.1	29,168,879.4	29,617,212.3
	Max	26,950,652.2	28,643,707.9	34,650,791.3	33,655,833.1	25,724,355.6	28,071,072.2	31,055,738.6	32,509,922.5	25,761,940.1	28,172,425.7	31,122,417.1	32,589,498.5

approach consumed time repositioning machines and required shifting costs, which increased the total cost. For re-layout, the total cost in terms of the NMM was higher than the moving distance cost. Whether the layout was robustly designed or redesigned to another layout in the next period depended on the shifting cost and number of machines and product types (datasets). However, it should be noted that the shifting costs considered in this work were excluded and other costs related to the shutting down of the manufacturing line were also omitted.

With breakdown maintenance, the lowest mean total cost for each %BMM is highlighted in Table 3. The total cost (TC*) of robust layout was lower than re-layout based on both types of shifting cost for 10M5N, 20M10N, 30M15N and 40M40N. For all %BMM, the difference in mean total cost between robust layout and re-layout for 10M5N and 20M10N were statistically significant with a 95 % confidence interval (since the P values obtained from the Student's t-test were less than 0.05). The total cost (TC*) of robust layout was higher than the TC* of re-layout in terms of MMD on the dataset of 20M40N but there were statistically insignificant differences.

The robust layout was more effective than re-layout for some datasets and some types of shifting costs. Within both types of shifting cost for re-layout, the total cost in terms of the number of machines moved was higher than in terms of moving distance. This confirmed that the shifting cost and number of machines and product types (datasets) have an influence on machine layout design. However, the appropriate % BMM determined whether the layout should be robust or re-laid out.

5   Conclusions

This paper has presented the application of Genetic Algorithm for quantifying the affect of manufacturing machine layout design with/without maintenance consideration for stochastic demand environments. The analysis considered scenarios where 10 %, 20 %

and 30 % of machines had breakdown maintenance. The material handling distances for both re-layout and robust layout increased when some machines were maintained during each period. This was caused by changes in routings due to the use of alternative machines. The experimental results indicated that the material handling distance for re-layout was shorter than for robust layout. However, redesigning the machine layout according to demand levels generated shifting costs.

The total costs of the robust layout designs that did not consider maintenance were lower than those that involved re-layout for almost all the datasets. Robust layout designs also produced lower cost in breakdown maintenance situations. This was because re-layout caused machines to be moved which caused shifting costs. However, shifting costs may have according to the machine movements, which has an influence on total cost for re-layout designs. It can be beneficial for companies to consider both demand and machine uncertainty when designing layouts, providing that the future demand and availability of machines are properly forecast and planned. Investors should make decisions based on a trade-off between rearrangement cost and material handling cost. Future research should be focused on designing the machine layout without considering preventive machine maintenance and both of preventive and breakdown maintenance.

Acknowledgement. This work was part of the research project supported by the Naresuan University Research Fund under the grant number R2559C230.

References

1. Poole, D., Mackworth, A., Goebel, R.: Computational Intelligence: A Logical Approach. Oxford University Press, New York (1998)
2. Engelbrecht, A.P.: Computational Intelligence: An Introduction, 2nd edn. Wiley, Hoboken (2007)
3. Tsang, E.: Forecasting - where computational intelligence meets the stock market. Front. Comput. Sci. China **3**, 53–63 (2009)
4. Wu, S.X., Banzhaf, W.: The use of computational intelligence in intrusion detection systems: a review. Appl. Soft Comput. **10**, 1–35 (2010)
5. Vipsita, S., Rath, S.K.: Sequence-based protein superfamily classification using computational intelligence techniques: a review. Int. J. Data Min. Bioinform. **11**, 424–457 (2015)
6. Bandyopadhyay, S., Maulik, U., Roy, D.: Gene identification: classical and computational intelligence approaches. IEEE Trans. Syst. Man Cybern. Part C-Appl. Rev. **38**, 55–68 (2008)
7. Araghi, S., Khosravi, A., Creighton, D.: A review on computational intelligence methods for controlling traffic signal timing. Expert Syst. Appl. **42**, 1538–1550 (2015)
8. Dapa, K., Loreungthup, P., Vitayasak, S., Pongcharoen, P.: Bat algorithm, genetic algorithm and shuffled frog leaping algorithm for designing machine layout. In: Ramanna, S., Lingras, P., Sombattheera, C., Krishna, A. (eds.) MIWAI 2013. LNCS (LNAI), vol. 8271, pp. 59–68. Springer, Heidelberg (2013). doi:10.1007/978-3-642-44949-9_6
9. Karray, F., Zaneldin, E., Hegazy, T., Shabeeb, A., Elbeltagi, E.: Computational intelligence tools for solving the facilities layout planning problem. In: Proceedings of the 2000 American Control Conference, vol. 1–6, pp. 3954–3958. IEEE, New York (2000)

10. Chansombat, S., Musikapun, P., Pongcharoen, P., Hicks, C.: A modified bat algorithm for production scheduling in the capital goods industry. In: 22nd International Conference on Production Research, ICPR 2013 (2013)

11. Lutuksin, T., Pongcharoen, P.: Best-worst ant colony system parameter investigation by using experimental design and analysis for course timetabling problem. In: 2nd International Conference on Computer and Network Technology, ICCNT 2010, Bangkok, pp. 467–471 (2010)

12. Chainate, W., Pongcharoen, P., Thapatsuwan, P.: Clonal selection of artificial immune system for solving the capacitated vehicle routing problem. J. Next Gener. Inf. Technol. **4**, 167–179 (2013)

13. Pongcharoen, P., Chainate, W., Pongcharoen, S.: Improving artificial immune system performance: inductive bias and alternative mutations. In: Bentley, P.J., Lee, D., Jung, S. (eds.) ICARIS 2008. LNCS, vol. 5132, pp. 220–231. Springer, Heidelberg (2008). doi:10. 1007/978-3-540-85072-4_20

14. Thapatsuwan, P., Sepsirisuk, J., Chainate, W., Pongcharoen, P.: Modifying particle swarm optimisation and genetic algorithm for solving multiple container packing problems. In: 2009 International Conference on Computer and Automation Engineering, ICCAE 2009, Bangkok, pp. 137–141 (2009)

15. Chen, G.Y.-H.: A new data structure of solution representation in hybrid ant colony optimization for large dynamic facility layout problems. Int. J. Prod. Econ. **142**, 362–371 (2013)

16. Kulturel-Konak, S.: Approaches to uncertainties in facility layout problems: perspectives at the beginning of the 21(st) century. J. Intell. Manuf. **18**, 273–284 (2007)

17. Sethi, A., Sethi, S.: Flexibility in manufacturing: a survey. Int. J. Flex. Manuf. Syst. **2**, 289–328 (1990)

18. Moslemipour, G., Lee, T.S.: Intelligent design of a dynamic machine layout in uncertain environment of flexible manufacturing systems. J. Intell. Manuf. **23**, 1849–1860 (2012)

19. Goel, L., Gupta, D., Panchal, V.K., Abraham, A.: Taxonomy of nature inspired computational intelligence: a remote sensing perspective. In: 4th World Congress on Nature and Biologically Inspired Computing, NaBIC 2012, Mexico City, pp. 200–206 (2012)

20. Yang, X.-S., Cui, Z., Xiao, R., Gandomi, A.H., Karamanoglu, M.: Swarm Intelligence and Bio-Inspired Computation: Theory and Applications. 1st edn. (2013)

21. Gen, M., Cheng, R., Lin, L.: Network Models and Optimization: Multiobjective Genetic Algorithm Approach (Decision Engineering). Springer, Heidelberg (2008)

22. Pongcharoen, P., Hicks, C., Braiden, P.M., Stewardson, D.J.: Determining optimum Genetic Algorithm parameters for scheduling the manufacturing and assembly of complex products. Int. J. Prod. Econ. **78**, 311–322 (2002)

23. Chaudhry, S.S., Luo, W.: Application of Genetic Algorithms in production and operations management: a review. Int. J. Prod. Res. **43**, 4083–4101 (2005)

24. Lenin, N., Kumar, M.S., Islam, M.N., Ravindran, D.: Multi-objective optimization in single-row layout design using a genetic algorithm. Int. J. Adv. Manuf. Technol. **67**, 1777–1790 (2013)

25. Kia, R., Khaksar-Haghani, F., Javadian, N., Tavakkoli-Moghaddam, R.: Solving a multi-floor layout design model of a dynamic cellular manufacturing system by an efficient genetic algorithm. J. Manuf. Syst. **33**, 218–232 (2014)

26. Rose, C.D., Coenen, J.M.G.: Comparing four metaheuristics for solving a constraint satisfaction problem for ship outfitting scheduling. Int. J. Prod. Res. **53**, 5782–5796 (2015)

27. Tompkins, J.A., White, J.A., Bozer, Y.A., Tanchoco, J.M.A.: Facilities Planning, 4th edn. Wiley, Hoboken (2010)

28. Loiola, E.M., de Abreu, N.M.M., Boaventura-Netto, P.O., Hahn, P., Querido, T.: A survey for the quadratic assignment problem. Eur. J. Oper. Res. **176**, 657–690 (2007)
29. Blackstone, J.H., Phillips, D.T., Hogg, G.L.: A state-of-the-art survey of dispatching rules for manufacturing job shop operations. Int. J. Prod. Res. **20**, 27–45 (1982)
30. Xiong, J., Xing, L.-N., Chen, Y.-W.: Robust scheduling for multi-objective flexible job-shop problems with random machine breakdowns. Int. J. Prod. Econ. **141**, 112–126 (2013)
31. Vitayasak, S., Pongcharoen, P.: Identifying optimum parameter setting for layout design via experimental design and analysis. Adv. Mater. Res. **931–932**, 1626–1630 (2014)
32. Leechai, N., Iamtan, T., Pongcharoen, P.: Comparison on rank-based ant system and shuffled frog leaping for design multiple row machine layout. SWU Eng. J. **4**, 102–115 (2009)
33. Vitayasak, S., Pongcharoen, P.: Genetic Algorithm based robust layout design by considering various demand variations. In: Tan, Y., Shi, Y., Buarque, F., Gelbukh, A., Das, S., Engelbrecht, A. (eds.) ICSI 2015. LNCS, vol. 9140, pp. 257–265. Springer, Heidelberg (2015). doi:10.1007/978-3-319-20466-6_28
34. Vitayasak, S., Pongcharoen, P.: Backtracking Search Algorithm for designing a robust machine layout. WIT Trans. Eng. Sci. **95**, 411–420 (2014)
35. Ousterhout, J.K.: Tcl and Tk tookit, 2nd edn. Addison Wesley, Boston (2010)
36. Starkweather, T., McDaniel, S., Mathias, K., Whitley, D., Whitley, C.: A comparison of genetic sequencing opeartors. In: Proceedings of the Third International Conference on Genetic Algorithms, pp. 69–76 (1991)
37. Murata, T., Ishibuchi, H.: Performance evaluation of genetic algorithms for flow shop scheduling problems. In: Proceedings of the First IEEE Conference on Evolutionary Computation, pp. 812–817 (1994)
38. Pongcharoen, P., Stewardson, D.J., Hicks, C., Braiden, P.M.: Applying designed experiments to optimize the performance of genetic algorithms used for scheduling complex products in the capital goods industry. J. Appl. Stat. **28**, 441–455 (2001)
39. Eklund, N.H.W., Embrechts, M.J., Goetschalckx, M.: Efficient chromosome encoding and problem-specific mutation methods for the flexible bay facility layout problem. IEEE Trans. Syst. Man Cybern. Part C-Appl. Rev. **36**, 495–502 (2006)
40. Hu, M.H., Wang, M.J.: Using genetic algorithms on facilities layout problems. Int. J. Adv. Manuf. Technol. **23**, 301–310 (2004)

A Random Forest-Based Self-training Algorithm for Study Status Prediction at the Program Level: *minSemi-RF*

Vo Thi Ngoc Chau$^{(\boxtimes)}$ and Nguyen Hua Phung

Ho Chi Minh City University of Technology,
Vietnam National University – HCMC, Ho Chi Minh City, Vietnam
{chauvtn, phung}@cse.hcmut.edu.vn

Abstract. Educational data mining aims to provide useful knowledge hidden in educational data for better educational decision making support. However, a large set of educational data is not always ready for a data mining task due to the peculiarities of the academic system as well as the data collection time. In our work, we focus on a study status prediction task at the program level where the data are collected and processed once a year in the time frame of the program of interest in an academic credit system. When there are little educational data labeled for the task, the effectiveness of the task might be affected and thus, the task should be considered in a semi-supervised learning process instead of a conventional supervised learning process to exploit a larger set of unlabeled data. In particular, we define a random forest-based self-training algorithm, named minSemi-RF, for the study status prediction task at the program level. The minSemi-RF algorithm is designed as a combination of Tri-training and Self-training styles in such a way that we turn a random forest-based self-training algorithm to be a parameter-free variant of the Tri-training algorithm. This algorithm produces a final classifier that can inherit the advantages of a random forest model. Based on the experimental results from the experiments conducted on the real data sets, our algorithm is proved to be effective and practical for early in-trouble student detection in an academic credit system as compared to some existing semi-supervised learning methods.

Keywords: Self-training · Random forest · Tri-training · Educational data mining · Study status prediction

1 Introduction

In recent times, educational data mining (EDM) has been regarded as a very active research area where the outcomes of data mining techniques are utilized. Indeed, a review on 240 related works has been conducted in [12]. Although stated in [12] as "EDM is living its spring time and preparing for a hot summer season.", we believe that EDM is always a significant research area to discover useful information and knowledge hidden in educational data for better educational decision making support.

As examined in [12], the classification task is among the most popular educational data mining tasks. However, just a few related works took this task into account at the program level in an academic credit system for detecting in-trouble undergraduate

© Springer International Publishing AG 2016
C. Sombattheera et al. (Eds.): MIWAI 2016, LNAI 10053, pp. 219–230, 2016.
DOI: 10.1007/978-3-319-49397-8_19

students as soon as possible along their study path. In addition, considering unlabeled data in the learning process to obtain a more effective classifier for student classification has not yet received much attention. In our work, a study status prediction task at the program level in an academic credit system is defined as a multi-class student classification task for early in-trouble student identification based on their study performance in a certain number of study years. We address this task in the aforementioned situation where there is a limitation of labeled data for model construction of the task and meanwhile, a larger set of unlabeled data is available for status prediction and able to be exploited in the learning process of the task.

As compared to the existing works in the educational data mining area, our work has the following different points. First of all, the context of our task is different from that in the related works such as [3, 5, 7, 9, 11, 13, 14, 16] where no consideration on unlabeled data in the learning process existed and there was no mention of an academic credit system. Indeed, [3] has used the existing supervised learning methods in Weka and social network analysis to predict unsuccessful students based on their student records and social behavior data. Also using the existing supervised learning algorithms, [5] obtained and examined 3 interpretable rule sets based on the student's course satisfaction which was not considered at the program level. [7] has enriched the characteristics of each student using multiple data sources for a student classification task. In [9], the success or failure of the students before the end of the course in the Moodle platform was predicted using the existing supervised learning methods based on the demographics and course attributes of each student. Instead of undergraduate students, the work in [11] was dedicated to high school students. In [11], the authors defined an evolutionary algorithm for high school student failure prediction based on general survey, specific survey, and current marks of each student. [13] has employed the existing supervised learning algorithms to classify the students with their similar final marks depending on the activities in web-based courses. [14] has handled data sparseness for a performance prediction task which was not viewed as a classification task in an academic credit system. Finally, [16] has performed a 4-class mark prediction task based on student's performance and demographic information by means of the existing supervised learning algorithms in Weka. In short, none of the aforementioned related works has exploited unlabeled data to enhance the performance of their mining tasks with the only use of the supervised learning algorithms.

As an exception, [8] is much related to our work because both take advantages of unlabeled data to make a final classifier more effective in a semi-supervised mechanism. Different from [8], our work examines the data of the full-time regular students in an academic credit system only based on their study performance. In addition, our work defines a new approach to the task and compares the resulting algorithm with the existing semi-supervised learning algorithms to emphasize its effectiveness.

In particular, a random forest-based self-training algorithm, named minSemi-RF, is proposed to build an effective classifier in a semi-supervised learning process. This algorithm is designed in a combined Tri-training [21] and Self-training [20] style with a global convergence point by optimizing the prediction on the given original set of labeled data. Besides, it can make the most of the random forest model [4] for the diversity of the random trees in the resulting ensemble model. Moreover, its design is made in a parameter-free configuration so that minSemi-RF can be more practical.

With the proposed solution, we expect that a study status prediction task can help forecasting the unpleasant cases in our students early and providing them with proper consideration and support for their final success with the program they enrolled.

2 Study Status Prediction at the Program Level

2.1 Task Definition

Study status prediction of an undergraduate student in an academic credit system aims to determine what his/her study status at the end point in time in the studying time frame permitted by the program. A study status label (e.g., "studying", "graduating", "study_stop", "first_warned", or "second_warnded") is given depending on how well the student has studied with respect to the extent that the program is accomplished.

As conducted in an academic credit system, our study status prediction task incurs several challenges inherent from the peculiarities of this academic system at the data and structure levels. They come from the flexibility of the system. Moreover, as witnessed in the existing works, no benchmark educational data set is available for the task. This would be more challenging when the data collection speed is slower and the check for real effectiveness of a task's result is time consuming at a unit of time which is academic year. Thus, the task needs to be considered in a different way.

In this paper, we define the study status prediction task as a multi-class classification task at the program level. In contrast to the existing related works [3, 5, 7, 9, 11, 13, 14], our work takes into account this task in such a different situation that educational data collected with available study status labels for supporting the task are limited while many students with unknown study status need to be examined. In this situation, a classification task should be formulated with a semi-supervised learning process rather than a traditional supervised learning process, in order to exploit the data from the students with unknown study status for the learning process. The effectiveness of the study status prediction task is expected to be enhanced significantly.

First of all, we establish the data space where each student is represented as an object in the space. In our work, the data space is a vector space whose dimensions correspond to the subjects in the program. These subjects are required for each student to successfully study to reach the degree of the program. As a result, each student is characterized as a vector in that space reflecting his/her study performance. A value of each vector at a dimension is a grade of the corresponding subject that the student has obtained. If the student has not yet taken a course of the subject, its grade is not available and its value at the corresponding dimension is zero (0) according to the minimum level of the accomplishment of the program for graduation. In addition, each student has a study status at the end point of study time for the enrolled program. If the final study status of a student is known, his/her data will be used in a training set of the learning phase. Otherwise, the student is a current student whose final study status will be determined in the classification phase. Besides, the data of such a student will also be used in the semi-supervised learning process as early mentioned.

In the following, we formally define the input and output of the study status prediction task. The input includes D and UD and the output is the prediction result on UD.

As for the input, D is a training set of n data vectors in a p-dimensional data space: $D = \{X_1, X_2, \ldots, X_n\}$ where $X_j = (x_{j,1}, x_{j,2}, \ldots, x_{j,p}, y_j)$ for $x_{j,d} \in [0, 10]$, $y_j \in S$ where S is a given set of study status labels, $j = 1..n$, $d = 1..p$. In the scope of this work, D is a small set of labeled data and S includes five study status labels: graduating, studying, first_warned, second_warned, and study_stop. UD is a set of un data vectors in a p-dimensional data space: $UD = \{X_1, X_2, \ldots, X_{un}\}$ where $X_j = (x_{j,1}, x_{j,2}, \ldots, x_{j,p}, y_j)$ for $x_{j,d} \in [0, 10]$, y_j which is an unknown class value to be determined in the task, $j = 1..un$, $d = 1..p$.

The output of the task includes a set Y of all the values predicted for all y_j's for $j = 1..un$ in UD that can be a study status label such as: "graduating", "studying", "first_warned", "second_warned", or "study_stop".

Therefore, the task is formally stated with two phases as follows:

Phase 1 – Learning phase for classifier construction in a semi-supervised process

$$C = L(D, UD) \qquad (1)$$

Where C is a resulting classifier and L is a semi-supervised learning process on both labeled and unlabeled data sets, D and UD, respectively.

Phase 2 – Classification phase for predicting a study status of each unknown student

$$Y = C(UD) \qquad (2)$$

Where Y is a set of predicted study status labels of all the students represented in UD.

2.2   Performing the Task

Nowadays, many learning algorithms are widely used for educational data mining. Specifically for an educational data classification task, several algorithms such as Naïve Bayes, Neural Network, Support Vector Machine, K-Nearest Neighbor (K-nn), Decision Tree C4.5, and Random Forest, etc. together with their variants have been utilized in a supervised learning process. As for a semi-supervised learning process, several existing well-known semi-supervised learning algorithms such as Self-training, Tri-training, Co-training, De-Tri-training, and so on have been employed for student dropout prediction in distance higher education in [8]. Different from the work in [8], our work defines a new solution to the task in a parameter-free manner so that the task can identify in-trouble students more effectively and practically. Indeed, a novel random forest-based self-training approach is proposed in the next section.

3   A Random Forest-Based Self-training Approach

As a solution to our aforementioned study status prediction task, a random forest-based self-training approach is determined. It is a semi-supervised learning approach using random forests in a Self-training style. It is also designed in a Tri-training style for no parameter setting requirement. An analysis is given to highlight its advantages.

3.1   The minSemi-RF Algorithm

As one of the simplest semi-supervised learning algorithms originally introduced in [20], the Self-training approach is well-known for the learning process that can exploit both labeled and unlabeled data to construct a prediction model. Like Tri-training proposed in [21], it asks for neither two sufficient and redundant views nor many different classifiers in its learning process. However, Self-training is less complicated than Tri-training. Based on the achievement of learning from noisy samples in [2], Tri-training is more advanced as a parameter-less semi-supervised learning approach that can handle the number of the most confidently predicted instances for an iterative enhancement on its training data set. In addition, Self-training is unfortunately inconvenient in practice as it requires an appropriate base classifier, how to define the so-called most confidently predicted instances, and the number of the selected instances for enhancement on the training set to be pre-determined by the users.

In our work, we would like to define a random forest-based self-training approach in the Tri-training style so that the resulting semi-supervised learning approach can deal with the disadvantages of Self-training and thus, become not only effective but also practical for the users. In particular, we use a random forest model of three random trees with $(\lfloor \log(p) \rfloor + 1)$ random features. This number of the random features is from the theory of random forest models in [4]. Besides, these three random trees play a role of the three classifiers in Tri-training. Different from the three classifiers in Tri-training, the random trees in a random forest model are based on bootstrap sampling and their diversity is ensured for a majority voting of an ensemble model due to the randomness in constructing a random forest model [4, 10]. Following the agreement of all the base classifiers in an ensemble for class prediction and the disagreement of the three classifiers in Tri-training, the probability threshold is automatically set; in order to select the most confidently predicted instances from a current set of unlabeled data in our approach. As a result, the proposed approach is free of parameter settings, inheriting the advantages of the random forest model for effectiveness and robustness.

Moreover, the optimization of our algorithm is based on the generalization of the final random forest model over the original set of labeled data containing true labels that we certainly know and the final classifier does too. It is different from the optimization in the semi-supervised learning process of Tri-training because it exhausts the selection of the most confidently predicted instances until no more instance is selected or no more instance is predicted. Tri-training performs an incremental examination on a sequence of decreasing error rates on the original labeled data set. This examination might lead to an early convergence, in spite of the less number of iterations.

For details, the pseudo-code of our minSemi-RF algorithm is given in Fig. 1. As described in an iterative manner, the minSemi-RF algorithm performs as follows.

In each iteration from line (5) to line (10), constructing a current random forest model is done on a current set *clSet* of labeled data. This current classifier is then evaluated on the original set *lSet* of labeled data. If its error rate is less than the minimum error rate, i.e. its prediction power is better than that of the best-so-far classifier, then the minimum error rate and the minimum classifier are updated with the new current ones.

Input:

    lSet: a labeled set which is originally given in the *p*-dimension vector space

    uSet: an unlabeled set which is originally given in the *p*-dimension vector space

Output:

    C: a resulting classifier

Process: **minSemi-RF**

(1).   Set a minimum error rate *Min_error_rate* to Double.MAX_VALUE

(2).   Assign *lSet* as a current set *clSet* which contains all instances with known labels

(3).   Assign *uSet* as a current set *cuSet* which contains all instances with unknown labels

(4).   Repeat until the termination conditions are met:

(5).   Build a current random forest *Current_RF* of three random trees with $(\lfloor \log(p) \rfloor + 1)$ random features on *clSet*

(6).   Compute a current error rate *Current_error_rate* by evaluating *Current_RF* on *lSet*

(7).   If *Min_error_rate* > *Current_error_rate* then

(8).   *Min_error_rate* = *Current_error_rate*

(9).   Save the current random forest *Current_RF* as a minimum random forest *Minimum_RF*

(10).   End If

(11).   If *cuSet* is not empty then

(12).   For each instance X^* in *cuSet* do

(13).   Predict a label of the current instance X^* in *cuSet* using *Current_RF* according to Equation (5)

(14).   Calculate a corresponding prediction score for the current instance X^* according to Equation (6)

(15).   If its prediction score = 1 then

(16).   Add this current instance X^* along with its predicted label into *sSet* which is *a set of the most confidently predicted instances* from *cuSet*

(17).   End If

(18).   End For

(19).   If *sSet* is not empty then

(20).   Update *clSet* by including *sSet*

(21).   Update *cuSet* by excluding *sSet*

(22).   Else

(23).   Return the minimum random forest *Minimum_RF* as a resulting classifier *C*

(24).   End If

(25).   Else

(26).   Return the minimum random forest *Minimum_RF* as a resulting classifier *C*

(27).   End If

(28). End Repeat

Fig. 1. The Proposed Semi-Supervised Learning Algorithm **minSemi-RF** on both Labeled and Unlabeled Data in the *p*-dimension Vector Space

From line (11) to line (18), exploiting unlabeled data is considered if the current set of unlabeled data is not empty. Otherwise, the best-so-far classifier with the minimum error rate is returned as a resulting classifier C. In these steps, all the instances of the non-empty unlabeled set are predicted according to Eq. (3) and their prediction scores are calculated according to Eq. (4). A set $sSet$ of the most confidently predicted instances with prediction scores = 1 is selected from the non-empty unlabeled set.

From line (19) to line (24), if the $sSet$ set is not empty, the current set $clSet$ of labeled instances is updated to enlarge the training set in the next iteration and the current set $cuSet$ of unlabeled instances is shrunk by removing those chosen instances. Otherwise, we return the best-so-far classifier with the minimum error rate as a resulting classifier C.

As specified in Fig. 1, a resulting classifier C is obtained with two termination conditions: no instance of the current set of unlabeled data is selected as the most confidently predicted instance in line (23) or no instance exists in the current set of unlabeled data in line (26). These termination conditions are based on the general rationale behind the semi-supervised learning approach which aims to exploit unlabeled instances in the learning process; that is the learning process will end if there is no unlabeled instance for the exploitation. These two termination conditions make our proposed algorithm converged in an optimized state.

In the rest of this subsection, we present the calculation of the prediction score of a current instance X^* and the *most confidently predicted instance selection* scheme in our algorithm. Let m be the number of classes and t be the number of random trees in the random forest model. In our algorithm, t is equal to three.

- Each random tree j performs a prediction on X^* and provides a class distribution score of each class C_i for $i = 1..m$ for X^* which is:

$$P_j(C_i|X^*) = \frac{k}{N} \qquad (3)$$

  where k is the number of instances in class C_i out of N instances in the training set of the tree j at the leaf node. It can be seen that such a class distribution score from each random tree is based on the purity of the instance set at the leaf node.

- A class distribution score of each class C_i at prediction for X^* by the random forest model is calculated as a total sum of all the class distribution scores of each class C_i for $i = 1..m$ by all t trees:

$$P(C_i|X^*) = \Sigma_{j=1..t}P_j(C_i|X^*) \qquad (4)$$

  These class distribution scores $P(C_i|X^*)$ for $i = 1..m$ are then normalized into the range [0, 1] such that: $\forall i = 1..m, 0 \le P(C_i|X^*) \le 1$ and $\Sigma_{i=1..m}P(C_i|X^*) = 1$.

- Based on the majority voting scheme, the final prediction score of X^* is determined as the maximum prediction score $P(C_i|X^*)$ for $i = 1..m$ and its predicted class is C_i corresponding to the maximum prediction score $P(C_i|X^*)$:

$$\text{Predicted class}(X^*) = \text{argmax}_{C_i}\{P(C_i|X^*) \text{ for } i = 1..m\} \qquad (5)$$

$$\text{Prediction score}(X^*) = \max\{P(C_i|X^*) \text{ for } i = 1..m\} \qquad (6)$$

Once a prediction score is computed for the instance X^*, it will be selected and added into the resulting set if and only if its prediction score is 1. Its predicted label is now considered true label once X^* becomes an element of the training data set. The reason for the threshold value of 1 in our selection scheme stems from the assumption about the prediction of each random tree in an ensemble and also about the final prediction of that ensemble. In our work, the final classifier is a random forest model.

3.2 Discussion

In short, our proposed semi-supervised learning algorithm, minSemi-RF, has been designed in a combined self-training and tri-training style. It is applicable to classifier construction from a small labeled data set and able to make use of a larger unlabeled data set to provide a resulting random forest model more effective for classification in practice. Above all, it is simple and practical with a restriction on neither the number of classes nor parameter configurations.

Compared to the works [6, 10, 15, 18, 20, 21], ours has the different points below:

Firstly, we have defined a new convergence point in order to avoid an early convergence as based on the original labeled data set. We believe that a global minimum error rate on the original labeled data set can be observed in a finite loop till no enhancement is made on the training data set of the learning phase. As of that moment, the most effectiveness of the resulting classifier can be examined and obtained.

Secondly, agreement/disagreement of the base classifiers in our resulting model is set in the majority voting scheme of an ensemble so that the most confidently predicted instance selection scheme in our algorithm can select the smallest but finest set of the instances that are truly labeled. This is because the prediction score of each selected instance is one, implying that all the random trees make the same decision on its label. Thus, a chance for such a decision to be correct is high with respect to an ensemble, especially to a random forest model.

Finally, our algorithm is free from setting the probability threshold required by the Self-training approach. Consequently, it is a parameter-less Self-training algorithm for practical use.

4 Experimental Results

For further evaluation, this section presents an empirical study using the real data sets of the regular students at Faculty of Computer Science and Engineering, Ho Chi Minh city University of Technology, Vietnam National University – HCMC, Vietnam, [1]. These students followed the program in Computer Science in 2005–2008.

Three data sets stemming from their study performance were prepared corresponding to three years of study from year 2 to year 4, namely, "Year 2" for the 2nd-year students, "Year 3" for the 3rd-year students, and "Year 4" for the 4th-year students. Table 1 describes the distribution of students over study status classes. The features characterizing each student are 43 attributes corresponding to 43 subjects and 1 class attribute for his/her study status at the end point of permitted study time. Thus, each student is represented as a 43-dimension vector in the vector space. Besides, each data set has 1334 instances corresponding to 1334 students that will be examined.

Table 1. Distribution of study status of the students at the program level

Study status	Graduating	Studying	First_warned	Second_warned	Study_stop	Total
Number of students	725	451	11	6	141	**1334**
Percentage (%)	54.35	33.81	0.82	0.45	10.57	**100**

As for the algorithms, we used the supervised learning algorithms J48 (i.e., C4.5) and Random Forest in Weka [19]. Thanks to the authors of [10, 21] for the availability of the source codes of Tri-training and Co-Forest, Tri-training [21] with C4.5 and Co-Forest [10] with default parameter settings are used for comparison. In addition, Random Forest [4] with 3 random trees and default random feature numbers and Self-training [20] with C4.5 and 2/3 for its probability threshold are included. A choice of Tri-training with C4.5 and Self-training with C4.5 is based on the empirical study in [17] while a choice of Co-Forest is made due to the same base classifier as ours, which is the random forest model in the semi-supervised learning process.

Besides the methods in the related works, the variants of the proposed minSemi-RF algorithm are considered to highlight the effectiveness of the characteristics of minSemi-RF. They are Semi-RF_2/3 and Semi-RF. Semi-RF_2/3 is a Self-training method with a random forest model of 3 random trees using the probability threshold of 2/3 like the agreement check among the three classifiers in a Tri-training style. Different from Semi-RF_2/3, Semi-RF is a Self-training method with a random forest model of 3 random trees using the probability threshold of 1. More advanced than Semi-RF, our proposed minSemi-RF algorithm uses the probability threshold of 1 and the minimization on the error rate over the original labeled data set for convergence.

Regarding performance measures, each value in Tables 2, 3 and 4 is the classification accuracy (%) reached by each algorithm. The larger number is the better. For reliable accuracy estimation, we use the k-fold cross validation scheme in the semi-supervised learning context where a data set is divided into k folds in a stratified sampling scheme. One fold is then used as a labeled data set and the other $(k - 1)$ folds are used as an unlabeled data set. The number of folds corresponds to the percentage of labeled data available for the task, i.e., k = 2..10 corresponds to 50 %..10 % labeled data.

The experimental results in Tables 2, 3 and 4 show that our minSemi-RF algorithm always outperforms the other algorithms regardless of the various data characteristics

Table 2. Accuracy (%) on the Year 2 data set

Fold#	Labeled data (%)	Random forest	Self-training	Tri-training	Co-forest	Semi-RF_2/3	Semi-RF	minSemi-RF
2	50.00	66.27	62.89	63.27	61.99	66.57	**69.04**	**69.04**
3	33.33	65.29	63.34	63.46	59.45	65.63	67.47	**70.46**
4	25.00	64.69	61.72	62.27	58.97	64.54	66.72	**68.72**
5	20.00	65.03	64.77	61.73	59.54	65.74	67.95	**69.12**
6	16.67	64.45	61.21	60.73	59.72	65.23	66.64	**67.36**
7	14.29	65.19	61.57	60.64	57.87	65.45	67.30	**68.17**
8	12.50	64.10	62.11	60.81	59.61	64.35	66.64	**68.14**
9	11.11	63.36	61.24	62.07	58.53	63.91	65.42	**67.66**
10	10.00	63.91	62.44	61.87	58.84	64.69	65.99	**67.34**

Table 3. Accuracy (%) on the Year 3 data set

Fold#	Labeled data (%)	Random forest	Self-training	Tri-training	Co-forest	Semi-RF_2/3	Semi-RF	minSemi-RF
2	50.00	72.11	69.04	66.57	67.99	72.19	73.39	**75.86**
3	33.33	71.89	68.52	68.97	65.93	72.26	73.24	**74.55**
4	25.00	70.64	65.97	67.79	63.52	71.01	72.59	**73.59**
5	20.00	69.90	67.88	68.12	64.47	70.01	71.50	**73.78**
6	16.67	69.54	67.21	66.19	65.05	69.93	71.02	**72.11**
7	14.29	69.00	67.08	65.84	63.53	69.60	71.55	**72.89**
8	12.50	68.97	66.13	64.79	62.91	69.78	71.02	**72.99**
9	11.11	68.68	67.05	68.61	62.34	68.68	70.78	**71.58**
10	10.00	67.32	65.95	66.33	63.94	67.89	69.12	**71.55**

Table 4. Accuracy (%) on the Year 4 data set

Fold#	Labeled data (%)	Random forest	Self-training	Tri-training	Co-forest	Semi-RF_2/3	Semi-RF	minSemi-RF
2	50.00	74.44	72.04	72.11	68.22	74.89	77.21	**77.74**
3	33.33	72.41	69.57	71.06	66.12	72.41	74.48	**74.93**
4	25.00	72.16	70.26	70.09	66.99	72.31	73.11	**75.86**
5	20.00	71.55	69.45	69.40	65.69	71.87	72.77	**74.27**
6	16.67	70.72	70.13	69.22	65.35	71.21	73.55	**74.20**
7	14.29	72.45	68.80	68.18	65.14	72.68	73.98	**75.07**
8	12.50	69.98	68.15	67.74	65.31	70.15	73.15	**73.89**
9	11.11	70.77	69.30	69.09	64.25	70.80	71.51	**72.75**
10	10.00	70.51	67.77	66.38	65.22	70.98	71.83	**73.16**

of each data set. Specific for the "Year 2" data set, minSemi-RF can improve the prediction accuracy of the classifiers from the other algorithms from about 1 % up to about 11 % except for Semi-RF. Besides, minSemi-RF can make use of unlabeled data well to enhance the final classifier as this resulting classifier is about 2 %–4 % better than the Random Forest model constructed in a supervised learning mechanism. The same experimental results are achieved for the "Year 3" and "Year 4" data sets showing that the final classifier from minSemi-RF provides better predictions than those from the other algorithms. However, the maximum improvements are around 10 % for the "Year 3" and "Year 4" data sets, which is a little bit less than one for the "Year 2" data set, implying that our minSemi-RF algorithm performs well on sparser data.

It is also worth noting that the accuracy of Semi-RF is always better than that of Semi-RF_2/3, implying that the setting of the probability threshold of one in Semi-RF and minSemi-RF is reasonable. Furthermore, the accuracy of minSemi-RF is always better than that of Semi-RF, implying that the use of the global convergence point is appropriate. Thus, the design of our proposed minSemi-RF algorithm is effective. Regarding the percentages of labeled data in each experiment group, it is realized that the more labeled data, the higher prediction accuracy is reached with our minSemi-RF algorithm. Therefore, once the unlabeled data are predicted, adding those newly predicted data into the training data set and utilizing them in the learning process is significant for a better classifier. This approach is also meaningful in practice as unlabeled data arrive and need to be predicted over the time.

5   Conclusions

In this paper, we have considered the study status prediction task at the program level for detecting in-trouble undergraduate students in an academic credit system. We formulate this task as a multi-class classification task in a practical situation where labeled data available for the task are limited and many unlabeled data that need to be predicted in the task exist. A new approach is addressed for the task to overcome the limitation of labeled data and exploit the richness of unlabeled data. In particular, a random forest-based self-training algorithm, minSemi-RF, is defined with a global convergence point. As a result, it provides an effective classifier to make better predictions on the real data sets than the one from the other existing algorithms. Above all, our proposed algorithm is parameter-free and thus, helpful for practical uses.

In the future, we plan to build a decision making support model based on the resulting classifier in practice to enhance a knowledge-driven educational decision support system. In addition, more data characteristics such as imbalance, overlapping, sparseness, etc. will be considered in the proposed approach for more effectiveness.

Acknowledgments. This research is funded by Vietnam National University Ho Chi Minh City under grant number C2016-20-16.

References

1. Academic Affairs Office, Ho Chi Minh City University of Technology, Vietnam. http://www.aao.hcmut.edu.vn. Accessed 29 June 2015
2. Angluin, D., Laird, P.: Learning from noisy examples. Mach. Learn. **2**(4), 343–370 (1988)

3. Bayer, J., Bydzovska, H., Geryk, J., Obsivac, T., Popelinsky, L.: Predicting drop-out from social behaviour of students. In: Proceedings of the 5th International Conference on Educational Data Mining, pp. 103–109 (2012)
4. Breiman, L.: Random forests. Mach. Learn. **45**(1), 5–32 (2001)
5. Dejaeger, K., Goethals, F., Giangreco, A., Mola, L., Baesens, B.: Gaining insight into student satisfaction using comprehensible data mining techniques. Eur. J. Oper. Res. **218**, 548–562 (2012)
6. Dong, A., Chung, F., Wang, S.: Semi-supervised classification method through oversampling and common hidden space. Inf. Sci. **349–350**, 216–228 (2016)
7. Koprinska, I., Stretton, J., Yacef, K.: Predicting student performance from multiple data sources. Artif. Intell. Educ. **9112**, 678–681 (2015)
8. Kostopoulos, G., Kotsiantis, S., Pintelas, P.: Estimating student dropout in distance higher education using semi-supervised techniques. In: Proceedings of the 19th Panhellenic Conference on Informatics, pp. 38–43 (2015)
9. Kravvaris, D., Kermanidis, K.L., Thanou, E.: Success is hidden in the students' data. Artif. Intell. Appl. Innovations **382**, 401–410 (2012)
10. Li, M., Zhou, Z.H.: Improve computer-aided diagnosis with machine learning techniques using undiagnosed samples. IEEE Trans. Syst. Man Cybern. Part-A: Syst. Hum. **37**(6), 1088–1098 (2007)
11. Márquez-Vera, C., Cano, A., Romero, C., Ventura, S.: Predicting student failure at school using genetic programming and different data mining approaches with high dimensional and imbalanced data. Appl. Intell. **38**, 315–330 (2013)
12. Peña-Ayala, A.: Educational data mining: a survey and a data mining-based analysis of recent works. Expert Syst. Appl. **41**, 1432–1462 (2014)
13. Romero, C., Espejo, P.G., Zafra, A., Romero, J.R., Ventura, S.: Web usage mining for predicting final marks of students that use Moodle courses. Comput. Appl. Eng. Educ. **21**, 135–146 (2013)
14. Saarela, M., Karkkainen, T.: Analysing student performance using sparse data of core bachelor courses. J. Educ. Data Min. **7**(1), 3–32 (2015)
15. Tanha, J., Someren, M., Afsarmanesh, H.: Semi-supervised self-training for decision tree classifier. Int. J. Mach. Learn. Cyber. 1–16 (2015). doi:10.1007/s13042-015-0328-7
16. Taruna, S., Pandey, M.: An empirical analysis of classification techniques for predicting academic performance. In: Proceedings of the IEEE International Advance Computing Conference, pp. 523–528 (2014)
17. Triguero, I., Garíca, S., Herrera, F.: Self-labeled techniques for semi-supervised learning: taxonomy, software and empirical study. Knowl. Inf. Syst. **42**(2), 245–284 (2015)
18. Triguero, I., Garíca, S., Herrera, F.: SEG-SSC: a framework based on synthetic examples generation for self-labeled semi-supervised classification. IEEE Trans. Cybern. **45**(4), 622–634 (2015)
19. Weka 3, Data Mining Software in Java. http://www.cs.waikato.ac.nz/ml/weka. Accessed 12 Dec 2015
20. Yarowsky, D.: Unsupervised word sense disambiguation rivaling supervised methods. In: Proceedings of the 33rd Annual Meeting of the Association for Computational Linguistics, pp. 189–196 (1995)
21. Zhou, Z.H., Li, M.: Tri-training: exploiting unlabeled data using three classifiers. IEEE Trans. Knowl. Data Eng. **17**, 1529–1541 (2005)

From Preference-Based to Multiobjective Sequential Decision-Making

Paul Weng[1,2]($\boxtimes$)

[1] School of Electronics and Information Technology,
SYSU-CMU Joint Institute of Engineering,
Guangzhou 510006, People's Republic of China
[2] SYSU-CMU Shunde Joint Research Institute,
Shunde 528300, People's Republic of China
paweng@cmu.edu
http://weng.fr/

Abstract. In this paper, we present a link between preference-based and multiobjective sequential decision-making. While transforming a multiobjective problem to a preference-based one is quite natural, the other direction is a bit less obvious. We present how this transformation (from preference-based to multiobjective) can be done under the classic condition that preferences over histories can be represented by additively decomposable utilities and that the decision criterion to evaluate policies in a state is based on expectation. This link yields a new source of multiobjective sequential decision-making problems (i.e., when reward values are unknown) and justifies the use of solving methods developed in one setting in the other one.

Keywords: Sequential decision-making · Preference-based reinforcement learning · Multiobjective markov decision process · Multiobjective Reinforcement Learning

1 Introduction

Reinforcement learning (RL) [27] has proved to be a powerful framework for solving sequential decision-making problems under uncertainty. For instance, RL has been used to build an expert backgammon player [28], an acrobatic helicopter pilot [1], a human-level video game player [15]. RL is based on the Markov decision process model (MDP) [21]. In the standard setting, both MDP and RL rely on scalar numeric evaluations of actions (and thus histories and policies). However, in practice, those evaluations may not be scalar or may even not be available.

Often actions are rather valued on several generally conflicting dimensions. For instance, in a navigation problem, these dimensions may represent duration, cost and length. This observation has led to the extension of MDP and RL to multiobjective MDP (MOMDP) and RL (MORL) [24]. In multiobjective optimization, it is possible to distinguish three interpretations for objectives.

© Springer International Publishing AG 2016
C. Sombattheera et al. (Eds.): MIWAI 2016, LNAI 10053, pp. 231–242, 2016.
DOI: 10.1007/978-3-319-49397-8_20

The first one corresponds to single-agent decision-making problems where actions are evaluated on different criteria, like in the navigation example. The second comes up when the effects of actions are uncertain and one then also wants to optimize objectives that correspond to probability of success or risk for instance. The last interpretation is in multiagent settings where each objective represents the payoff received by a different agent. Of course, in one particular multiobjective problem, one may encounter objectives with different interpretations.

More generally, sometimes no numerical evaluation of actions is available at all. In this case, inverse reinforcement learning (IRL) [16] has been proposed as an approach to learn a reward function from demonstration provided by a human expert who is assumed to use an optimal policy. However, this assumption may be problematic in practice as humans are known not to act optimally. A different approach, qualified as preference-based, takes as initial preferential information the comparisons of actions or histories instead of a reward function. This direction has been explored in the MDP setting [12] and the RL setting where it is called preference-based RL (PBRL) [2,9].

This theoretic paper presents a short overview of some recent work on multi-objective and preference-based sequential decision-making with the goal of relating those two research strands. The contribution of this paper is threefold. We build a bridge between preference-based RL and multiobjective RL, and highlight new possible approaches for both settings. In particular, our observation offers a new interpretation of an objective, which yields a new source of multi-objective problems.

The paper is organized as follows. In Sect. 2, we recall the definition of standard MDP/RL, their extensions to the multiobjective setting and their generalizations to the preference-based setting. In Sect. 3, we show how MORL can be viewed as a PBRL problem. This then allows the methods developed for PBRL to be imported to the MORL setting. Conversely, in Sect. 4, we show how some structured PBRL can be viewed as an MORL, which then justifies the application of MORL techniques on those PBRL problems. Finally, we conclude in Sect. 5.

2   Background and Related Work

In this section, we recall the necessary definitions needed in the next sections while presenting a short review of related work. We start with the reinforcement learning setting (Sect. 2.1) and then present its extension to the multiobjective setting (Sect. 2.2) and to the preference-based setting (Sect. 2.3).

2.1   Reinforcement Learning

A reinforcement learning problem is usually defined using the Markov Decision Process (MDP) model. A *standard MDP* [21] is defined as a tuple $\langle S, A, T, R \rangle$ where:

- S is a finite set of states,
- A is a finite set of actions,

– $T : S \times A \times S \rightarrow [0,1]$ is a transition function with $T(s, a, s')$ being the probability of reaching state s' after executing action a in state s,
– $R : S \times A \rightarrow \mathbb{R}$ is a reward function with $R(s, a)$ being the immediate numerical environmental feedback received by the agent after performing action a in state s.

In this framework, a *t-step history* h_t is a sequence of state-action:

$$h_t = (s_0, a_1, s_1, \ldots, s_t)$$

where $\forall i = 0, 1, \ldots, t, s_i \in S$ and $\forall i = 1, 2, \ldots, t, a_i \in A$. The value of such a history h_t is defined as:

$$R(h_t) = \sum_{i=1}^{t} \gamma^{i-1} R(s_{i-1}, a_i)$$

where $\gamma \in [0, 1)$ is a discount factor. A *policy* specifies how to choose an action in every state. A *deterministic* policy $\pi : S \rightarrow A$ is a function from the set of states to the set of actions, while a *randomized* policy $\pi : S \times A \rightarrow [0, 1]$ states the probability $\pi(s, a)$ of choosing an action a in a state s.

The *value function* of a policy π in a state s is defined as:

$$v^{\pi}(s) = \mathbb{E}\Big[\sum_{t \geq 0} \gamma^t \mathcal{R}_t\Big] \tag{1}$$

where $\mathcal{R}_t$ is a random variable defining the reward received at time t under policy π and starting in state s. Equation (1) can be computed iteratively as the limit of the following sequence: $\forall s \in S$,

$$v_0^{\pi}(s) = 0 \tag{2}$$

$$v_{t+1}^{\pi}(s) = R(s, \pi(s)) + \gamma \sum_{s' \in S} T(s, \pi(s), s') v_t^{\pi}(s'). \tag{3}$$

In a standard MDP, an *optimal* policy can be obtained by solving the Bellman's optimality equations: $\forall s \in S$,

$$v^{\pi}(s) = \max_{a \in A} R(s, a) + \gamma \sum_{s' \in S} T(s, a, s') v^{\pi}(s'). \tag{4}$$

Many solution methods can be used [21] to solve this problem exactly: for instance, value iteration, policy iteration, linear programming. Approaches based on approximating the value function for solving large-sized state space have also been proposed [27].

Classically, in reinforcement learning (RL), it is assumed that the agent does not know the transition and reward functions. In that case, an optimal policy has to be learned by interacting with the environment. Two main approaches can be distinguished here [27]. The first (called indirect or model-based method), tries to first estimate the transition and reward functions and then use an MDP solving method on the learned environment model (e.g., [26]). The second (called direct or model-free method), searches for an optimal policy without trying to learn a model of the environment.

The preference model that describes how policies are compared in standard MDP/RL is defined as follows. A history is valued by the discounted sum of rewards obtained along that history. Then, as a policy in a state induces a probability distribution over histories, it also induces a probability distribution over discounted sums of rewards. The decision criterion used to compare policies in standard MDP is then based on the expectation criterion.

Both MDP and RL assume that the environmental feedback from which the agent plans/learns a (near) optimal policy is a scalar numeric reward value. In many settings, this assumption does not hold. The value of an action may be determined over several often conflicting dimensions. For instance, in the autonomous navigation problem, an action lasts a certain duration, has an energy consumption cost and travels a certain length. To tackle those situations, MDP and RL have been extended to deal with vectorial rewards.

2.2   Multiobjective RL

Multiobjective MDP (MOMDP) [24] is an MDP $\langle S, A, T, \boldsymbol{R} \rangle$ where the reward function is redefined as $\boldsymbol{R} : S \times A \to \mathbb{R}^d$ with d being the number of objectives. The value function $\boldsymbol{v}^\pi$ of a policy π is now vectorial and can be computed as the limit of the vectorial version of (2) and (3): $\forall s \in S$,

$$\boldsymbol{v}_0^\pi(s) = (0, \ldots, 0) \in \mathbb{R}^d \tag{5}$$

$$\boldsymbol{v}_{t+1}^\pi(s) = \boldsymbol{R}(s, \pi(s)) + \gamma \sum_{s' \in S} T(s, \pi(s), s') \boldsymbol{v}_t^\pi(s'). \tag{6}$$

In MOMDP, the value function of policy π *Pareto-dominates* that of another policy π' if in every state s, $\boldsymbol{v}^\pi(s)$ is not smaller than $\boldsymbol{v}^{\pi'}(s)$ on every objective and $\boldsymbol{v}^\pi(s)$ is greater than $\boldsymbol{v}^{\pi'}(s)$ on at least one objective. By extension, we say that π Pareto-dominates π' if value function v^π Pareto-dominates value function $v^{\pi'}$. A value function (resp. policy) is *Pareto-optimal* if it is not Pareto-dominated by any other value function (resp. policy). Due to incomparability of vectorial value functions, there are generally many Pareto-optimal value functions (and therefore policies), which constitutes the main difficulty of the multiobjective setting.

Similarly to standard MDP, MOMDP can be extended to multiobjective reinforcement learning (MORL), in which case the agent is not assumed to know the transition function, neither the vectorial reward function.

In multiobjective optimization, four main families of approaches can be distinguished. One first natural approach is to determine the set of all Pareto-optimal solutions (e.g., [14,33]). However, in practice, searching for all the Pareto-optimal solutions may not be feasible. Indeed, it is known [20] that this set can be exponential in the size of the state and action spaces. A more practical approach is then to determine an ϵ-cover of it [7,20], which is an approximation of the set of Pareto-optimal solutions.

Definition 1. *A set $C \subseteq \mathbb{R}^d$ is an ϵ-cover of a set $P \subseteq \mathbb{R}^d$ if*

$$\forall v \in P, \exists v' \in C, (1 + \epsilon)v' \geq v$$

where $\epsilon > 0$.

Another approach related to the first one is to consider refinements of Pareto dominance, such as Lorenz dominance (which models a certain notion of fairness) or lexicographic order [10,34]. In fact, with Lorenz dominance, the set of optimal value functions may still be exponential in the size of the state and action spaces. Again, one may therefore prefer to determine its ϵ-cover [20] in practice.

Still another approach to solve multiobjective problems is to assume the existence of a scalarizing function $f : \mathbb{R}^d \to \mathbb{R}$, which, given a vector $v \in \mathbb{R}^d$, returns a scalar valuation. Two cases can be considered: f can be either linear [3] or nonlinear [17–19].

The scalarizing function can be used at three different levels:

– It can be directly applied on the vectorial reward function leading to the definition of a scalarized reward function. This boils down to defining a standard MDP/RL from a MOMDP/MORL, which can then be tackled with standard solving methods.
– It can also aggregate the different objectives of the vector values of histories and then a policy in a state can be valued by taking the expectation of those scalarized evaluation of histories.
– It can be applied on the vectorial value functions of policies in order to obtain scalar value functions.

For linear scalarizing functions, those three levels lead to the same solutions. However, for nonlinear scalarizing functions, they generally lead to different solutions. In practice, it generally only makes sense to use a nonlinear scalarizing function on expected discounted sum of vector rewards (i.e., vector value functions), as the scalarizing function is normally defined to aggregate over the final vector values. To the best of our knowledge, most previous work has applied a scalarizing function in this fashion. In Sect. 3, we describe a setting where applying a nonlinear scalarizing function on vector values of histories could be justified.

A final approach to multiobjective problem assumes an interactive setting where a human expert is present and can provide additional preferential information (i.e., how to trade-off between different objectives). This approach loops between the following two steps until a certain criterion is satisfied (e.g., the expert is satisfied with a proposed solution or there is only one solution left):

– show potential solutions or ask query to the expert
– receive a feedback/answer from the expert

The feedback/answer from the expert allows to guide the search for a preferred solution among all Pareto-optimal ones [25], or elicit unknown parameters of user preference model [22].

In both standard MDP/RL and MOMDP/MORL, it is assumed that numeric environmental feedback is available. In fact, this may not be the case in some situations. For instance, in the medical domain, it may be difficult and even impossible to value a treatment of a life-threatening illness in terms of patient well-being or death with a single numeric value. Preference-based approaches have been proposed to handle these situations.

2.3   Preference-Based RL

A preference-based MDP (PBMDP) is an MDP where possibly no reward function is given. Instead, one assumes that a preference relation $\gtrsim$ is defined over histories. In the case where the dynamics of the system is not known, this setting is referred to as preference-based reinforcement learning (PBRL) [2,4,6,9]. Due to this ordinal preferential information, it is not possible to directly use the same decision criterion based on expectation like in the standard or multiobjective cases. Most approaches in PBRL [4,6,9] relies on comparing policies with *probabilistic dominance*, which is defined as follows:

$$\pi \gtrsim \pi' \iff \mathbf{P}[\pi \gtrsim \pi'] \geq \mathbf{P}[\pi' \gtrsim \pi] \tag{7}$$

where $\mathbf{P}[\pi \gtrsim \pi']$ denotes the probability that policy π generates a history preferred or equivalent to that generated by policy π'. Probabilistic dominance is related to Condorcet methods (where a candidate is preferred to another if more voters prefers the former than the latter) in social choice theory. This is why the optimal policy for probabilistic dominance is often called a *Condorcet winner*.

The difficulty with this decision model is that it may lead to preference cycles (i.e., $\pi \succ \pi' \succ \pi'' \succ \pi$) [12]. To tackle this issue, three approaches have been considered. The first approach simply consists in assuming some consistency conditions that forbid the occurence of preference cycles. This is the case in the seminal paper [35] that proposed the framework of dueling bandits. This setting is the preference-based version of multi-armed bandit, which is itself a special case of reinforcement learning. The second approach consists in considering stronger versions of (7). Drawing from voting rules studied in social choice theory, refinements such as Copeland's rule or Borda's rule for instance, have been considered [5,6]. The last approach, which was proposed recently [8,12], consists in searching for an optimal mixed[1] policy instead of an optimal deterministic policy, which may not exist. Drawing from the minimax theorem in game theory, it can be shown that an optimal mixed policy is guaranteed to exist.

---

[1] The randomization is over policies and not over actions, like in randomized policies.

3   MORL as PBRL

An MOMDP/MORL problem can obviously be seen as a PBMDP/PBRL problem. Indeed, the preference relation $\succsim$ over histories can simply be taken as the preference relation induced over histories by Pareto dominance. Then probabilistic dominance (7) in this setting can be interpreted as follows. A policy π is preferred to another policy π' if the probability that π generates a history that Pareto-dominates a history generated by π' is higher than the probability of the opposite event. A minor issue in this formulation is that incomparability is treated in the same way as equivalence.

More interestingly, when a scalarizing function f is given, scalarized values of histories can then be used and compared in (7), leading to:

$$\pi \succsim \pi' \iff \mathbf{P}[f(\boldsymbol{R}(H_\pi)) \geq f(\boldsymbol{R}(H_{\pi'}))] \geq \mathbf{P}[f(\boldsymbol{R}(H_{\pi'})) \geq f(\boldsymbol{R}(H_\pi))]$$

where H_π (resp. $H_{\pi'}$) is a random history generated by policy π (resp. π') and $\boldsymbol{R}(H_\pi)$ (resp. $\boldsymbol{R}(H_{\pi'})$) is its vectorial value. Notably, this setting motivates the application of a nonlinear scalarizing function on vector values of histories, which has not been investigated before [24].

More generally, viewing MOMDP/MORL as a PBMDP/PBRL, one can import all the techniques and solving methods that have been developed in the preference-based settings [6,9,12]. As far as we know, both cases above (with Pareto dominance or with a scalarizing function) have not been investigated. We expect that efficient solving algorithms exploiting the additively decomposable vector rewards could possibly be designed by adapting PBMDP/PBRL algorithms.

When transforming a multiobjective into a preference-based problem, the decision criterion has generally to be changed from one based on expectation to one based on probabilistic dominance. This change may be justified for different reasons. For instance, when it is known in advance that an agent is going to face the decision problems only a limited number times, the expectation criterion may not be suitable because it does not take into account notions of variability and risk attitudes. Besides, when the decision problem really corresponds to a competitive setting, probabilistic dominance is particularly well-suited.

4   PBRL as MORL

While viewing MOMDP/MORL as PBMDP/PBRL is quite natural, the other way around may be less obvious and more interesting. We therefore develop in more details this direction by focusing on one particular case of PBMDP/PBRL where the preference relation over histories is assumed to be representable by an additively decomposable utility function and the decision criterion is based on expectation (e.g., as assumed in inverse reinforcement learning [16]). This amounts to assuming the existence of a reward function $\hat{R} : S \times A \to \{x_1, \dots, x_d\}$ where the x_i's are unknown scalar numeric reward values. Exploiting this assumption, we present two cases where PBMDP/PBRL can be transformed into

MOMDP/MORL, and justifies the use of one scalarizing function, the Chebyshev norm, on the MOMDP/MORL model obtained from a PBMDP/PBRL model.

4.1 From Unknown Rewards to Vectorial Rewards

The first transformation assumes that an order over unknown rewards is known, while the second assumes more generally that an order over some histories are known.

Ordered Rewards. In the first case, we assume that we know the order over the x_i's. Without loss of generality, we assume that $x_1 < x_2 < \ldots < x_d$.

Following previous work [29,30], it is possible to transform a PBMDP into an MDP with vector rewards by defining the following vectorial reward function $\bar{R}$ from $\hat{R}$:

$$\forall s \in S, \forall a \in A, \bar{R}(s, a) = \mathbf{1}_i \text{ if } \hat{R}(s, a) = x_i \tag{8}$$

where $\mathbf{1}_i$ is the i-th canonical vector of $\mathbb{R}^d$. Using $\bar{R}$, one can compute the vector value function of a policy by adapting (5) and (6). The i-th component of a vector value function of a policy π in a state can be interpreted as the expected discounted count of reward x_i obtained when applying policy π. However, note that because of the preferential order over components, two vectors cannot be directly compared with Pareto dominance. Another transformation is needed to obtain a usual MOMDP.

Given a vector v, we define its decumulative $v^\downarrow$ as follows:

$$\forall k = 1, \ldots, d, v_k^\downarrow = \sum_{j=k}^{d} v_j$$

A PBMDP/PBRL can be reformulated as the following MOMDP/MORL where the reward function is defined by:

$$\forall s \in S, \forall a \in A, \boldsymbol{R}(s, a) = \mathbf{1}_i^\downarrow \text{ if } \hat{R}(s, a) = x_i \tag{9}$$

Using this reward function, the vector value function $\boldsymbol{v}^\pi$ of a policy π can be computed by adapting (5) and (6). One may notice that $\boldsymbol{v}^\pi(s)$ is the decumulative vector computed from $\bar{v}^\pi$.

The relations between the standard value function v^π, the vectorial value functions $\bar{v}^\pi$ and $\boldsymbol{v}^\pi$ are stated in the following lemma.

Lemma 1. *We have:*

$$\forall s \in S, v^\pi(s) = (x_1, x_2, \ldots, x_d) \cdot \bar{v}^\pi(s) = (x_1, x_2 - x_1, \ldots, x_d - x_{d-1}) \cdot \boldsymbol{v}^\pi(s)$$

where $x \cdot y$ denotes the inner product of vector x and vector y.

It is then easy to see that if $\boldsymbol{v}^\pi(s)$ Pareto-dominates $\boldsymbol{v}^{\pi'}(s)$ then $v^\pi(s) \geq v^{\pi'}(s)$ thanks to the order over the x_i's.

Ordered Histories. In some situations, the order over unknown rewards may not be known and may not be easily determined. For instance, in a navigation problem, it may not be obvious how to compare each action locally. However, comparing trajectories may be more natural and easier to perform for the system designer. Note that although vectorial reward function $\bar{R}$ in (8) can be defined, without the order over rewards x_i's, vectorial reward function $\boldsymbol{R}$ in (9) (and thus the corresponding MOMDP/MORL) cannot be defined anymore.

In those cases, if sufficient preferential information over histories is given, the previous trick can be adapted using simple linear algebra. We now present this new transformation from PBMDP/PBRL to MOMDP/MORL. We assume that the following comparisons are available:

$$h_1 \prec h_2 \prec \ldots \prec h_d \tag{10}$$

where the h_i's are histories. Using the vector reward $\bar{R}$, one can compute the vector value of each history, i.e., $\forall i = 1, 2, \ldots, d$, if $h_i = (s_0, a_1, s_1, \ldots, s_t)$ then its value is defined by:

$$\bar{r}_i = \sum_{j=1}^{t} \gamma^{j-1} \bar{R}(s_{j-1}, a_j) \in \mathbb{R}^d.$$

We assume that $\{\bar{r}_1, \ldots, \bar{r}_d\}$ form an independent set, which implies that the matrix H whose columns are composed of $\{\bar{r}_1, \ldots, \bar{r}_d\}$ is invertible. Recall H is the basis change matrix from basis $\{\bar{r}_1, \ldots, \bar{r}_d\}$ to the canonical basis $\{\mathbf{1}_1, \ldots, \mathbf{1}_d\}$ and its inverse matrix H^{-1} is the basis change matrix in the other direction. Rewards x_i's represented by the canonical basis can then be expressed in the basis formed by the independent vectors $\{\bar{r}_1, \ldots, \bar{r}_d\}$ using the basis change matrix H^{-1}. Now, let us define a new vector reward function $\boldsymbol{R}_H$ by:

$$\forall s \in S, \forall a \in A, \boldsymbol{R}_H(s, a) = H_i^{-1\downarrow} \text{ if } \hat{R}(s, a) = x_i \tag{11}$$

where $H_i^{-1\downarrow}$ is the decumulative of the i-th column of matrix H^{-1}. Using this new reward function, one can define vector value function $\boldsymbol{v}^\pi$ of a policy π by adapting (5) and (6).

Lemma 2. *We have:*

$$\forall s \in S, v^\pi(s) = (r_1, r_2 - r_1, \ldots, r_d - r_{d-1}) \cdot \boldsymbol{v}^\pi(s)$$

where r_i is the value of history h_i, i.e., $r_i = (x_1, \ldots, x_d) \cdot \bar{r}_i$.

As the value of the r_i's is increasing with i, if $\boldsymbol{v}^\pi(s)$ Pareto-dominates $\boldsymbol{v}^{\pi'}(s)$, then π should be preferred.

Applying MORL Techniques to PBRL. We have seen two cases where a PBMDP/PBRL problem can be transformed into an MOMDP/MORL problem. As a side note, one may notice that the second case is a generalization of the first one. Thanks to this transformation, the multiobjective approaches that we recalled in Sect. 2.2 can be applied in the preference-based setting. We now mention a few cases that would be interesting to investigate in our opinion.

Here, a Pareto-optimal solution corresponds to a policy that is optimal for admissible reward values that respects the order known over rewards or histories. Like in MOMDP/MORL, it may not be feasible to determine the set of all Pareto optimal solutions. A natural approach [20] is then to compute its ϵ-cover to obtain a representative set of solutions that are approximately optimal.

Another approach is to use a non-linear scalarizing function like the Chebyshev distance to an ideal point. A policy π^* is *Chebyshev-optimal* if it minimizes:

$$\pi^* = \underset{\pi}{argmin} \max_{i=1,\ldots,d} \mathbf{I}_i - \sum_{s \in S} \mu(s) \boldsymbol{v}_i^\pi(s) \tag{12}$$

where $\mathbf{I}_i = \max_\pi \sum_{s \in S} \mu(s) \boldsymbol{v}_i^\pi(s)$ defines the i-th component of the ideal point $\mathbf{I} \in \mathbb{R}^d$, μ is a positive probability distribution over initial states and $\boldsymbol{v}_i^\pi$ is the i-th component of the vector value function of an MOMDP/MORL obtained from a PBMDP/PBRL. It is possible to show that a Chebyshev-optimal policy is a *minimax-regret-optimal* policy [23], whose definition can be written as follows:

$$\pi^* = \underset{\pi}{argmin} \max_{x \in \mathcal{R}} \max_{\pi'} \sum_{s \in S} \mu(s) x \cdot \boldsymbol{v}^{\pi'}(s) - \sum_{s \in S} \mu(s) x \cdot \boldsymbol{v}^\pi(s) \tag{13}$$

where $\mathcal{R} \subset [0,1]^d$ is the set of nonnegative values representing differences of consecutive reward values.

Lemma 3. *A policy is Chebyshev-optimal if and only if it is minimax-regret-optimal.*

It is easy to see that the maximum (over x) in (13) is attained by choosing x as a canonical vector and equal to the maximum (over i) in (12). This simple property justifies the application of one simple non-linear scalarizing function used in multiobjective optimization in the preference-based setting.

The interactive approach mentioned in Sect. 2.2 has been already exploited for eliciting the unknown rewards in interactive settings where comparison queries can be issued to an expert by interleaving optimization/learning phases with elicitation phases in PBMDP with value iteration [11,31] and PBRL with Q-learning [32]. It would be interesting to use an interactive approach to elicit the reward values by comparing the element of an ϵ-cover of the Pareto optimal solutions. This technique may help reduce the number of queries.

5   Conclusion

In this paper, we highlighted the relation between two sequential decision-making settings: preference-based MDP/RL and multiobjective MDP/RL. In particular, we showed that multiobjective problems can also arise in situations of unknown reward values. Based on the link between both formalisms, one can possibly import techniques designed for one setting to solve the other. To illustrate our points, we also listed a few interesting cases.

Besides, in our translation of a PBMDP/PBRL to an MOMDP/MORL, we assumed that rewards were Markovian, which may not always be true in practice. It would be interesting to extend our translation to the non-Markovian case [13].

References

1. Abbeel, P., Coates, A., Ng, A.Y.: Autonomous helicopter aerobatics through apprenticeship learning. Int. J. Rob. Res. **29**(13), 1608–1639 (2010)
2. Akrour, R., Schoenauer, M., Sebag, M.: APRIL: active preference learning-based reinforcement learning. In: Flach, P.A., Bie, T., Cristianini, N. (eds.) ECML PKDD 2012. LNCS (LNAI), vol. 7524, pp. 116–131. Springer, Heidelberg (2012). doi:10.1007/978-3-642-33486-3_8
3. Barrett, L., Narayanan, S.: Learning all optimal policies with multiple criteria. In: ICML (2008)
4. Busa-Fekete, R., Szörenyi, B., Weng, P., Cheng, W., Hüllermeier, E.: Preference-based reinforcement learning. In: European Workshop on Reinforcement Learning, Dagstuhl Seminar (2013)
5. Busa-Fekete, R., Szörenyi, B., Weng, P., Cheng, W., Hüllermeier, E.: Top-k selection based on adaptive sampling of noisy preferences. In: International Conference on Marchine Learning (ICML) (2013)
6. Busa-Fekete, R., Szorenyi, B., Weng, P., Cheng, W., Hüllermeier, E.: Preference-based reinforcement learning: evolutionary direct policy search using a preference-based Racing algorithm. Mach. Learn. **97**(3), 327–351 (2014)
7. Chatterjee, K., Majumdar, R., Henzinger, T.A.: Markov decision processes with multiple objectives. In: Durand, B., Thomas, W. (eds.) STACS 2006. LNCS, vol. 3884, pp. 325–336. Springer, Heidelberg (2006). doi:10.1007/11672142_26
8. Dudík, M., Hofmann, K., Schapire, R.E., Slivkins, A., Zoghi, M.: Contextual dueling bandits. In: COLT (2015)
9. Fürnkranz, J., Hüllermeier, E., Cheng, W., Park, S.: Preference-based reinforcement learning: a formal framework and a policy iteration algorithm. Mach. Learn. **89**(1), 123–156 (2012)
10. Gábor, Z., Kalmár, Z., Szepesvári, C.: Multicriteria reinforcement learning. In: Proceedings of International Conference of Machine Learning (1998)
11. Gilbert, H., Spanjaard, O., Viappiani, P., Weng, P.: Reducing the number of queries in interactive value iteration. In: Walsh, T. (ed.) ADT 2015. (LNAI), vol. 9346, pp. 139–152. Springer, Heidelberg (2015). doi:10.1007/978-3-319-23114-3_9
12. Gilbert, H., Spanjaard, O., Viappiani, P., Weng, P.: Solving MDPs with skew symmetric bilinear utility functions. In: IJCAI, pp. 1989–1995 (2015)
13. Gretton, C., Price, D., Thiebaux, S.: Implementation and comparison of solution methods for decision processes with non-Markovian rewards. In: UAI, vol. 19, pp. 289–296 (2003)
14. Lizotte, D.J., Bowling, M., Murphy, S.A.: Efficient reinforcement learning with multiple reward functions for randomized controlled trial analysis. In: ICML (2010)
15. Mnih, V., Kavukcuoglu, K., Silver, D., Rusu, A.A., Veness, J., Bellemare, M.G., Graves, A., Riedmiller, M., Fidjeland, A.K., Ostrovski, G., Petersen, S., Beattie, C., Sadik, A., Antonoglou, I., King, H., Kumaran, D., Wierstra, D., Legg, S., Hassabis, D.: Human-level control through deep reinforcement learning. Nature **518**, 529–533 (2015)

16. Ng, A., Russell, S.: Algorithms for inverse reinforcement learning. In: ICML. Morgan Kaufmann (2000)
17. Ogryczak, W., Perny, P., Weng, P.: On minimizing ordered weighted regrets in multiobjective Markov decision processes. In: Brafman, R.I., Roberts, F.S., Tsoukiàs, A. (eds.) ADT 2011. LNCS (LNAI), vol. 6992, pp. 190–204. Springer, Heidelberg (2011). doi:10.1007/978-3-642-24873-3_15
18. Ogryczak, W., Perny, P., Weng, P.: A compromise programming approach to multiobjective Markov decision processes. Int. J. Inf. Technol. Decis. Making **12**, 1021–1053 (2013)
19. Perny, P., Weng, P.: On finding compromise solutions in multiobjective Markov decision processes. In: Multidisciplinary Workshop on Advances in Preference Handling (MPREF) @ European Conference on Artificial Intelligence (ECAI) (2010)
20. Perny, P., Weng, P., Goldsmith, J., Hanna, J.: Approximation of Lorenz-optimal solutions in multiobjective Markov decision processes. In: International Conference on Uncertainty in Artificial Intelligence (UAI) (2013)
21. Puterman, M.: Markov Decision Processes: Discrete Stochastic Dynamic Programming. Wiley, Hoboken (1994)
22. Regan, K., Boutilier, C.: Eliciting additive reward functions for Markov decision processes. In: IJCAI, pp. 2159–2164 (2011)
23. Regan, K., Boutilier, C.: Robust online optimization of reward-uncertain MDPs. In: IJCAI, pp. 2165–2171 (2011)
24. Roijers, D., Vamplew, P., Whiteson, S., Dazeley, R.: A survey of multi-objective sequential decision-making. J. Artif. Intell. Res. **48**, 67–113 (2013)
25. Steuer, R., Choo, E.U.: An interactive weighted Tchebycheff procedure for multiple objective programming. Math. Program. **26**, 326–344 (1983)
26. Strehl, A.L., Littman, M.L.: Reinforcement learning in finite MDPs: PAC analysis. J. Mach. Learn. Res. **10**, 2413–2444 (2009)
27. Sutton, R., Barto, A.: Reinforcement Learning: An Introduction. MIT Press, Cambridge (1998)
28. Tesauro, G.: Temporal difference learning and TD-Gammon. Commun. ACM **38**(3), 58–68 (1995)
29. Weng, P.: Markov decision processes with ordinal rewards: Reference point-based preferences. International Conference on Automated Planning and Scheduling (ICAPS), vol. 21, pp. 282–289 (2011)
30. Weng, P.: Ordinal decision models for Markov decision processes. In: European Conference on Artificial Intelligence (ECAI), vol. 20, pp. 828–833 (2012)
31. Weng, P., Zanuttini, B.: Interactive value iteration for Markov decision processes with unknown rewards. In: IJCAI (2013)
32. Weng, P., Busa-Fekete, R., Hüllermeier, E.: Interactive Q-learning with ordinal rewards and unreliable tutor. In: ECML/PKDD Workshop Reinforcement Learning with Generalized Feedback, September 2013
33. White, D.: Multi-objective infinite-horizon discounted Markov decision processes. J. Math. Anal. Appls. **89**, 639–647 (1982)
34. Wray, K.H., Zilberstein, S., Mouaddib, A.I.: Multi-objective MDPs with conditional lexicographic reward preferences. In: AAAI (2015)
35. Yue, Y., Broder, J., Kleinberg, R., Joachims, T.: The k-armed dueling bandits problem. J. Comput. Syst. Sci. **78**(5), 1538–1556 (2012)

A Comparison of Domain Experts and Crowdsourcing Regarding Concept Relevance Evaluation in Ontology Learning

Gerhard Wohlgenannt[(✉)]

Vienna University of Economics and Business,
Welthandelsplatz 1, 1200 Wien, Austria
gerhard.wohlgenannt@wu.ac.at
http://www.wu.ac.at

Abstract. Ontology learning helps to bootstrap and simplify the complex and expensive process of ontology construction by semi-automatically generating ontologies from data. As other complex machine learning or NLP tasks, such systems always produce a certain ratio of errors, which make manually refining and pruning the resulting ontologies necessary. Here, we compare the use of domain experts and paid crowdsourcing for verifying domain ontologies. We present extensive experiments with different settings and task descriptions in order to raise the rating quality the task of relevance assessment of new concept candidates generated by the system. With proper task descriptions and settings, crowd workers can provide quality similar to human experts. In case of unclear task descriptions, crowd workers and domain experts often have a very different interpretation of the task at hand – we analyze various types of discrepancy in interpretation.

Keywords: Ontology learning · Evaluation · Crowdsourcing · Human computation

1 Introduction

With the emergence of Web technologies, *knowledge creation* has evolved into a distributed process that integrates groups of users with different levels of expertise [5]. Recent approaches further broaden the knowledge creation process to include large populations of non-experts by using crowdsourcing techniques [6]. Crowdsourcing, in the form of gamification, and esp. in the form of paid micro-task crowdsourcing, has become a popular means to solve tasks that computers cannot solve yet. It is often used to create training data for supervised machine learning, or for annotation and evaluation tasks.

Ontology engineering is a crucial knowledge acquisition process in the area of the Semantic Web. Ontologies are the vocabulary and therefore the backbone of the Semantic Web. Ontology construction is a complex and expensive task, therefore ontology learning systems, which (semi-)automatically generate

© Springer International Publishing AG 2016
C. Sombattheera et al. (Eds.): MIWAI 2016, LNAI 10053, pp. 243–254, 2016.
DOI: 10.1007/978-3-319-49397-8_21

ontologies from existing data (eg. unstructured domain text corpora), have been
proposed. As the automatically generated ontological constructs need re-design
and pruning, we apply crowdsourcing and domain experts for evaluating various
parts of the ontologies.

More specifically, in our ontology learning system [13,16], we have applied
both domain expert evaluation, as well as paid crowd workers to rate the domain
relevance of domain concept candidates generated by the system. The ontology
learning system learns lightweight ontologies from scratch in monthly intervals
– in various knowledge domains. Therefore, the system has accumulated a lot
of data which we will use in this publication to compare the characteristics and
quality of domain expert judgments versus ratings by crowd workers.

We want to give some insights and lessons learned about the following questions: what are the quality and characteristics of crowd worker judgments in a
task setting like judging the domain relevance of concept candidate labels? What
are the differences in task setup and task description between crowd workers and
domain experts? What influence do task description and settings in the crowdsourcing platform have on the resulting quality? And, in general, how well suited
is crowdsourcing for domain specific knowledge acquisition jobs?

To address the research questions, we first compared the original ratings of
crowdsourcing and domain experts for the data collected between 2013–2016.
Then, we had another domain expert evaluation using a clearer task description
in order to create a gold standard. We also repeated the crowdsourcing rating
process with an extended task description and a careful selection of crowd workers, with the goal to improve the quality of the crowdsourcing results as far as
possible – with regards to the gold standard. In a nutshell, our experiments show
that crowd workers can provide quality similar to domain experts, if measures
to raise quality are taken. But, obviously, the more specialized and complex the
domain and the tasks, the harder it is to maintain good quality.

2   Related Work

There are three main types of crowdsourcing methods: paid-for crowdsourcing,
games with a purpose, and altruism. Games with a purpose include human
computation tasks as a side effect into playing (online) games [1,7]. In paid-for
crowdsourcing, more precisely *Mechanized Labor*, contributors carry out small
tasks for a small amount of money, this is also call micro-task crowdsourcing.
Two popular marketplaces that bring together crowd workers and customers are
Amazon Mechanical Turk and CrowdFlower.

In the realm of ontology engineering, paid crowdsourcing has been used for
various tasks. Eckert et al. [4] build concept hierarchies in the philosophy domain
using Amazon Mechanical Turk. They use crowdsourcing to judge the related-
ness of concept pairs, and to find taxonomic structures. An important aspect of
ontology creation is taxonomy building, Noy et al. [10] verify the correctness of
taxonomic relations with paid micro-task crowdsourcing. Wohlgenannt et al. [15]
build and evaluate a crowdsourcing plugin for the Protege ontology editor.

The authors focus on the tight integration of crowdsourcing (paid crowdsourcing, and games with a purpose) into the knowledge engineering workflow, and analyze the benefits of crowdsourcing in terms of cost, time and scalability as compared to domain experts.

A closely related field of research is ontology alignment, where Sarasua et al. [11] use crowd workers to evaluate the correctness of *sameAs* relations, and to choose relations between terms. ZenCrowd [3] verifies the output of automatic entity linking algorithms. For a given term, crowd workers select the best fitting DBpedia URL that represents the entity. Rather recently, Amazon Mechanical Turk was used to generate Semantic Web benchmark data in the Conference track of the Ontology Alignment Evaluation Initiative (OAEI).

Most previous work which studies the quality of annotations generated by crowd workers in the field of knowledge acquisition comes to the conclusion that the quality delivered by crowd workers is similar to domain experts, esp. when the complexity of the task is moderate [2]. Here, we want to provide more detailed insights into differences between domain expert and crowd worker judgements, and on task setup and task description for improved crowd worker quality.

3 System Description

This section includes a short description of the ontology learning system used to generate the concept candidates. More details about the system can be found for example in [9,13]. The goal of the system is to learn lightweight ontologies from so-called sources of evidence. These evidence sources include (domain-filtered) text collected from news media Web sites, social sources such as text from Twitter and Facebook, and structured sources like DBpedia and WordNet. Figure 1 provides an overview of the system. The starting point is a seed ontology. The seed ontology typically contains 2–3 root concepts in the respective domain.

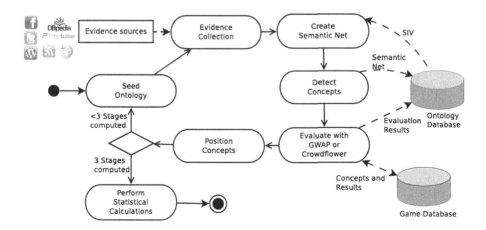

Fig. 1. The ontology learning process.

For these seed concepts, the system collects new evidence for related terms from the evidence sources. All this new evidence is stored into a Semantic Network. The neural networking algorithm of spreading activation then helps to select the most important concept candidates from the Semantic Network [14]. These concept candidates are then manually verified for domain relevance by either domain experts or crowd workers. Finally, the verified concepts are positioned in the existing ontology, resulting in an *extended ontology*. The extended ontology now serves as new seed ontology, and the extension process starts over. Usually we do three extension runs.

For this work, the most important step is the selection of concept candidates–as these will be evaluated in the remainder of the paper. From the plethora of terms in the Semantic Network (typically many thousands of terms), our system selects the 25 most promising concept candidates, according to their activation levels from the spreading activation algorithm.

4   Evaluation Setup

All data used in the experiments in this paper stems from ontology learning experiments conducted from October 2013 to December 2015. In every month the system [13] computes ontologies in various domains, in each for various system settings, from scratch. Each month we only use the corpus data (for example from news media sites) collected in that respective month, which leads to an evolution of ontologies.

The ontology system generates (among other things) concept candidate labels, which are then manually evaluated for domain relevance using either domain experts (DE) or crowd workers (CW). In this research we evaluate 4 domains, but only one domain (*Tennis*) has concept candidate relevance ratings by both DE and crowdsourcing. In crowdsourcing, we collected five votes for each concept candidate label, and used majority voting to make a decision. Figure 2 shows a screenshot of what a crowd worker is presented when doing the

Fig. 2. Screenshot of an evaluation question on CrowdFlower.

evaluation task. We did not include the task instructions in the screenshot, as they would take up too much space.

5   Results

This section presents the evaluation results. In Sect. 5.1 we give an overview of the evaluation data collected, and then compare the original crowdsourcing and domain expert evaluations. In Sect. 5.2 we do a re-evaluation of the concept candidates with domain experts – using more precise task descriptions. This gives us a gold standard evaluation, which we then compare to the original evaluation of DE and CW. Section 5.3 presents the results from repeating the CW evaluations with improved task descriptions and settings, and compares them to the gold standard. And finally, in Sect. 5.4 we compare the original to the repeated CW evaluations.

5.1   Analysis of Original Data

Firstly, we present an aggregated view on the evaluation data. Table 1 shows the ratio of concept candidates judged as *relevant* in our four domains. In this table the statistics are referring to *distinct* concept labels. The concepts are counted only once, even if the concept candidates are occurring in various ontologies for the specific domain over time. We distinguish between results from *domain experts* and *crowd workers*.

Table 1. Concept candidate labels judged relevant compared to total number of candidates automatically generated by the system in the domain; this table uses distinct concept labels, disregarding repeated occurrence of a label.

Domain	Domain experts	Crowd workers
Tennis	137 of 647 (21.17 %)	157 of 291 (53.95 %)
Climate change	304 of 889 (34.19 %)	
NOAA	147 of 358 (41.06 %)	
Middle-east crisis (DE)		322 of 570 (56.49 %)

The most obvious observation in Table 1 is that CW were by far less strict with judging concept candidates. Crowd workers rated between 53 %–57 % of distinct concepts as relevant, whereas the ratio was only between 21 % and 41 % for DE, depending on the domain. So, domain experts naturally tend to have a stricter view on the world, and CW are more likely to accept a term if in doubt – esp. if they are not given precise task instructions.

Table 2 is similar to Table 1, but now we take multiple occurrences of the same concept candidates in different ontologies into account. As the underlying

Table 2. Ratio of concept candidate labels judged relevant; this table takes into account repeated occurrence of a concept label.

Domain	Domain experts	Crowdworkers
Tennis	1675 of 4165 (40.22 %)	632 of 1030 (61.36 %)
Climate change	11508 of 16332 (70.46 %)	
NOAA	578 of 1124 (51.42 %)	
Middle-east crisis (DE)		592 of 1025 (57.75 %)

data changes, the light-weight ontologies evolve, but obviously many domain concept candidates often re-occur.

When we take repeated occurrence of concepts into account, the ratio of relevant concepts raises drastically for both judgments from DE and CW. The reason is the following: the concepts that re-occur over time and under different settings are typically the concepts which the ontology learning system regards the most relevant, whereas the concept candidates that are generated very rarely over time are likely to be not as important to the domain or the result of rather random appearance in the underlying text corpora. In domains where we have ratings both from CW and DE, ie. *Tennis*, we still have the situation of DE being more strict with their judgments.

The percentage of concepts ranked positive in the *climate change* domain is very high, although the concept labels have been ranked by DE. Our interpretation is the following: In the *climate change* domain we apply high quality corpora, not only mirrored from general news media sites, but also from domain-specific Websites of environmental NGOs. The corpora are larger and of better quality than for the other domains. (ii) The *climate change* domain is more stable than the other domains, and has a number of relevant concepts which re-occur in most generated ontologies, such as "global warming" or "climate". Savenkov and Wohlgenannt [12] evaluate the ontology learning data, which is also underlying this work, regarding ontological volatility, and find that the *Tennis* domain is more volatile than *climate change*.

In order to better interpret the observation that DE are much stricter in judgment, we analyzed the concept candidates that are overlapping, ie. which both appear in ontologies in the *Tennis* domain – which were evaluated using either by DE or CW. In total we have 691 distinct concept candidates in the *Tennis* domain, 247 of these overlapping between the evaluation methods (DE, CW). Taking into account repeated occurrence, the numbers change to 5195 total candidate concepts, and 4380 overlapping.

Similarly to a confusion matrix, Table 3 separates the overlapping concepts into four classes: (a) where both CW as well as DE judged the label as relevant to the domain, (b) DE say the concept is relevant, CW say it is not, (c) DE rate as non-relevant, but CW rate as relevant, (d) where both evaluator types come to the same conclusion that the concept candidate is not relevant to the domain (of *Tennis*).

Table 3. Relevance judgments for overlapping concept labels between Domain Experts (DE) and Crowd Workers (CW). Gives the numbers for repeated occurrence of a concept label in different ontologies, and in parenthesis counts for distinct concept labels.

	CW: relevant	CW: non-relevant
DE: relevant	1889 (69)	108 (6)
DE: non-relevant	966 (76)	1417 (96)

Taking multiple occurrence of concept candidates into account, we can see the DE and CW agree in most cases (1889 plus 1417), which is an agreement on about 75 % of candidates. But there is also a big group of concept candidates where crowd workers are less strict with their judgment (966). Only 108 of the 4380 overlapping concept labels DE judge as relevant, while CW rate as non-relevant.

Looking closer at the data, we can see that many of the terms in group *CW: relevant/DE: non-relevant* fall in one of three categories: (a) terms which have some relevance, but where the DE were more strict, eg. baseline, field, loss or summer, (b) wrong judgments by CW, for example test, table tennis, or dog, (c) terms which are not proper English words or phrases, such as world no or ausopen.

In order get a clearer picture of the differences between DE and CW, we had another domain expert re-evaluate (rate) all the 691 distinct concept candidates in the *Tennis* domain. This time we gave very clear instructions on how to measure domain relevance, most importantly: (a) only accept proper English phrases and abbreviations as relevant. For example ATP or us open is relevant, but not usopen; (b) if in doubt if a term is (closely) related to the domain rate as relevant.

5.2   Re-evaluation with an Additional Domain Expert

In this section, we used an additional evaluation made by a domain expert over all 691 concept candidate labels as a gold standard – in order to analyze and interpret the results from the original DE and CW judgments.

First, we evaluated the accuracy, ie. the ratio of judgments, where CW and DE agree with ratings given in the gold standard. Table 4 shows the percentage of agreement, split into three categories: concept candidates rated *relevant* by

Table 4. Accuracy of DE and CW ratings with regards to the GS, for concept candidates incl. re-occurring candidates; distinct values in parenthesis

	Relevant	Non-relevant	Total
Accuracy DE	72.93 % (58.51 %)	96.46 % (94.09 %)	**84.00 %** (83.72 %)
Accuracy CW	85.38 % (85.71 %)	66.12 % (58.72 %)	**77.37 %** (69.75 %)

the gold standard evaluation, concepts rated not relevant, and the sum of the two (*total*). In the table we distinguish, again, between taking the number of occurrences of concept candidates into account, and using distinct concepts only (in parenthesis).

We see some interesting facts, for example that most concepts rated non-relevant by the gold standard evaluation were also rated non-relevant by DE evaluations, with 96.46 %. So there is a strong agreement on non-relevant concepts, but for concepts rated *relevant* by the gold standard (GS) evaluation, agreement is much lower. This shows that the original DE were a lot stricter with accepting concept candidates as relevant. For CW evaluations the data shows the opposite effect, strong overlap for rating concepts as relevant, but only for 66.12 % of the concepts rated as negative by the gold standard evaluator the CW agreed.

In total, as expected, the ratio of overlap is higher between DE evaluations and the gold standard evaluations than for CW versus GS, although the differences are rather small (84.00 % versus 77.37 %).

We also applied Cohen's kappa as prominent measure to compute inter-annotator agreement. The kappa value for CW and GS is 0.53 (again taking the number of occurrences into account), for DE and GS it is 0.68. According to the interpretation of Landis and Koch [8] we see *substantial* agreement between DE and GS, and moderate agreement between CW and GS.

A closer look at the data reveals some of the causes for the observation made. In a number of cases, GS evaluations rated concepts as relevant, whereas DE did not. Examples are: `draw, lawn, history, baseline`. All these examples are at least remotely relevant for the *Tennis* domain. The relatively high number of disagreement about *relevant* concepts clearly stems from the instructions given to the gold standard evaluator, namely to judge as relevant when in doubt. In contrast to DE versus GS, for CW compared to the GS, there was a lot of disagreement on concepts judged to be non-relevant by the GS evaluations. As in the last section, this often concerns concepts labels with are not English terms (eg. `womenstennis, andymurray`), plainly wrong ratings by CW (eg. `hair, garden, inning`), or terms too remotely relevant for the domain (eg. `foot, qualifier, livescore`).

5.3   Repetition of Crowdsourcing with New Settings

With the lessons learned from our original experiments, in May 2016, we repeated the CrowdFlower evaluations for all the 291 (overlapping) concept candidates in the Tennis domain. In this new crowdsourcing job, we gave preciser *task instructions*, which included a number of examples. The instructions were similar to the instructions given to the gold standard domain expert evaluator, but a bit more detailed. Furthermore, we tried to get high quality results by only accepting the best workers (*level 3 workers*), and restricting worker residence to English speaking countries such as UK or US. Finally we used carefully designed test questions, which are called "gold units" in CrowdFlower, and crowd workers needed to pass at least 80 % of the gold units.

Table 5. Accuracy of repeated (modified) CW evaluation with regards to the GS, for concept candidates including re-occurring candidates; distinct values in are given parenthesis.

	Relevant	Non-relevant	Total
Number of terms	2157 of 2631	1591 of 1801	3748 of 4432
Number of terms (distinct)	93 of 119	153 of 172	246 of 291
Accuracy CW	81.98 % (78.15 %)	88.34 % (88.95 %)	**84.56 %** (84.53 %)

Again, we use the GS evaluation as a baseline, and compare the newly gathered crowdsourcing results (*CW-new*) to the gold standard. Table 5 presents the results, distinguishing between agreement on *relevance* resp. *non-relevance* of candidate concepts for the Tennis domain.

As we can see, the accuracy of the new CW evaluation is substantially higher than for the original CW data. Cohen's kappa is now 0.68 for distinct candidates, and 0.69 when taking the number of occurrences into account – as compared to 0.53 in the original CW evaluation. We attribute the improvement mainly to the updated task instructions. In *CW-new* we now rarely see unwanted terms such as `andymurray` (which is not a proper entity), but obviously there are still problems. For example, *CW-new* rated `quarterback` as relevant, which hints at domain knowledge missing. Another example is `world no`, which is a bi-gram fragment, and should not be rated as relevant. On the other hand, some terms which are relevant according to the GS evaluation, such as `ball games, fight, defender`, which were rated as non-relevant by *CW-new*.

Improved task instruction will never solve uncertainty with borderline cases regarding domain relevance, but should help to reduce other reasons for wrong judgements. We analyzed our two main reasons for wrong judgements: (i) concept candidates which are clearly not relevant to the domain, and (ii) terms which are not proper English concept labels (eg. hashtags, etc.). For group (i) the number of errors was reduced from *21* to *3*, and for group (ii) from *14* to *3*. This clearly shows that the improved task descriptions and settings helped with these sources of errors.

Despite the occasional errors, the quality of *CF-new* has the same level of agreement to the GS as the original DE evaluation. In an attempt to further improve quality, we did another set of CrowdFlower evaluations of the 291 terms, where we only accepted workers that had at least 99 % of the test questions (gold units) correct. From this additional crowdsourcing evaluation we expected an increase in accuracy, but actually it stayed about the same. We archived a total accuracy of 84.2 % on distinct candidates, and 82.1 % when taking the number of occurrences into account. Cohen's kappa was 0.66 and 0.65, respectively. This shows the limits of human evaluation, which are caused by two main reasons: Most importantly, for some terms domain relevance is not clear, they are border-line cases. Furthermore, humans make mistakes in judgment, either because they did not understand the task instructions in all details, or they lack knowledge of parts of the domain.

5.4   Comparing the Crowdsourcing Evaluations

Finally, we investigate the differences between the results of the original CW evaluation, and the new CW evaluation. Table 6 shows the agreement on relevant and non-relevant concept candidates, as well as the differences between the two evaluations.

Table 6. Relevance judgements comparing the original CW evaluation (*CW-orig*) and the re-evaluation with new settings and task descriptions (*CW-new*). Given are the numbers counting repeated occurrence of a concept label in different ontologies, and in parenthesis the numbers for distinct concept labels.

	CW-orig: relevant	CW-orig: non-relevant
CW-new: relevant	1785 (98)	582 (49)
CW-new: non-relevant	430 (81)	1635 (145)

There is a large discrepancy between *CW-orig* and *CW-new*, the agreement is only *moderate*, with a Cohen's kappa value of 0.54. For distinct terms, the kappa is even lower, at 0.30. Again, the differences mostly come from the improved and clearer task descriptions. The original CW evaluation led to a more open view on the domain, and as already mentioned, included many terms which are not proper phrases or entities as relevant.

5.5   Discussion

One of the key learnings regarding *task setup* is that crowd workers will have a different interpretation of task definitions than domain experts. Therefore task definitions for crowd workers must be very precise and be backed up by more examples on how to solve a task. So extensive and precise *task descriptions* are crucial when using crowd workers, in addition to traditional measures to improve worker quality, such as using gold units (with CrowdFlower), allowing only workers with the highest skill level (as recorded in previous tasks), and only native English speakers. A discussion of the detailed results for various evaluation setups is included in the sections above already.

All the data itself, examples of task descriptions, gold units, and the code we used to analyze the data can be found online[1].

6   Conclusion and Future Work

In this paper, we compare micro-task crowdsourcing to the use of domain experts for the task of domain relevance assessment of concept candidates in ontology learning. First, we compared the data from our original crowdsourcing evaluation

---

[1] https://aic.ai.wu.ac.at/~wohlg/miwai2016.

to the original domain expert evaluation. Then, we repeated the domain expert evaluation with improved task descriptions to create a gold standard. We also repeated the crowdsourcing evaluations with improved task descriptions and settings with the goal to raise crowd worker quality, and then compared that data to the gold standard.

We found that a very precise task description, including a number of examples, as well as strict worker selection and the use of gold units are crucial to ensure high quality results from crowd workers. Using these measures allows to deliver quality similar to human experts. But esp. in complex domains, crowd worker quality will vary, so we advise to explore the results with experiments with different settings and task descriptions in such cases. A limitation of all evaluation types (crowdsourcing and domain experts) are cases which cannot be judged clearly – in our case concept candidates with only moderate domain relevance. Disagreement in judgment among crowd users helps to detect these cases.

The main contributions of this work are the following: (i) evaluating the suitability of crowdsourcing for a specific ontology learning task, (ii) comparing the quality of crowd worker assessment to domain experts – based on extensive experiments which used different settings and variations, (iii) doing a detailed analysis of the effects of evaluation strategies on the quality of results and the types of errors, (iv) giving guidelines on how to set up crowdsourcing tasks in order to improve the evaluation quality.

In future work we will have a closer look at other domains used in our system, such as *climate change*, and also re-evaluate its respective concept candidates with updated CrowdFlower settings and task description. Furthermore, our system keeps relevance judgements about concepts only for a given time period – in order to facilitate the evolution of the domain by data-driven change. Concept candidates which re-appear will be judged again in a 6 month interval. For the data collected we will study how concept relevance understanding changes over time.

Acknowledgments. The work presented in this paper was created based on results from project uComp. uComp received the funding support of EPSRC EP/K017896/1, FWF 1097-N23, and ANR-12-CHRI-0003-03, in the framework of the CHIST-ERA ERA-NET.

References

1. von Ahn, L., Dabbish, L.: Designing games with a purpose. Commun. ACM **51**(8), 58–67 (2008). http://doi.acm.org/10.1145/1378704.1378719
2. Cheatham, M., Hitzler, P.: Conference v2.0: An Uncertain Version of the OAEI Conference Benchmark. In: Mika, P., Tudorache, T., Bernstein, A., Welty, C., Knoblock, C., Vrandečić, D., Groth, P., Noy, N., Janowicz, K., Goble, C. (eds.) ISWC 2014. LNCS, vol. 8797, pp. 33–48. Springer, Heidelberg (2014). doi:10.1007/978-3-319-11915-1_3

3. Demartini, G., Difallah, D.E., Cudré-Mauroux, P.: ZenCrowd: leveraging probabilistic reasoning and crowdsourcing techniques for large-scale entity linking. In: Proceedings of the 21st International Conference on World Wide Web, pp. 469–478. ACM (2012)

4. Eckert, K., Niepert, M., Niemann, C., Buckner, C., Allen, C., Stuckenschmidt, H.: Crowdsourcing the assembly of concept hierarchies. In: Proceedings of the 10th Annual Joint Conference on Digital Libraries, JCDL 2010, pp. 139–148. ACM (2010)

5. Gil, Y.: Interactive knowledge capture in the new millennium: how the semantic web changed everything. Knowl. Eng. Rev. **26**(1), 45–51 (2011)

6. Howe, J.: Crowdsourcing: why the power of the crowd is driving the future of business (2009). http://crowdsourcing.typepad.com/

7. Krause, M., Smeddinck, J.: Human computation games: a survey. In: Proceedings of 19th European Signal Processing Conference (EUSIPCO 2011) (2011)

8. Landis, J., Koch, G.: The measurement of observer agreement for categorical data. Biometrics **33**(1), 159–174 (1977)

9. Liu, W., Weichselbraun, A., Scharl, A., Chang, E.: Semi-automatic ontology extension using spreading activation. J. Univ. Knowl. Manag. (1), 50–58 (2005)

10. Noy, N.F., Mortensen, J., Musen, M.A., Alexander, P.R.: Mechanical turk as an ontology engineer? Using microtasks as a component of an ontology-engineering workflow. In: Proceedings of the 5th Annual ACM Web Science Conference, WebSci 2013, pp. 262–271 (2013)

11. Sarasua, C., Simperl, E., Noy, N.F.: CROWDMAP: crowdsourcing ontology alignment with microtasks. In: Cudré-Mauroux, P., Heflin, J., Sirin, E., Tudorache, T., Euzenat, J., Hauswirth, M., Parreira, J.X., Hendler, J., Schreiber, G., Bernstein, A., Blomqvist, E. (eds.) ISWC 2012. LNCS, vol. 7649, pp. 525–541. Springer, Heidelberg (2012). doi:10.1007/978-3-642-35176-1_33

12. Savenkov, V., Wohlgenannt, G.: Similarity metrics in ontology evolution. In: Klinov, P., Mourmotsev, D. (eds.) KESW 2015, Posters and Position papers. Moscow, Russia, October 2015

13. Wohlgenannt, G.: Leveraging and Balancing Heterogeneous Sources of Evidence in Ontology Learning. In: Gandon, F., Sabou, M., Sack, H., dAmato, C., Cudré-Mauroux, P., Zimmermann, A. (eds.) ESWC 2015. LNCS, vol. 9088, pp. 54–68. Springer, Heidelberg (2015). doi:10.1007/978-3-319-18818-8_4

14. Wohlgenannt, G., Belk, S., Schett, M.: Computing Semantic Association: Comparing Spreading Activation and Spectral Association for Ontology Learning. In: Ramanna, S., Lingras, P., Sombattheera, C., Krishna, A. (eds.) MIWAI 2013. LNCS (LNAI), vol. 8271, pp. 317–328. Springer, Heidelberg (2013). doi:10.1007/978-3-642-44949-9_29

15. Wohlgenannt, G., Sabou, M., Hanika, F.: Crowd-based ontology engineering with the uComp protege plugin. Semantic Web J. (SWJ) p. Accepted/Scheduled for publication (2015)

16. Wohlgenannt, G., Weichselbraun, A., Scharl, A., Sabou, M.: Dynamic integration of multiple evidence sources for ontology learning. J. Inf. Data Manag. **3**(3), 243–254 (2012)

Evolutionary Analysis and Computing
of the Financial Safety Net

Ke Yang[1,2(✉)], Kun Yue[1], Hong Wu[1,3], Jin Li[1], and Weiyi Liu[1]

[1] School of Information Science and Engineering,
Yunnan University, Kunming, China
19732978@qq.com
[2] Head Office, Fudian Bank, Kunming, China
[3] School of Information Engineering, Qujing Normal University, Qujing, China

Abstract. Governments want to establish a financial safety net (FSN) to prevent financial crises from spreading. The FSN is a series of institutional arrangements to preserve financial stability. In the real FSN, it includes the central bank, deposit insurance institutions and their premium, commercial banks and their benefit rates, etc., and these parameters are interdependent and dynamic change. And thus, analysis and computing of the FSN is very challenging. Inspired by evolutionary game theory, in this paper, we first establish a network game model of the FSN to analyze the evolution of bank deposit insurance strategies, and further propose a method to measure the effectiveness of the FSN. Finally, we use computational experiments to simulate the operation of the FSN. In the experiments, an evolutionary computation method is employed to compute banks' decisions to reduce computing time. Experimental results show that our evolutionary approach is suitable for the FSN, and is able to provide suggestions of macro policy for regulators.

Keywords: Evolutionary computation · Evolutionary game · Financial safety net · Simulation · Effectiveness measurement

1 Introduction

The concept of the financial safety net (FSN) was first proposed by Bank for International Settlements in 1986. Later, it is a series of institutional arrangements to preserve financial stability in financial sector. It consists of prudential regulation, last resort of the central bank and deposit insurance system [1]. As financial crises break out in recent years, governments want to design and build an effective FSN in order to prevent financial crises from spreading. The design objectives of the FSN, on the one hand are to preserve financial stability, and on the other hand are to maximize social benefits of all members in financial system. This financial issue also attracts scholars' attentions on mathematics [2], management science [3], computer science [4] fields. In this paper, we propose an evolutionary analysis approach to understand the mechanism of the FSN, and use computational experiments to simulate the operation of the FSN. Experimental results can help regulators to design the optimal FSN.

The design of the FSN is a multi-objective optimization problem [2]. Research methods to this problem are broadly divided into the following categories. The first is

© Springer International Publishing AG 2016
C. Sombattheera et al. (Eds.): MIWAI 2016, LNAI 10053, pp. 255–267, 2016.
DOI: 10.1007/978-3-319-49397-8_22

the cost-benefit analysis, which is a widely used theoretical approach. By adjusting factors of regulation cost, this problem becomes a Pareto optimization process [5, 6]. Secondly, appropriate deposit insurance policies, such as: insurance limit and premium pricing, are designed to strengthen market disciplines [7, 8]. Finally, based on game analysis, an incentive-compatible mechanism is designed, so that the benefits objectives of banks and regulators are agreed [9, 10]. The analysis approach of this paper is similar to the last category, but we give special consideration to a challenge problem that there exists asymmetric information in real financial environment.

First, we develop a network game model to describe the FSN. Under the deposit insurance system, financial stability is a public good [9]. Because, if a bank takes a full-amount insurance strategy, after a crisis of this bank occurs, the crisis will not spread throughout financial system due to guarantee of deposit insurance institutions, so financial stability is preserved. On the contrary, if a bank does not take deposit insurance and the central bank will do not rescue after a crisis of this bank occurs, this situation may lead to crises spread in financial system and result in losses of other banks. So the effectiveness of a single bank's insurance strategy depends on other banks. Thus, a public goods game is formed, and the game equilibrium status is the steady-state of the FSN under operation. In addition, we introduce an ineffectiveness index of game theory to measure the effectiveness of the FSN.

Due to the information asymmetry in the FSN, a bank does not have full information, not knowing benefit function of other banks, not knowing the central bank's rescue policy. Therefore, a bank is unable to compute equilibrium strategy. To address this difficulty, we notice that the strategy of a bank is a dynamic tuning process. Inspired by evolutionary game theory, we propose updating rules for a bank seeking optimum strategy, which does not depend on the other bank's information. We prove that the evolution of banks' strategies can converge to equilibrium status.

Next, in order to simulate the FSN's operation, we need to solve computational complexity problems. Because computing of the FSN is related to the central bank, deposit insurance institutions and their premium, commercial banks and their benefit rates, etc., and these parameters are interdependent and dynamic change. For computing of complex networks, evolutionary computation is a suitable optimization approach based on Darwinian evolution, reducing computing time. Evolutionary computation approaches mainly include genetic algorithms, genetic programming, multi-objective evolutionary algorithms and evolutionary game method. These methods are widely employed for financial optimization issue [11]. For example, genetic algorithm is employed for portfolio optimization [12] and financial fraud detection [13] problems. Multi-objective evolutionary algorithms employed for related transactions mining [14] and credit portfolios [15] problems. Evolutionary game method is employed for market simulation [16] and crises spread [17] problems.

In this paper, we employ the best response algorithm, one of evolutionary game method, to compute approximate solution of the game equilibrium strategy. We give a theorem that the time complexity of our algorithm is bounded.

Finally, we make simulation experiments by using the real bank asset data as experimental parameters. Our computation method is employed to compute banks' decision in parallel, simulating the evolution of bank insurance strategies. After finite iteration computing, bank strategies are nearly invariable, and it means that the bank strategies reach a stable state. We find that there is a negative correlation between the effectiveness of the FSN and the deposit insurance premium. Experimental results show that our evolutionary analysis approach can be adapted to the optimization problem of the FSN.

The contributions of this paper are:

- We develop a network game model to describe the FSN, and analyze the evolution of bank insurance strategies.
- We give a measure method for the effectiveness of the FSN.
- We employ an evolutionary computation method to solve the approximate equilibrium of the game, reducing computing time.

The paper is organized as follows. In Sect. 2, we describe the FSN model and benefit functions. In Sect. 3, we analyze the operation of the FSN. In Sects. 4 and 5, an evolutionary computation method is applied to simulation experiments. In the last Section, we conclude this paper. The proofs of theorems of this paper are given in the Appendix.

2   The FSN Model

Considering a country's FSN, it includes 1 central bank, n commercial banks and m deposit insurance institutions. We model relationships of these institutions as Fig. 1. Below, we further illustrate their competitive or cooperative relationships by proposing their benefits functions below.

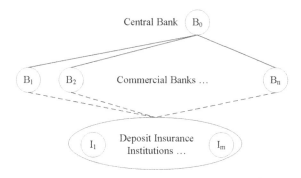

Fig. 1. A financial safety net model

First, from the perspective of commercial banks, they absorb funds from depositors, then invest. The operation objective is to maximize benefits. Denote the average success rate of investment by θ. Denote the average benefits rate of investment by π. Under interest rate market environment, there are 2 deposit rates, one is in guarantee of deposit insurance institution, and another is out of guarantee. Denote the deposit rate in guarantee by α, denote the deposit rate out of guarantee by β, and $\alpha < \beta$. Under the deposit insurance system, banks pay the premium if depositor funds are insured by insurance institutions. Denote the premium rate by μ. Denote the assets of bank i by A_i, $i \in \{1, \ldots, n\}$. Denote total assets of all banks in financial system by W. Thus, $W = \sum_i A_i$.

Under the deposit insurance system, banks can choose insurance or not. When a bank chooses the insurance, it first pays premiums. If its investment is successful, the benefits rate is $(\pi - \alpha)$, otherwise, an insurance institution compensates depositors' funds. Thus, if a bank chooses the insurance, its benefits rate v can be expressed as

$$v = \theta \cdot (\pi - \alpha) + (1 - \theta) - \mu \tag{1}$$

If a bank does not choose insurance, and its investment is successful, then the benefits rate of the bank is $(\pi - \beta)$, otherwise, it may get rescue from the central bank with probability q to compensate depositors' funds. Thus, if a bank does not chooses the insurance, its benefits rate u can be expressed as

$$u = \theta \cdot (\pi - \beta) + (1 - \theta) \cdot q \tag{2}$$

Next, from the perspective of the central bank, let H denote the total amount of risk assets in financial system. If the value of H is very small (i.e., $H = 0$), then the central bank can fully rescue (i.e., $q = 1$). Otherwise, the value of H reaches the maximum (i.e., $H = W$), then the central bank has not enough funds to rescue (i.e., $q = 0$). Therefore, the rescue probability of central bank q is a decreasing function of H, We have

$$q(H) = 1 - (H/W)^e \tag{3}$$

where e is the central bank's risk aversion degree, $e \geq 1$.

Let $\bar{r}$ denote the average benefits rate of all banks, then

$$\bar{r} = u \cdot H/W + v \cdot (1 - H/W) \tag{4}$$

Let $a = \theta \cdot (\pi - \beta)$, $b = (1 - \theta)$, and $c = \theta \cdot (\pi - \alpha) + (1 - \theta) - \mu$. We have $v = c$, $u = a + b \cdot q$.

In the FSN, there are competitive relationships between banks. There are rescue relationships between each bank and the central bank because the central rescue of bank is the last resort to financial crises. If a bank chooses one of deposit insurance, then there are insurance relationships between them.

3   Evolutionary Analysis on FSN

3.1   The Game Equilibrium

We first analyze the mixed strategies of the public goods game in the FSN. Denote the action space for each bank by $S = \{0, 1\}$. Let action (1) and action (0) denote a bank chooses to insure or not. Let p_i denote the mixed strategy of bank i, $0 \leq p_i \leq 1$. It does not choose to insure with probability p_i, and chooses to insure with probability $(1 - p_i)$. Denote the mixed strategies combination of n banks by $p = (p_1, p_2, ..., p_n)$. Then, the expected value of uninsured assets of bank i is $p_i \cdot A_i$, and the expected value of uninsured assets of all banks is $\sum_i p_i \cdot A_i$. Thus, the total amount of risk assets H is a function of p.

$$H(p) = \sum_{i=1}^{n} p_i \cdot A_i \tag{5}$$

Together with Eq. (3), we have

$$q(p) = 1 - \left(\frac{1}{W} \cdot \sum_{i=1}^{n} p_i \cdot A_i \right)^e \tag{6}$$

When the mixed strategies combination p is chosen, denote the expected benefits rate of bank i by r_i.

$$r_i(p) = p_i \cdot (a + b \cdot q(p)) + (1 - p_i) \cdot c \tag{7}$$

Denote the expected benefits of bank i by F_i, and its function is

$$F_i(p) = r_i(p) \cdot A_i \tag{8}$$

From Nash existence theorem, Nash equilibrium of this public goods game exists.

Furthermore, we explain below that this bank game is also a potential game. Potential game obeys a potential function, which is an objective function of common benefits for all game participants. Potential game processes have some nice properties. Pure equilibrium always exists. After a finite number of iterations of participants' strategies, the optimal solution of the objective function can be found [18]. Potential game is widely used in network model for real-world problems. Here we introduce the definition of a potential game.

Definition 1 [18]. A game is a potential game, if there exists a continuous differentiable scalar function, which gradient is equal to a vector of this game's utility function, namely, $\partial \Phi(p)/\partial p_i = \partial F_i(p)/\partial p_i$, where p_i is the strategy of participant i, p is the strategy combination of all participants, $F_i(p)$ is the utility function of participant i, $\Phi(p)$ is called the potential function of this game.

Referring to construction method of potential function [19], we give our theorems as below and their proofs in Appendix.

Theorem 1. The bank game is a potential game in the FSN, and Eq. (9) is the potential function of this game.

$$\Phi(p) = \int_0^H u(x)dx + (W - H) \cdot v \tag{9}$$

There is another important property of potential game. The strategies are Nash equilibrium when potential function reaches extreme values [20]. Thus, the equilibrium strategies of our bank game process the following property.

Theorem 2. Let $p^* = (p_1^*, p_2^*, \cdots, p_n^*)$ be a Nash equilibrium of the bank game in the FSN, and Eq. (10) holds.

$$\sum_{i=1}^{n} p_i^* \cdot A_i = ((a + b - c)/b)^{1/e} \cdot W \tag{10}$$

3.2 Computing of Effectiveness

Since the game equilibrium is not the optimal solution of the system, there is the most popular measure of the inefficiency of equilibrium, namely, the *Price of Anarchy* (PoA), which is defined as the ratio between the worst objective function value of an equilibrium of the game and that of an optimal outcome [20]. Therefore, we also use the PoA to measure the effectiveness of the FSN.

By deposits and loans, banks obtain benefits while promoting the development of the real economy. Thus, we define that, the social welfare of the FSN is the sum of all commercial banks' benefits. Denote the social welfare function of the FSN is

$$R(p) = \sum_{i=1}^{n} F_i(p) \tag{11}$$

Denote the effectiveness value by I, and we define

$$I = R(p^*) / R(p^{OPT}) \tag{12}$$

where p^* is the Nash equilibrium strategies of the game, $p^{OPT} = \arg\max_p R(p)$. We give our theorem for computing the effectiveness as below and its proof in Appendix.

Theorem 3. The formula of the effectiveness value of the FSN is Eq. (13)

$$I = c \cdot \left(\left((a+b-c)^2 \cdot m \big/ 2b \cdot (m+1) \right) + c \right)^{-1} \tag{13}$$

3.3 Evolution of Strategies

When making decisions, banks do not know the benefits function of other banks, and they can not compute the exact equilibrium strategies. However, since the operation of banks is a long-term process, the behavior of making decisions is also a dynamic tuning process. Therefore, in order to depict such evolution of banks' strategies, we propose updating rules for banks seeking optimum decisions as follow.

We set an accounting period as a time unit, and analyze the evolution of banks insurance strategies. Denote a strategies combination in the k-th accounting period by p [k], and $p[k] = (p_1[k], p_2[k], \ldots, p_n[k])$. Banks do settlements at the end each period, and know the benefits rates $u[k]$ and $v[k]$. Then, the strategy of bank i in the $(k + 1)$-th period is adjusted according to the following updating rules.

$$p_i[k+1] = p_i[k] + l[k] \cdot d[k] \cdot p_i[k] \cdot (1 - p_i[k]) \tag{14}$$

where $l[k]$ is the learning rate of banks on surrounding environment in the k-th period, d [k] is the difference between two benefits rates, namely, $d[k] = (u[k] - v[k])/(a + b - c)$. From Eq. (14), we know that strategies of current period only depend on strategies and results of last period.

Sastry et al. [21] have proved that the discrete probability sequence $p_i[k]$ in Eq. (14) converges to $p_i(t)$ for any i when $l[k] \to 0$, where $p_i(t)$ is the solution of the following continuous time ordinary differential equation.

$$\dot{p}_i(t) = \lambda \cdot p_i(t) \cdot (1 - p_i(t)) \cdot [u(p_i(t)) - v(p_i(t))] \qquad (15)$$

By using *Lyapunov* stability method, we give our convergence theorem for Eq. (15). The detailed proof is in Appendix.

Theorem 4. For the bank game in the FSN, p_0 is an initial strategies combination. If $\xi(p_0, t)$ is a solution of ordinary differential Eq. (15), then $\xi(p_0, t)$ converges towards Nash equilibrium and each bank's benefits rate is equal as $t \to \infty$.

However, because the strategy of bank is a continuous value, it is impossible to reach the accurate strategy of Nash equilibrium according to the discrete Eq. (14). Therefore, we propose a concept of approximate equilibrium.

Definition 2. $\tilde{p}$ is a strategy combination of the bank game in the FSN. Let $I_\delta = \{i \in \{1, \ldots, n\} | r_i(\tilde{p}) < (1 - \delta) \cdot \bar{r}\}$, $A_\delta = \sum_{i \in I_\delta} A_i$. The strategy combination $\tilde{p}$ is $\varepsilon - \delta$ approximate equilibrium iff $A_\delta / W \le \varepsilon$.

4   Evolutionary Computation

In this section we employ an evolutionary computation method to compute the approximate equilibrium of our bank game.

Updating rules of banks' decision behavior, Eq. (14), is a class of dynamic model in evolutionary game theory. And our bank game is a potential game from Theorem 1. Therefore, we can employ a best response algorithm, which always converge to Nash equilibrium [21]. More specially, the strategy of each bank is computed in each period in parallel. The adjustment direction of strategies is close to the exact solution in next period. Then, such computation is repeated until conditions of approximate equilibrium are reached. The main steps of our algorithm are shown in Table 1.

The control parameter $l[k]$ in Eq. (14) is the learning rate of banks on surrounding environment. Because operation principles of banks are smooth and orderly development, the adjustment of insurance strategies does not have large fluctuations. Through experiments, we observe that it can ensure $p[k]$ orderly convergence to equilibrium by setting $l[k] = \bar{r}[k]$, where $\bar{r}[k]$ is the average benefits rate in the k-th period. We give our theorem for computing time of this algorithm as below and its proof in Appendix.

Theorem 5. According to the algorithm of Table 1, banks' strategies converge toward the $\varepsilon - \delta$ approximate equilibrium by iterative computation. The number of iterations is bounded by

$$\mathcal{O}\left(\ln(R_{\max}/R_{\min})\big/\varepsilon\delta^2\right)$$

where $R_{\min}$ and $R_{\max}$ are the minimum and maximum of social welfare, respectively.

5 Experiments

In this section, we use computational experiment to simulate the operation of the FSN. By using the real bank asset data as experimental parameters, we employ the algorithm in Table 1 to compute the evolution process of banks' insurance strategies. Moreover, we compute the effectiveness of the FSN, and analyze the relevance between the effectiveness and financial indicators.

We run experiments on a PC with Windows operating system, which is configured as a 4G memory and a 1.6 GHz CPU.

The input of experiments is a FSN. We use assets data of 238 Chinese banks in 2014 [22]. The total asset size $W = 1550419$ hundred million. According to uniform distribution on the interval [0, 1], random numbers are selected as an initial strategy of each bank. In addition, there is only one deposit insurance institution in the FSN. We also use financial indicators as experimental parameters. Referring to banks non-performing loans ratio of 1.75 %, set $\theta = 0.9825$ (the average success rate of investment). Referring to year bank lending rate of 4.75 %, set $\pi = 0.0475$ (the average benefits rate of investment). Under interest rate market environment, there are two kinds of deposit rates. One is in guarantee of deposit insurance institution, and another is out of guarantee. Referring to year bank fixed deposit interest rate of 1.5 %, set $\alpha = 0.015$ (the deposit rate in guarantee). Referring to floating benchmark interest rate 1.1 times, set $\beta = 0.0165$ (the deposit rate out guarantee). Set $\mu = 0.2$ % (the deposit insurance premium rate). Set $e = 1.2$ (the central bank's risk aversion degree).

The output of experiments is the evolution process of the banks' insurance strategies and their benefits.

We set $\varepsilon = 0.1$ and $\delta = 0.05$ as conditions of experiment termination.

Table 1. The evolutionary algorithm

Procedure Appro_Equilibrium

Initialize $p_i[0]$ for each bank i, and $k=0$

while (stopping criterion is not satisfied)

    Compute: $u[k]$, $v[k]$, $d[k]$ and $l[k]$

    in_parallel on each bank i

        $p_i[k+1] \leftarrow p_i[k] + l[k] \cdot d[k] \cdot p_i[k] \cdot (1 - p_i[k])$

    end in_parallel

    $k \leftarrow k + 1$

    Compute stopping criterion with $p[k]$

end while

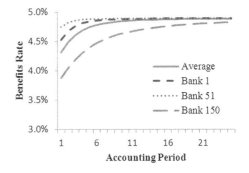

Fig. 2. Evolution of banks' benefits rate

5.1   Convergence of Strategies' Evolution

The purpose of this experiment is to simulate the evolution of banks' strategies by computing benefits rate and decisions of each bank in each period.

The experiment results are shown in Fig. 2. After computing in 25 iterations, termination condition is reached. It illustrates the convergence of our algorithm, which matches Theorem 4. Further, the number of iterations is bounded, which matches Theorem 5. Secondly, lines show changes of banks' benefits rate, where we select bank 1, bank 51 and bank 151, respectively. Benefits rates are orderly increasing with period, and there are no fluctuations. It illustrates that control parameters of the algorithm are reasonably set. In addition, when the approximate equilibrium state is reached in the last period, banks' benefits rates are close to the average rate, which matches Theorem 4.

5.2   Computing of the Effectiveness of the FSN

The effectiveness of the FSN can be directly computed from experimental results. Below we set different financial indicators and execute repeatedly experiments. By changing input parameters, we analyze the relevance between the effectiveness and indicators.

Figure 3 shows the effectiveness value based on 6 experiments. By setting different deposit insurance premium rates in each experiment, we observe that there is a negative correlation between the deposit insurance premium and the effectiveness of the FSN, which matches Eq. (13).

Figure 4 shows the calculated effectiveness value based on other 6 experiments. By setting different the central bank's risk aversion degree in each experiment, we observe that there is also a negative correlation between the central bank's risk aversion and the effectiveness of the FSN, which also matches Eq. (13).

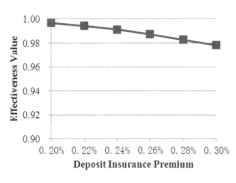 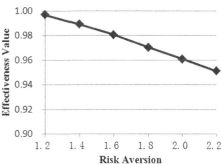

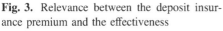

Fig. 3. Relevance between the deposit insurance premium and the effectiveness

Fig. 4. Relevance between the central bank's risk aversion and the effectiveness

6 Conclusion

This paper develops a network game model to describe the FSN. Based on evolutionary game theory, updating rules of bank insurance strategies are proposed. Such analysis can help to really understand the operation of the FSN. We verify the convergence of the evolution of banks' strategies by theoretical proof and computational experiments. In addition, we give a measure method for the effectiveness of the FSN. Such an approach can provide suggestions of macro policy for regulators.

For the research on the optimal FSN, there are still many problems to answer in the future. In this paper, we define that social welfare is the sum of all commercial banks' benefit. In fact we should also consider rescue costs of the central bank and payment of deposit insurance institutions. But this makes the problem more difficult due to more optimization objective. In addition, we do not consider dynamic changes of external factors, such as deposit interest rates and bad debt rate. How these factors synchronously affect the FSN, is also a difficult problem.

Acknowledgments. This paper is supported by the National Natural Science Foundation of China (Nos.61562091), Natural Science Foundation of Yunnan Province (Nos. 2014FA023), and the Research Foundation of the Education Department of Yunnan Province (Nos. 2014C134Y, 2016ZZX013).

Appendix

We give proofs of theorems in this paper as below.

Theorem 1: We understand that the bank game is a routing game for funds seeking a better investment path. There are 2 alternative paths, and the funds amounts are H and $(W - H)$, respectively. Therefore, the potential function of the bank game is

$$\Phi(p) = \int_0^H u(x)dx + \int_0^{W-H} v(x)dx = \int_0^H u(x)dx + (W - H) \cdot v$$

∎

Theorem 2: Since strategies are Nash equilibrium when potential function reaches extreme values, compute its first and second order derivative of Φ.

$$\Phi'(H) = u(H) - v = a + b - c - b \cdot (H/W)^e, \qquad \Phi''(H) = -b \cdot H^{e-1} \cdot W^{-e} < 0.$$

Thus, Φ reaches the maximum when $\Phi'(H) = 0$. And due to $H = \sum_i p_i \cdot A_i$, we get this theorem. ∎

Theorem 3: Compute the first and second order derivative of social welfare function.
$R(H) = u(H) \cdot H + v \cdot (W - H) = -b \cdot H \cdot (H/W)^e + (a + b - c) \cdot H + c \cdot W$,   then
$R'(H) = (a + b - c) - b \cdot (e + 1) \cdot (H/W)^e$,
$R''(H) = -b \cdot e \cdot (e + 1) \cdot H^{-1} \cdot (H/W)^e < 0$, $R(H)$ reaches the maximum when
$R'(H) = 0$. Then, $R^{OPT} = \left(\left((a + b - c)^2 \cdot m / 2b \cdot (m + 1) \right) + c \right) \cdot W$. In addition,

$R(p^*) = c \cdot W$ from Theorem 2, together with the Eq. (12), we finally get the formula of this theorem.   ∎

Theorem 4: Let *Lyapunov* function be $L(p) = \Phi^* - \Phi(p)$, where $\Phi(p)$ is given by Eq. (9), Φ^* is the maximum of $\Phi(p)$. Thus, $L(p)$ is positive definite.

$$\dot{L}(p) = -\dot{\Phi}(p) = -(u(p) - v) \cdot \dot{H} = -(u(p) - v) \cdot \sum_i A_i \cdot \dot{p}_i(t)$$

$$= -(u(p) - v) \cdot \sum_i A_i \cdot \lambda(t) \cdot p_i(t) \cdot (1 - p_i(t)) \cdot (u(p) - v)$$

$$= -(u(p) - v)^2 \cdot \sum_i A_i \cdot \lambda(t) \cdot p_i(t) \cdot (1 - p_i(t)) \leq 0$$

According to *Lyapunov* stability theorem, the solution $\xi(p_0, t)$ of Eq. (15) is asymptotic stable. Iff $u(p) = v$, $\dot{L}(p) = 0$ holds. Therefore, $\xi(p_0, t)$ converges to the Nash equilibrium as $t \to \infty$.   ∎

Theorem 5: Let $\Psi(p(t)) = \Phi(p(t)) + R_{\min}$, and its first order derivative with respect time t of Ψ is $\dot{\Psi}(p(t)) = \dot{\Phi}(p(t)) \geq 0$. Iff $u(p) = v$, it holds $\dot{\Psi} = 0$, and $r_i = \bar{r}$ for each i.

According to Definition 2, if $p(t)$ is not $\varepsilon - \delta$ equilibrium, then, it at least has $\varepsilon \cdot W$ part of funds amount, and its benefits rate is $(1 - \delta) \cdot \bar{r}$ at most. We use *Jensen* inequality in the following case. Suppose: the benefits rate of $\varepsilon \cdot W$ part is equal to $(1 - \delta) \cdot \bar{r}$, the rate of the rest $(1 - \varepsilon) \cdot W$ part is equal to $\hat{r}$. Therefore, $\bar{r}$ is given as below.

$$\bar{r} = \varepsilon \cdot (1 - \delta)\bar{r} + (1 - \varepsilon) \cdot \hat{r} \tag{16}$$

Jensen inequality is employed again, we have:

$$\dot{\Psi}(p(t)) \geq \lambda \left(\varepsilon W \cdot ((1 - \delta)\bar{r})^2 + (1 - \varepsilon)W \cdot \hat{r}^2 - W \cdot \bar{r}^2 \right) \tag{17}$$

We get $\hat{r} = \bar{r} \cdot (1 - \varepsilon + \varepsilon\delta)/1 - \varepsilon$ from Eq. (16), and substitute it into Eq. (17), then

$$\dot{\Psi}(p(t)) \geq \lambda \varepsilon W \delta^2 \bar{r}^2 / (1 - \varepsilon) = \varepsilon W \delta^2 \bar{r} / (1 - \varepsilon) \geq \varepsilon W \delta^2 \bar{r} = \varepsilon \delta^2 R$$

Next, $\Psi = \Phi + R_{\min} \leq (2e + 3)R/(e + 1)$, Substitute it into above inequality, we have

$$\dot{\Psi}(p(t)) \geq \varepsilon \delta^2 \cdot \Psi(p(t)) \cdot (e + 1)/(2e + 3)$$

Thus, if $p(t)$ is not $\varepsilon - \delta$ equilibrium in period $(t + \Delta t)$, it holds that

$$\Psi(p(t + \Delta t)) \geq \Psi(p(t)) \cdot \exp\left(\varepsilon \delta^2 \cdot (e + 1)/(2e + 3) \cdot \Delta t\right)$$

These periods include $(t_0 + \Delta t_0)$, $(t_1 + \Delta t_1)$, ..., $(t_m + \Delta t_m)$, ...

Let $T = \sum_m \Delta t_m$, When m goes to infinity, T includes the total periods when p does not reach equilibrium, we have

$$exp\left(\frac{e+1}{2e+3}\varepsilon\delta^2 \cdot T\right) \leq \frac{\Psi(p(t_i + \Delta t_i))}{\Psi(p(t_0))} \leq \frac{\Psi_{max}}{\Psi_{min}} \leq \frac{\Phi_{max} + R_{max}}{R_{min}} \leq \frac{2e+3}{e+1} \cdot \frac{R_{max}}{R_{min}}$$

$$T \leq (e+1)\ln((2e+3)R_{max}/(e+1)R_{min})/(2e+3)\varepsilon\delta^2 \sim \mathcal{O}\left(\ln(R_{max}/R_{min})/\varepsilon\delta^2\right)$$

∎

References

1. Schich, S.: Financial crisis: deposit insurance and related financial safety net aspects. OECD J.: Finan. Mark. Trends **2008**, 1–39 (2009)
2. Pan, I., Das, S., Das, S.: Multi-objective active control policy design for commensurate and incommensurate fractional order chaotic financial systems. Appl. Math. Model. **39**, 500–514 (2015)
3. Gan, L., Wang, G.W.Y.: Partial deposit insurance and moral hazard in banking. Int. J. Commer. Manag. **23**, 8–23 (2013)
4. Fischer, T.: News reaction in financial markets within a behavioral finance model with heterogeneous agents. Algorithmic Finan. **1**(2), 123–139 (2011)
5. An, H., Gu, Y., Zhong, H.: The study of effectiveness of U.S. financial regulation based on the financial market efficiency. In: International Conference on Artificial Intelligence, Management Science and Electronic Commerce, pp. 843–846 (2011)
6. Freixas, X.: Systemic risk and prudential regulation in the global economy. In: Globalization and Systemic Risk, pp. 145–167. World Scientific (2012)
7. Wang, X., Li, L., Feng, J.: A model build and empirical test on efficiency evaluation about deposit insurance system in China based on variance decomposition. Inf. Technol. J. **12**, 2708–2711 (2013)
8. Jyh-Horng, L., Chieh, W.-H., Wang, C.-C.: Actuarially fair deposit insurance premium during a financial crisis: a barrier-capped barrier option framework. Int. J. Innovative Comput. Inf. Control **10**, 2067–2085 (2014)
9. Blum, A., Morgenstern, J., Sharma, A., Smith, A.: Privacy-preserving public information for sequential games. In: Proceedings of the 2015 Conference on Innovations in Theoretical Computer Science, pp. 173–180. ACM, Rehovot (2015)
10. Milano, M., O'Sullivan, B., Gavanelli, M.: Sustainable policy making: a strategic challenge for artificial intelligence. Ai Mag. **35**, 22–35 (2014)
11. Aguilar-Rivera, R., Valenzuela-Rendón, M., Rodríguez-Ortiz, J.J.: Genetic algorithms and Darwinian approaches in financial applications: a survey. Expert Syst. Appl. **42**, 7684–7697 (2015)
12. Gupta, P., Mehlawat, M.K., Mittal, G.: Asset portfolio optimization using support vector machines and real-coded genetic algorithm. J. Glob. Optim. **53**, 297–315 (2012)
13. Lu, J.: Data modeling for searching abnormal noise in stock market based on genetic algorithm. In: International Symposium on Computational Intelligence and Design, pp. 129–131 (2010)
14. Ganghishetti, P., Vadlamani, R.: Association rule mining via evolutionary multi-objective optimization. In: Murty, M.N., He, X., Chillarige, R.R., Weng, P. (eds.) MIWAI 2014. LNCS (LNAI), vol. 8875, pp. 35–46. Springer, Heidelberg (2014). doi:10.1007/978-3-319-13365-2_4

15. Ponsich, A., Jaimes, A.L., Coello, C.A.C.: A survey on multiobjective evolutionary algorithms for the solution of the portfolio optimization problem and other finance and economics applications. IEEE Trans. Evol. Comput. **17**, 321–344 (2013)
16. Li, H., Wu, C., Yuan, M.: An evolutionary game model of financial markets with heterogeneous players. Procedia Comput. Sci. **17**, 958–964 (2013)
17. Caporale, G.M., Serguieva, A., Wu, H.: A mixed-game agent-based model for simulating financial contagion. In: IEEE Congress on Evolutionary Computation, CEC 2008, 1–6 June 2008, Hong Kong, China, pp. 3421–3426 (2008)
18. Monderer, D., Shapley, L.S.: Potential games. Games Econ. Behav. **14**, 124–143 (1996)
19. Fischer, S., Vöcking, B.: On the evolution of selfish routing. In: Albers, S., Radzik, T. (eds.) ESA 2004. LNCS, vol. 3221, pp. 323–334. Springer, Heidelberg (2004). doi:10.1007/978-3-540-30140-0_30
20. Roughgarden, T.: Potential functions and the inefficiency of equilibria. In: Proceedings of the International Congress of Mathematicians (ICM), pp. 1071–1094 (2006)
21. Sastry, P.S., Phansalkar, V.V., Thathachar, M.: Decentralized learning of Nash equilibria in multi-person stochastic games with incomplete information. IEEE Trans. Syst. Man Cybern. **24**, 769–777 (1994)
22. NetEase Finance. http://money.163.com/special/2014f500bank/

Short Papers

Learning to Navigate in a 3D Environment

Nurulhidayati Haji Mohd Sani[1(✉)], Somnuk Phon-Amnuaisuk[1,2],
Thien Wan Au[1], and Ee Leng Tan[3]

[1] School of Computing and Informatics, Universiti Teknologi Brunei,
Jalan Tungku Link, Gadong BE1410, Brunei Darussalam
nurulhidayati.sani@gmail.com
[2] Centre for Innovative Engineering, Universiti Teknologi Brunei,
Jalan Tungku Link, Gadong BE1410, Brunei Darussalam
[3] Sesame World Technology, Berakas, Brunei Darussalam

Abstract. In this paper, we investigate the knowledge acquisition and
the learning ability of an agent in a three-dimensional (3D) environment
using data mining techniques. We apply three data mining techniques:
naïve Bayes, decision tree and apriori; to a human-controlled navigation
and then investigate the characteristic of knowledge discovered from each
of these techniques. The results shows that the agent is able to learn to
navigate automatically in the environment but with different outcomes
and limitations.

Keywords: Navigation · Machine learning · Apriori · Decision tree ·
Naïve bayes

1 Introduction

Our work is motivated by the goal of building a machine capable of learning to
automatically navigate in an unknown environment. A self-navigating machine,
can be useful, especially in a hazardous situation for example. This paper investigates the knowledge acquisition and learning abilities of a goal-directed agent
using data mining techniques. We model an agent that is able to automatically
navigate its way in a 3D environment using knowledge discovered by mining
human-controlled navigation dataset.

Initially, humans are responsible in teaching the agent on how to navigate
through the environment by controlling and selecting the best action for the
agent in respect to the state of the environment. The players actions that navi-
gates the agent to the goal are recorded over 4090 runs. This forms a substantial
dataset that we apply data mining (DM) techniques (naïve Bayes [1], decision
tree [2], apriori [3]) and examine emerging patterns which will formulate the
knowledge for the agent. This paper investigates (i) whether the agent can navigate in a new environment using knowledge learned from the human-controlled
navigation dataset, and (ii) the nature of the agents performance based on these
three knowledge discovery techniques.

The rest of the paper is organized as follows. In Sect. 2, we present the related
works. Section 3 defines and formulates the problem and provides an overview

© Springer International Publishing AG 2016
C. Sombattheera et al. (Eds.): MIWAI 2016, LNAI 10053, pp. 271–278, 2016.
DOI: 10.1007/978-3-319-49397-8_23

of the environment and agent. Section 4 outlines the experimental design. Our experimental results and discussions are reported in Sect. 5. Finally, we conclude our findings and propose our future work in Sect. 6.

2   Related Works

A self-navigated agent must be able to learn from their experiences; how to avoid obstacles and to find the optimal path to its destination. Many intelligent computing techniques have been investigated e.g., reinforcement learning [4], Dempster-Shafers theory of evidence [5], fuzzy logic rules [5], etc.

There are many types of techniques and approaches involved in machine learning that have been studied by researchers in the field of Artificial Intelligence (AI). Inclusively, many of these works have shown that knowledge can be learned from observed data. A previous work used history replays of Real-Time Strategy (RTS) games as the dataset to learn the gamers behavior [6]. In addition to learning behavior, Derezynski et al. [7] identified future actions as well as simulate their possible future actions and/or identified novel strategies based on the dataset obtained from logs of past games. The study [8] that built a probabilistic model that uses the historic behavior of gamers in a commercial social videogame as their dataset and provides a real-time estimation of next action expected. Although these works offer useful information to the research fields, few others modeled this knowledge into an automated agent capable of proficiency gaming. Such study is made by Weber et al. [9] where he developed an autonomous agent using a Goal-Driven Autonomy (GDA) model that can learn to perform tasks based on the demonstration in a RTS game. Similarly, Gemine et al. [10] used set of recorded games and applied supervised learning to teach an agent to learn building-production strategies. More recently, the application of Artificial Neural Network (ANN) in learning to play a Tetris game is investigated [11]. The ANN is also exploited in [12] to create an autonomous and adaptive first person shooter agent that uses reinforcement learning based on an observed environment.

One method of extracting knowledge is DM [13]. Knowledge can be described as a relationship between stimuli and response, e.g., a proposition describing antecedent and consequence. Many data mining techniques can find patterns in this shape. For example, a study is made where the state of filled and unfilled Tetris board was used as the condition to decide the actions of where to place the next tetromino in a Tetris gameplay [14]. Through several gameplays, and once enough data is collected, apriori algorithm is applied to mine stimuli and response patterns. While this technique has been shown to work extremely well in applications, the resulting predicted outcome relies entirely on the knowledge discovered based on previous selected actions performed which in turn yields a single outcome. The resulting outcome in [6,7], however, uses probabilistic models which varies based on the possibility of occurrences and provides a variety of selections which in turn might give a better performance to the application.

In this work, we consider the agent to learn to navigate based on knowledge and probabilistic approaches and to investigate the differences of these approaches in a new environment.

3 Problem Formulation

In this section, we discuss the components used to formulate the problem described. The environment used in this study is formulated first. This is followed by an explanation of the agent used to navigate the environment.

3.1 3D Virtual Environment

The environment setup for this experiment is a real world human setting modeled in a 3D virtual world representation using a Unity3D application. The environment is a virtual 3D space, populated with m game objects O_m. An object may be an agent A, a goal G, or obstacle such as wall W. The agent would navigate its way to the goal while avoiding obstacles. For every state, the agent is able to perform k actions, which in this case is a movement of one unit in any of the eight directions (N, NE, E, SE, S, SW, W, NW) in the environment but cannot pass through obstacles. The game ends when the agent reaches the goal.

3.2 Agent

The agent is modeled similarly to a human, that can perceive its environment through sensors and react accordingly through actions [15]. In this experiment, the agent is assumed to be able to: (i) analyse the environment through perceptors; (ii) acquire knowledge through learning (iii) perform actions based on the agent's knowledge.

4 Experimental Design

Two experiments will be carried out for this study as outlined in the systems architecture (see Fig. 1). In the first experiment, we (human) manually choose the actions and control the agent based on the current state of the environment. During this experiment, data is being collected for analysis. The collected data is analysed using DM techniques to obtain the knowledge for the agent. The

Fig. 1. Learning to navigate in a 3D environment system architecture

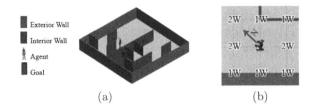

(a)   (b)

Fig. 2. (a) The outline of the environment: An environment, an agent, a goal and obstacles (exterior and interior walls); (b) Recording the environment as a data, taken from the agents north direction. The data reads: 2W, 1W, 1W, 2W, 1W, 1W, 1W, 2W;

result based on the first experiment should provide the basic knowledge for the agent. In the second experiment, the agent applies the knowledge learned through interacting with the environment. Figure 2(a) shows the outline of the environment (top view). The area of the environment is a 3D space with a size of $10 \times 10 \times 2$. For every new gameplay, the exterior 20 walls are positioned stationery to form a squared-shaped room with five walls each side. One goal, one agent and 15 interior walls are randomly positioned in the environment are randomly positioned within the squared room. If a goal is reached, the position of the walls remains and while the agent and the goal is relocated randomly. This state can be repeated number of times until the player decides to end or play a new game. This parameter setup is shown in Table 1.

Table 1. Parameter settings for the experiment

Parameter settings	Values	Remarks
Grid size	$10 \times 10 \times 2$	Default Unity3D unit
Number of agent, A	1	
Number of goal, G	1	
Number of walls, W	35	20 Walls surrounding the environment 15 randomly generated walls
Number of actions, k	8	Each action is a movement to one of the eight directions (N, NE, E, SE, S, SW, W, NW)

Below are summary of the process involved in the knowledge acquisition stage which focuses on obtaining the knowledge to train the agent outlined in Fig. 3(a):

i Agent percepts the environment using its sensors.
ii Human selects and controls the actions performed by the agent using one of the directional buttons (*up*, *left* or *right*) based on the state of the environment. For example, if the *right* key is pressed once, the agent will face the *NE* direction, twice *E* direction, thrice *SE* direction, etc. But data will only be captured when the forward (*up*) button is pressed.
iii The selected actions are recorded along with the state of the environment.

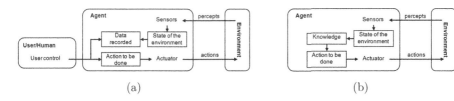

Fig. 3. (a) A Human Control Agent. It performs actions according to human control. State of environment and Human/user control information is recorded to be analysed for knowledge discovery; (b) A Knowledge-based Agent. Agent uses state of environment matched with policies to select which action to perform in environment.

The recorded data is analysed using DM techniques. This experiment is done separately. Patterns emerging from the analysis is extracted as a knowledge for the agent to perform in a new environment.

The knowledge application experiment (outlined in Fig. 3(b)) requires the agent to use the knowledge learned in a new environment without human intervention. This experiment was repeated three times, one for each of the three different DM techniques. Below summarizes the processes in the experiment:

i Agent percepts the environment using its sensors.
ii The agent matches the percepted environment to the knowledge, and uses the given projected outcome to perform its actions.

4.1 Recorded Data

We decided to record (i) the environment as viewed by the agent in 360 degrees around him with division analysis according to the cardinal and ordinal directions of a compass (i.e. eight basic directions in a compass), that consist of one of the following data: $1W$, $2W$, $1G$, $2G$ (where 1 = near, 2 = far, W = wall and G = goal); and (ii) the k action taken by the agent.

4.2 Association and Classification Rule Mining

The following DM techniques were applied to the collected data for analysis:

i Association: Apriori
ii Classification: J48 C4.5 decision tree and naïve Bayes

Weka [13] explorer toolkit is implemented to perform these DM analysis. A ten-fold cross validation is used in our classification techniques and the confidence level of 0.75 is set for our association technique while all other options are set to default settings. The algorithms were conducted at different times throughout the study. The Weka simulation is conducted separately and all the three DM techniques are generated using the same dataset.

4.3   Evaluation Criteria

The evaluation was conducted by observing the agents performance based on two parameters: (i) the number of times the agent avoids an obstacle W; and (ii) The number of times the agent A approaches goal G if it is visible.

5   Results and Discussions

A reasonable pattern emerged from all the three DM techniques from a total of 4090 recorded training instances from 40 game sessions. In order to apply each of these knowledge to the agent, it requires an interpretation of the results. Some of the raw results as well as the translated results are summarized in Table 2.

Table 2. Sample of analysed raw and interpreted rule for apriori, decision tree and naive bayes

Technique	Analysis Result	Interpreted Result
Apriori	• $angle\_N = 2G$ 518 $==>$ $action = N$ 518 $conf : (1)$ • $angle\_N = 1G$ $angle\_S = 2W$ 513 $==>$ $action = N$ 513 $conf : (1)$	• IF $angle\_N == 2G$ THEN $action = N$ • IF $angle\_N == 1G$ AND $angle\_S = 2W$ THEN $action = N$
Decision Tree	$angle\_N = 1W$ | $angle\_NE = 1W$ || $angle\_NW = 1W$ ||| $angle\_E = 1W$ |||| $angle\_SE = 1W$ ||||| $angle\_W = 1W$: $S(13.0/1.0)$ ||||| $angle\_W = 2W$ |||||| $angle\_S = 1W$: $W(9.0)$	IF $angle\_N == 1W$ IF $angle\_NE == 1W$ IF $angle\_NW == 1W$ IF $angle\_E == 1W$ IF $angle\_SE == 1W$ IF $angle\_W == 1W$ THEN $action = S$ ELSE IF $angle\_W == 2W$ IF $angle\_S == 1W$ THEN $action = 2W$
Naive Bayes	$angle\_N$ 1W 12.0 64.0 91.0 1.0 109.0 55.0 16.0 29.0 2W 27.0 84.0 298.0 644.0 243.0 62.0 28.0 48.0 1G 1.0 1.0 1.0 213.0 1.0 1.0 1.0 1.0 2G 1.0 1.0 1.0 283.0 1.0 1.0 1.0 1.0 [total] 41.0 150.0 391.0 1141.0 354.0 119.0 46.0 79.0	

Apriori and decision tree algorithms directly translate the results into knowledge. Naïve Bayes, however provides a more probabilistic approach, which requires us to use a random variable R to determine the actions to be performed for the agent. For example, suppose the analysed environment for N, NE, E, SE, S, SW, W, NW are $2W$, $1W$, $2W$, $1W$, $1W$, $1W$, $1W$, $1W$, respectively. The calculated probability and cumulative probability for each of the possible outcome is sampled as illustrated in Table 3 and we decide that the order of these

Table 3. Sample probability computed for the agent to select an action

Action	N	NE	E	SE	S	SW	W	NW
Probability	0.525	0.000	0.431	0.005	0.031	0.001	0.006	0.001
Cumulative probability	0.525	0.526	0.957	0.962	0.993	0.993	0.999	1

Table 4. Summary of experimental results

Technique	Criteria	Correctly performed	Incorrectly performed
Tree diagram	(i)	654 (100 %)	0 (0 %)
	(ii)	7 (100 %)	0 (0 %)
Nave Bayes	(i)	611 (98.5 %)	9 (1.5 %)
	(ii)	66 (93 %)	5 (7 %)

actions are stationary. If R, lets assume, equals to 0.61, based on the cumulative probability, then an action $k = E$ is selected.

Although the results given by apriori algorithm shows a reasonable knowledge to the agent, the limitation to this is that the knowledge is incomplete. Some of the environment percepted by the agent are unknown which results in action k to be *null*. Because of this, experiment to run the simulator is impracticable.

Table 4 shows summary of experiments. In a sample of 661 instances that has been recorded to run the interpreted result of decision tree, 654 (100 %) actions are performed as expected with 0 % or no incorrect performed action made. Although it shows an excellent result, the problem with this technique is that the agent will only perform the action stated repeatedly when logically, there are multiple choices. This behavior sometimes lead the agent into going in circles and end up in a loop. The seven actions stated in criteria (ii) where it found the goal most likely happened because the goal is already nearby when the game started.

The agent also performed rather well (>90 %) by using naïve Bayes technique with a collected sample of 691 instances. The 1.5 % and 7 % incorrect action is, by any chance, selected because although the probability of that action is close to 0, there is still a possibility for that action to be selected. The advantage of using this technique is that it gives more flexibility on the action chosen compared to that of apriori and decision tree but the drawback is that some of the actions are performed incorrectly.

6   Conclusions and Future Work

In this paper, we demonstrated a DM approach to study the learning ability of an agent to navigate in a 3D Environment. We collected a total of 4090 data and implemented apriori, decision tree and naïve Bayes by using Weka toolkit as the techniques used to acquire knowledge for the agent based on human controlling navigation dataset. Our result shows the agent can learn to navigate in the 3D Environment through this knowledge but with a few limitations.

As DM techniques allows knowledge discovery from dataset, possible future work is to include more actions to improve the immersion of the 3D environment and knowledge obtained by the agent can be further improvised, either by incorporating two or three techniques or by inserting extra knowledge to the learned knowledge.

References

1. Witten, I., Frank, E., Hall, M.: Data Mining: Practical Machine Learning Tools and Techniques. Morgan Kaufmann Publishers Inc., San Francisco (2011)
2. Quinlan, J.R.: C4.5: Programs for Machine Learning. Morgan Kaufmann Publishers Inc., San Francisco (1993)
3. Agrawal, R., Srikant, R.: Mining sequential patterns. In: Proceedings of the Eleventh International Conference on Data Engineering, pp. 3–14, March 1995
4. Muhammad, J., Bucak, I.O.: An improved q-learning algorithm for an autonomous mobile robot navigation problem. In: International Conference on Technological Advances in Electrical, Electronics and Computer Engineering (TAEECE) (2013)
5. Jaafar, J., McKenzie, E.: Autonomous virtual agent navigation in virtual environments. Int. J. Comput. Electr. Autom. Control Inf. Eng. **4**, 91–98 (2010)
6. Hsieh, J.L., Sun, C.T.: Building a player strategy model by analyzing replays of real-time strategy games. In: 2008 IEEE International Joint Conference on Neural Networks (IEEE World Congress on Computational Intelligence). IEEE (2008)
7. Dereszynski, E., Hostetler, J., Fern, A., Dietterich, T., Hoang, T.T., Udarbe, M.: Learning probabilistic behavior models in real-time strategy games. In: Proceedings of the Seventh AAAI Conference on Artificial Intelligence and Interactive Digital Entertainment (2011)
8. Baldominos, A., Albacete, E., Marrero, I., Saez, Y.: Real-time prediction of gamers behavior using variable order Markov and big data technology: a case of study. Int. J. Interact. Multimedia Artif. Intell. **3**, 44–51 (2016)
9. Weber, B.G., Mateas, M., Jhala, A.: Learning from demonstration for goal-driven autonomy. In: Proceedings of the Twenty-Sixth AAAI Conference on Artificial Intelligence (2012)
10. Gemine, Q., Safadi, F., Fonteneau, R., Ernst, D.: Imitative learning for real-time strategy games. In: 2012 IEEE Conference on Computational Intelligence and Games (CIG), pp. 424–429. IEEE (2012)
11. Phon-Amnuaisuk, S.: Learning to play tetris from examples. In: Phon-Amnuaisuk, S., Au, T.W. (eds.) Computational Intelligence in Information Systems. AISC, vol. 331, pp. 255–264. Springer, Heidelberg (2015)
12. Wang, D., Tan, A.H.: Creating autonomous adaptive agents in a real-time first-person shooter computer game. In: IEEE Transactions on Computational Intelligence and AI in Games, vol. 7, pp. 123–138. IEEE (2015)
13. Witten, I.H., Frank, E., Hall, M.A.: Data Mining: Practical Machine Learning Tools and Techniques. Morgan Kaufmann, San Francisco (2011)
14. Phon-Amnuaisuk, S.: Evolving and discovering tetris gameplay strategies. Procedia Comput. Sci. **60**, 458–467 (2015)
15. Russell, S., Norvig, P.: Artificial Intelligence: A Modern Approach. Pearson, New York (2010)

Analysis of Similarity/Dissimilarity of DNA Sequences Based on Pulse Coupled Neural Network

Xin Jin[1], Dongming Zhou[1(✉)], Shaowen Yao[2], Rencan Nie[1],
Quan Wang[1], and Kangjian He[1]

[1] Information College, Yunnan University, Kunming 650091, China
zhoudm@ynu.edu.cn
[2] School of Software, Yunnan University, Kunming 650091, China
yaosw@ynu.edu.cn

Abstract. To calculate the similarity or dissimilarity of DNA sequences, a new method is proposed based on pulse coupled neural network (PCNN) model. First, according to the characteristics of PCNN model, we encode DNA primary sequences using a simple coding method. Then we use PCNN model to extract the entropy sequence (ES) of the encoded DNA sequence; the ES expresses the features of the DNA sequences. At last, we calculate the similarity of the ES by Euclidean distance to get the similarity of DNA sequences. We take several sets of data to test our method. The experimental results demonstrate that our method is effective.

Keywords: DNA sequences · Pulse coupled neural network · Euclidean distance

1 Introduction

DNA sequences are composed of four bases over the four-letters alphabet A, C, G, T, which play a significant role in the determination of the functions of various biological structures. However, it is still very difficult to obtain the important information of DNA primary sequences [1]. In DNA similarity analysis field, scholars have proposed many methods to analyze the similarity of this kind of biological sequences [1–5]. In 1983, graphical representations of DNA sequences were firstly initiated by Hamori and Ruskin [1], and then the method was expanded by Nandy and Randic [2, 3]. Since this, many new graphical representation based methods have been proposed by researchers in two-dimension [4, 5] and three-dimension [6, 7]. Except graphical representation based methods, Wavelet transforms and Fourier transform was applied to DNA sequences similarity analysis and hierarchical clustering [8, 9]. Segmented K-mer applied on similarity analysis of mitochondrial genome sequences [10], however, it still did not completely solve the deficiency of the K-mer method in biological sequences analysis, and would lose same information when DNA sequence data is condensed into a vector of K-mer counts.

© Springer International Publishing AG 2016
C. Sombattheera et al. (Eds.): MIWAI 2016, LNAI 10053, pp. 279–287, 2016.
DOI: 10.1007/978-3-319-49397-8_24

In 1993, pulse coupled neural network (PCNN) based on Eckhorn research in cat's visual cortex was proposed [11–13]. It has been widely used in image processing [14, 15], features extraction [16], and other fields [17, 18]. The capture features of PCNN neurons will cause the capture and firing of the similar external neurons, by this way, it can automatically couple and transmit the information to other neurons [18]. Compared with other artificial neural network in the pattern recognition field, PCNN is no need sample to train and has an incomparable advantage over other traditional artificial neural network [11–18]. Our researches in PCNN model reveal that it can be used for extracting the features of DNA sequence, and the features will include not only the structural and functional information of DNA sequences, but also the connection among the bases in DNA sequences [19].

In this paper, we propose a new method of DNA sequences similarity calculation based on PCNN model. First, according to the different permutations of bases of triplet code, we encode DNA primary sequences using simple coding way, then the encoded DNA sequences are normalized into numerical sequences range from 0 to 1, which will make the code suit for PCNN model to extract the entropy sequence (ES) of DNA sequences. At last, Euclidean distance is used to calculate the similarities among ES of different species. In order to examine the validity of the new method, we analyzed the similarity of the first exon sequences of β-globin genes of 11 species. The experimental results show that our method is effective and consistent with the evolutionary relation of the species.

2 DNA Sequences Code

Since most of the methods cannot deal with A, G, C, T alphabet sequences directly, before processing we should recode the DNA sequences to make the letter sequences numeralization. By this way, DNA primary sequences can be converted into one-dimensional numerical sequences. There are many methods for transforming DNA sequences into numerical sequences, such as CMI code analysis [20], codon-based encoding [21]. However, most of the methods may lose the structural and functional information of DNA sequences. Therefore, these methods may not fully reflect the actual differences among DNA sequences.

We encode A, G, C, T by 1, 2, 3, 4, respectively, so that DNA sequences are translated into quaternary sequences, and it will hold all the structural and functional information of DNA sequences. Based on this simple method, four numerical descriptors of DNA sequences are obtained without complicated calculations. After coding, the coded sequences are normalized from 0 to 1, the rule is in accordance with (1). It will be consistent with the range of normalized gray image pixel value, which ranges from 0 to 1, and this range is suitable for PCNN to process DNA sequences code. Besides, the process does not change the numerical characters of coded DNA sequences. After this, we can use PCNN model to extract the features of the coded sequences of DNA primary sequences.

$$C = \frac{I_{\max} - i}{I_{\max} - I_{\min}}, \tag{1}$$

where C is the normalized value, i is the input quaternary code, I_{max} and I_{min} are the maximum and minimum of all input values.

3 DNA Sequences Features Extraction Based on PCNN

3.1 PCNN Model

PCNN model is described by five equations to, as (2)–(6).

$$F_{ij}(n) = V^F \sum_{kl} M_{ijkl} Y_{kl}(n-1) + e^{-\alpha^F} F_{ij}(n-1) + S_{ij}, \tag{2}$$

$$L_{ij}(n) = V^L \sum_{kl} W_{ijkl} Y_{kl}(n-1) + L_{ij}(n-1)e^{-\alpha^L}, \tag{3}$$

$$U_{ij}(n) = F_{ij}(n)[1 + \beta L_{ij}(n)], \tag{4}$$

$$\theta_{ij}(n) = e^{-\alpha^\theta} \theta_{ij}(n-1) + V_{ij}^\theta Y_{ij}(n-1), \tag{5}$$

$$Y_{ij}(n) = \begin{cases} 1, & U_{ij}(n) > \theta_{ij}(n) \\ 0, & otherwise \end{cases}. \tag{6}$$

In above equations, the subscripts i and j represent the neuron location, n denotes the current iteration. The receptive field is composed by L and F channels, and it will receive neighboring neurons' coupling input Y and external stimulus input S, which is described by (2) and (3). In L and F channels of the receptive field, the neuron links with its neighborhood neurons by the synaptic linking weights W and M, respectively, the decay exponentials are α^L and α^F, while the channel amplitudes are V^L and V^F. The total internal activity U is calculated according to L and F, and β is the linking strength of the neuron. Pulse generator is described by (5) and (6), which will generate one pulse would be output by the neuron when the threshold value θ is less than U. α^θ and V_{ij}^θ is decay coefficient and amplification coefficient, respectively.

When PCNN is used for DNA sequences processing, each code is connected to a unique neuron, and the number of the neurons is equal to the number of bases in the input code sequences. And the code value of a base is taken as the external input stimulus of the neuron in F channel, $S_{ij} = I(i,j)$. The output of each neuron results in two states, namely, non-pulse and pulse, which are represented by 0 and 1. As a result, the output status of the neurons is a binary sequence. And more information about PCNN will be found in [11–19].

3.2   DNA Sequences Linking Relations in PCNN Neurons

In PCNN model [14–19], the matrix W and M represent the synaptic connection weights. We usually let W equal to M, and it is a very important characteristic of PCNN model. For image processing, PCNN is a single layer two-dimensional array of laterally linked pulse coupled neurons, and all neurons are identical. As a result, the matrix W and M is two-dimensional. In this paper, the codes of the DNA sequences are served as the input of the PCNN, which are one-dimensional. A code bit is corresponding to a neuron, the matrix W and M are also one-dimensional, which makes the PCNN model can deal with one-dimensional DNA sequences. There are many setting methods for the matrix, and one simple setting way is that let the element of the matrix equal to the reciprocal of the distance between the central neuron and its adjacent neurons. For the neuron (i, j), the element of the matrix can be defined by (7).

$$W_{ijkl} = \frac{1}{(i - k)}, \tag{7}$$

where W_{ijkl} denotes the element of linking matrix, (i, j) is the central neuron, and (k, l) is adjacent neuron.

As we all know the information of the DNA sequences represent the composition and distribution of the bases in DNA sequences. However, normal analysis ways do not consider their linking relations. One base has a great relationship with its adjacent bases, so their linking relation does need a special consideration. One codon has three bases, which contacts with two codons. Therefore, one base at least has a close-knit linking relation with 8 bases, and the length of W in this paper is 8.

3.3   The Entropy Sequences of DNA Sequences

In image processing field, the output of PCNN is two-dimensional binary pulses map. The entropy of each two-dimensional binary pulses map is the description of the features in the image, which can translate the binary pulses maps into a number sequence. The number sequence is known as entropy sequence (ES). In DNA sequences processing field, we encode DNA primary sequences with a simple coding way to keep all structural and functional information of DNA primary sequences, then PCNN model is used to extract the features of the coded DNA primary sequences. According to PCNN model, different S_{ij} will have different pulse status at the same time, some neurons will output pulses, but others will not, namely, non-pulse (0) and pulse (1), so the output statuses of neurons comprise a binary sequence (0 or 1). These binary sequences contain the connection, composition, and distribution information of the bases in DNA sequences, and ES of the binary sequences can effectively express the features of the coded DNA primary sequences.

The compute mode of ES is described by (8) and (9). We calculate the entropy of the output binary sequences at each time of iteration using (12). We arrange the entropy according to its appearing in order to obtain an one-dimensional ES, which is denoted by ES(n), as shown in (9). ES is a comprehensive description of the code DNA sequences, and each code DNA sequences will have its own unique ES according to the characters of PCNN model. So it can be applied to DNA primary sequences pattern classification.

$$E[Y(n)] = -P_1 \log_2(P_1) - P_0 \log_2(P_0), \tag{8}$$

$$ES(n) = \sum_1^N E[Y(n)], \tag{9}$$

where $E[Y(n)]$ is the entropy of $Y[n]$ at nth iteration. P_1 is the probability of 1 status in binary sequences; P_0 is the probability of 0 states in binary sequences. ES(n) is ES of the DNA sequences, N is the total number of iterations, n varies from 1 to N.

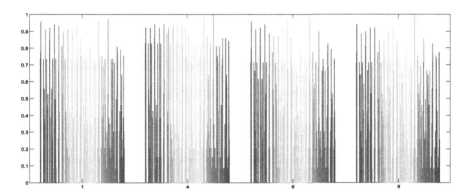

Fig. 1. The ES of sample 1, 4, 6 and 8.

From above analysis, we can know that ES of PCNN can be used to calculate the similarity of the primary DNA. ES not only reflects the composition and distribution of the bases in DNA sequences, but also embodies the connections among the bases. Figure 1 is ES bar chart of sample 1, 4, 6 and 8 in Table 1, respectively. In Fig. 1, it can be seen sample 4 is obviously different with others. We all know that sample 4 is poultry and the others are mammal. However, ES of sample 6 and 8 are very similar, they are murine. Besides, ES of three kinds of mammal also have a little difference, which depends on the evolutionary relationship among the species. Therefore, we can infer that the close species have similar ES, and distant species have different ES. So we can use ES of PCNN to measure the similarity of the primary DNA, it will express kinship among the species.

4   Algorithm Process and Experimental Analysis

To verify the validity of the presented algorithm, we use the first exon of β-globin gene in different species as experimental samples, which DNA sequences are often used, as shown in Table 1. It should be pointed out that we use Euclidean distance to calculate

DNA sequences similarity. The parameters of PCNN model are: $W = [1/5 \ 1/4 \ 1/3 \ 1/2 \ 1 \ 1/2 \ 1/3 \ 1/4 \ 1/5]$, $\beta = 0.01$, $\alpha^F = 0.54$, $\alpha^L = 0.2$, $\alpha^\theta = 0.2$, $V^L = 1$, $V^F = 0.1$, $V^\theta = 15$ and the iteration $N = 300$.

Table 2 is similarity matrix, it can be seen that the three kinds of primates (human, chimpanzee and gorilla), which are strongly similar to each other, for the smallest distances are associated with human–gorilla and gorilla–human, chimpanzee–human. Besides, the two kinds of muridaes' (mouse and rat) DNA sequences are similar to each other as well. And goat and bovine are the most similar with each other. On the other hand, Gallus is the only none mammal among them, so it shows great dissimilarity with others among 11 species. Lemur shows greater similarities to rabbit, chimpanzee, human, gorilla than to the other species. This is analogous results were reported by other authors [4, 5, 7], which proved the proposed method was feasible.

Table 1. The DNA sequences of the first exon of β-globin gene of 11 species

Species	Coding sequence
Human	ATGGTGCACCTGACTCCTGAGGAGAAGTCTGCCGTTACTGCCCTGTGGGGCAAGGTGAACGTGGATGA AGTTGGTGGTGAGGCCCTGGGCAG
Goat	ATGCTGACTGCTGAGGAGAAGGCTGCCGTCACCGGCTTCTGGGGCAAGGTGAAAGTG GATGAAGTTGGTGCTGAGGCCCTGGGCAG
Opossum	ATGGTGCACTTGACTTCTGAGGAGAAGAACTGCATCACTACCATCTGGTCTAAGGTGCAGGTTGACC AGACTGGTGGTGAGGCCCTTGGCAG
Gallus	ATGGTGCACTGGACTGCTCAGGAGAAGCAGCTCATCACCGGCCTCTGGGGCAAGGTCAATGTGGC CGAATGTGGGGCCGAAGCCCTGGCCAG
Lemur	ATGACTTTGCTGAGTGCTGAGGAGAATGCTCATGTCACCTCTCTGTGGGGCAAGGTGGATGTAGAGA AAGTTGGTGGCGAGGCCTTGGGCAG
Mouse	ATGGTTGCACCTGACTGATGCTGAGAAGTCTGCTGTCTCTTGCCTGTGGGCAAAGGTGAACCCCGA TGAAGTTGGTGGTGAGGCCCTGGGCAGG
Rabbit	ATGGTGCATCTGTCCAGTGAGGAGAAGTCTGCGGTCACTGCCCTGTGGGGCAAGGTGAATGTGGA AGAAGTTGGTGGTGAGGCCCTGGGC
Rat	ATGGTGCACCTAACTGATGCTGAGAAGGCTACTGTTAGTGGCCTGTGGGGAAAGTGAACCCTGATA ATGTTGGCGCTGAGGCCCTGGGCAG
Gorilla	ATGGTGCACCTGACTCCTGAGGAGAAGTCTGCCGTTACTGCCCTGTGGGGCAAGGTGAACGTGGAT GAAGTTGGTGGTGAGGCCCTGGGCAGG
Bovine	ATGCTGACTGCTGAGGAGAAGGCTGCCGTCACCGCCTTTTGGGGCAAGGTGAAAGTGGATGAAGT TGGTGGTGAGGCCCTGGGCAG
Chimpanzee	ATGGTGCACCTGACTCCTGAGGAGAAGTCTGCCGTTACTGCCCTGTGGGGCAAGGTGAACGTGGAT GAAGTTGGTGGTGAGGCCCTGGGCAGGTTGGTATCAAGG

Table 2. The similarity matrix of eleven different species

	Human	Goat	Opossum	Gallus	Lemur	Mouse	Rabbit	Rat	Gorilla	Bovine	Chimpanzee
Human	0	1.4910	1.4560	**1.6750**	1.5850	1.4077	1.2965	1.6530	**0.0761**	1.3057	0.4985
Goat	1.4910	0	1.6441	**1.9841**	1.8822	1.3144	1.4384	1.5509	1.4994	**0.7923**	1.4160
Opossum	1.4560	1.6441	0	1.9269	**2.1256**	**1.4288**	1.8076	1.6523	1.4762	1.6895	1.3973
Gallus	1.6750	1.9841	1.9269	0	**2.5710**	2.0517	2.2715	2.3574	1.6662	1.8955	1.7628
Lemur	1.5850	1.8822	2.1256	**2.5710**	0	1.6710	**1.3634**	1.5605	1.5954	1.7784	1.3910
Mouse	1.4077	1.3144	1.4288	2.0517	1.6710	0	1.2821	**1.1518**	1.4183	1.2353	1.1992
Rabbit	1.2965	1.4384	1.8076	**2.2715**	1.3634	1.2821	0	1.5014	1.3039	1.3352	**1.0783**
Rat	1.6530	1.5509	1.6523	**2.3574**	1.5605	**1.1518**	1.5014	0	1.6740	1.6296	1.4758
Gorilla	**0.0761**	1.4994	1.4762	1.6662	1.5954	1.4183	1.3039	**1.6740**	0	1.3060	0.4993
Bovine	1.3057	**0.7923**	1.6895	**1.8955**	1.7784	1.2353	1.3352	1.6296	1.3060	0	1.2258
Chimpanzee	**0.4985**	1.4160	1.3973	**1.7628**	1.3910	1.1992	**1.0783**	1.4758	0.4993	1.2258	0

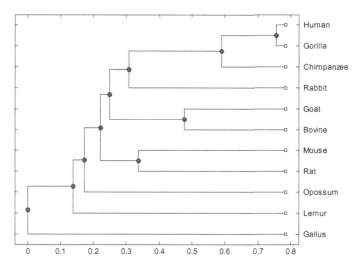

Fig. 2. Phylogenetic tree of the first exon of β-globin gene in Table 2.

UPGMA phylogenetic tree of Table 2 was built from the similarity-matrix using the proposed method, as shown in Fig. 2. We note that the classification of 11 mammals is basically consistent with the evolution, all the mammals are on a branch, and are far and gallus. Figure 2 and Table 2 are associated with the group of human, gorilla and chimpanzee; all the primates are closely related to each. On the contrary, the longest distance was observed between the group of primates and gallus. The pairs mouse and rat, goat and bovine. The method can almost distinguish the genetic relationship among 11 species. Although the method has some not perfect aspects, it should be a promising area for further research. The results basically conform to the evolution rule of the biological species.

5 Conclusions

In this paper, we proposed a novel method based on PCNN for DNA sequences similarity calculation. Our encoding method is simple, which can hold all the information of DNA sequences and be suitable for PCNN to extract ES of the species. ES is independent of the lengths of DNA sequences, so it can deal with different lengths DNA sequences. Using the characters of PCNN, ES can effectively express the features of DNA sequences. At last, we calculate the similarity by Euclidean distance to get the similarity matrix from DNA primary sequences. The experimental results show that the proposed method for the analysis of DNA sequences is relatively reasonable and effective.

Acknowledgements. Our work is supported by the National Natural Science Foundation of China (Grant 61365001, 61463052).

References

1. Hamori, E., Ruskin, J.: H-curves, A novel method of representation of nucleotide series especially suited for long DNA-sequences. J. Biol. Chem. **258**, 1318–1327 (1983). http://www.scopus.com/record/display.url?eid=2-s2.0-0020659327&origin=inward&txGid=9392 B6CE9565C5B0010C85CC7B4198E4.iqs8TDG0Wy6BURhzD3nFA%3a2

2. Randic, M., Vracko, M., Lers, N., Plavsic, D.: Novel 2-D graphical representation of DNA sequences and their numberical characterization. Chem. Phys. Lett. **368**, 1–6 (2003)

3. Nandy, A.: A new graphical representation and analysis of DNA-sequence structure. I: methodology and application to globin genes. Curr. Sci. Assoc. **66**, 309–314 (1994). http://www.scopus.com/record/display.url?eid=2-s2.0-84892519017&origin=inward&txGid= 9392B6CE9565C5B0010C85CC7B4198E4.iqs8TDG0Wy6BURhzD3nFA%3a6

4. Yao, Y.H., Nana, X.Y., Wang, T.M.: A new 2D graphical representation—Classification curve and the analysis of similarity/dissimilarity of DNA sequences. J. Mol. Struct.: THEOCHEM **764**, 101–108 (2006). http://ac.els-cdn.com/S0166128006000996/1-s2.0-S0166128006000996-main.pdf?_tid=884c5c5e-9ee6-11e4-b791-00000aab0f6c&acdnat= 1421567549_8424d048d1a4f3b4d24d00d3125520f8

5. Wang, J., Zhang, Y.: Characterization and similarity analysis of DNA sequences grounded on a 2-D graphical representation. Chem. Phys. Lett. **423**, 50–53 (2006)

6. Yuan, C., Liao, B., Wang, T.: New 3D graphical representation of DNA sequences and their numerical characterization. Chem. Phys. Lett. **379**, 412–417 (2003)

7. Liao, B., Wang, T.: Analysis of similarity/dissimilarity of DNA sequences based on 3-D graphical representation. Chem. Phys. Lett. **388**, 195–200 (2004)

8. Qi, Y., Jin, N., Ai, D.: Wavelet Analysis of DNA Walks on the Human and Chimpanzee MAGE/CSAG-palindromes. Genomics Proteomics Bioinform. **10**, 230–236 (2012)

9. Yin, C.C., Chen, Y., Yau, S.T.: A measure of DNA sequence similarity by Fourier Transform with applications on hierarchical clustering. J. Theoret. Biol. **359**, 18–28 (2014). http://ac.els-cdn.com/S0022519314003324/1-s2.0-S0022519314003324-main.pdf?_tid= f1c4c6ae-9ee5-11e4-92d8-00000aab0f27&acdnat=1421567296_b82053ab3d6f075b85060a 94972ab4f8

10. HongJie, Yu.: Segmented K-mer and its application on similarity analysis of mitochondrial genome sequences. Gene **518**, 419–424 (2013)

11. Eckhorn, R., Reitboeck, H.J., Arndt, M., Dicke, P.: Feature linking via synchronization among distributed assemblies: Simulation of result from cat visual cortex. Neutral Comput. **2**(3), 293–307 (1990). http://www.mitpressjournals.org/doi/abs/10.1162/neco.1990.2.3. 293#.VWvXBdKl8h1

12. Johnson, J.L., Ritter, D.: Observation of periodic waves in a pulse-coupled neural network. Opt. Lett. **18**(15), 1253–1255 (1993). https://www.osapublishing.org/ol/abstract.cfm?uri=ol-18-15-1253&origin=search

13. Johnson, J.L.: Pulse-Coupled Neural Nets: Translation, rotation, scale, distortion, and intensity signal invariances for images. Appl. Opt. **33**(26), 6239–6253 (1994). https://www.osapublishing.org/ao/abstract.cfm?uri=ao-33-26-6239&origin=search

14. Zhao, C.H., Shao, G.F., Ma, L.J., et al.: Image fusion algorithm based on redundant-lifting NSWMDA and adaptive PCNN. Optik-Int. J. Light Electron Opt. **125**, 6247–6255 (2014). http://ac.els-cdn.com/S0030402614009632/1-s2.0-S0030402614009632-main.pdf?_tid=cda b64e0-9e44-11e4-86f4-00000aacb362&acdnat=1421498087_14cff5924b80014a4a3a6755 f33a9d7f

15. Fu, J.C., Chen, C.C., Chai, J.W., Wong, S.T.C., Li, I.C.: Image segmentation by EM-based adaptive pulse coupled neural networks in brain magnetic resonance imaging. Comput. Med. Imaging Graph. **34**, 308–320 (2010). http://ac.els-cdn.com/S0895611109001402/1-s2.0-S0895611109001402-main.pdf?_tid=ce5cdbc2-9ee3-11e4-9e15-00000aab0f27&acdnat=14 21566378_59724ed68bcf2b33c37c8cb8cf796504

16. Li, H., Jin, X., Yang, N., Yang, Z.: The recognition of landed aircrafts based on PCNN model and affine moment invariants. Pattern Recogn. Lett. **51**, 23–29 (2015). http://ac.els-cdn.com/S0167865514002463/1-s2.0-S0167865514002463-main.pdf?_tid=93310268-e375-11e4-9045-00000aacb35f&acdnat=1429105665_c013622eabb670d98b3c08af70b278f7

17. Wang, Z., Ma, Y.D., Cheng, F.Y., Yang, L.Z.: Review of pulse-coupled neural networks. Image Vis. Comput. **28**, 5–13 (2010). http://ac.els-cdn.com/S0262885609001346/1-s2.0-S0262885609001346-main.pdf?_tid=7b89e29e-9ee0-11e4-a1ed-00000aab0f6c&acdnat=14 21564951_78ec6b8d24a162782ada72347989bcf8

18. Subashini, M.M., Sahoo, S.K.: Pulse coupled neural networks and its applications. Expert Syst. Appl. **41**, 3965–3974 (2014). http://ac.els-cdn.com/S0957417413010026/1-s2.0-S0957417413010026-main.pdf?_tid=adcb248a-9ee4-11e4-849a-00000aab0f01&acdnat=14 21566753_21936162bc6fb2185b736b58e0d338e0

19. Jin, X., Nie, R.C., Zhou, D.M., et al.: A novel DNA sequence similarity calculation based on simplified pulse-coupled neural network and Huffman coding. Phys. A: Stat. Mech. Appl. **461**, 325–338 (2016)

20. Hou, W.B., Pan, Q.H., He, M.F.: A novel representation of DNA sequence based on CMI coding. Phys. A **409**, 87–96 (2014). http://ac.els-cdn.com/S0378437114003410/1-s2.0-S0378437114003410-main.pdf?_tid=246ac810-9ee6-11e4-8cb5-00000aab0f26&acdnat= 1421567381_40c798c611f860711709cb99ad0e5c03

21. Jeong, B.S., Bari, A.T.M.G., Reaz, M.R., Jeona, S., Lima, C.G., Choi, H.J.: Codon-based encoding for DNA sequence analysis. Methods **67**, 373–379 (2014). http://ac.els-cdn.com/S1046202314000267/1-s2.0-S1046202314000267-main.pdf?_tid=0f59e96a-9ee6-11e4-b20a-0 0000aab0f6b&acdnat=1421567346_78ae735f9469508666622130a9d52721

Computing Sentiment Scores of Adjective Phrases for Vietnamese

Thien Khai Tran$^{(\boxtimes)}$ and Tuoi Thi Phan

Faculty of Computer Science and Engineering,
Ho Chi Minh City University of Technology, Ho Chi Minh City, Vietnam
{thientk, tuoi}@cse.hcmut.edu.vn

Abstract. Building sentiment lexicons is an essential task that provides "material" for all sentiment analysis levels: document-based, sentence-based, and aspect-based. For Vietnamese researchers, this problem is still a hot issue and should be resolved because the Vietnamese sentiment corpus is not complete. In this paper, we propose a fuzzy language computation based on Vietnamese linguistic characteristics to provide an effective method for computing the sentiment polarity of adjective phrases. Then, from this base, we built a sentiment phrase dictionary for Vietnamese with fine-grained scores. In our experiments on a real data set, we show that our approach gives perfectly acceptable results.

Keywords: Sentiment lexicons · Valence shifter · Language computation · Linguistic variable · Vietnamese hedges · Fuzzy logic · Approximate reasoning

1 Introduction

In the task of sentiment analysis, researchers aim to identify positive, negative, and neutral opinions, emotions, and evaluations of target objects, such as products, individuals, or topics. Building sentiment lexicons is an essential task that provides "material" for all sentiment analysis levels: document-based, sentence-based, and aspect-based. One of the biggest English sentiment lexicons is SentiWordNet [8]. It contains sentiment terms extracted from WordNet with a semi-supervised learning method and is available for research purposes. In recent years, some researches on Vietnamese text have been published by Vu et al. [12], Nguyen et al. [2], and Trinh et al. [7] in which [7] is the newest one building a Vietnamese emotional dictionary that contains five sub-dictionaries of noun, verb, adjective, adverb, and proposed features. However, the way the authors calculated the sum of the emotional values of the hedges was based on feelings.

In this paper, based on Vietnamese linguistic characteristics and the fuzzy computation proposed by Zadeh [4, 13], we present an effective method for the computation of the sentiment polarity of adjective phrases, and from this we built a sentiment dictionary for Vietnamese. Zadeh developed the concept of fuzzy linguistic variables that modify the meaning and intensity of their operands, and we developed a modified fuzzy function suitable for use with the Vietnamese language. In our experiments, we showed that our system provides good results.

© Springer International Publishing AG 2016
C. Sombattheera et al. (Eds.): MIWAI 2016, LNAI 10053, pp. 288–296, 2016.
DOI: 10.1007/978-3-319-49397-8_25

With this paper, we describe our research contributions, as follows: (1) The mining of Vietnamese linguistic characteristics to propose sentiment computing rules for adjective phrases. (2) Proposal of modified fuzzy functions suitable for Vietnamese linguistic variables. (3) Taking of steps toward building a Vietnamese sentiment phrase dictionary with fine-grained scores.

We organized the rest of this paper as follows: in Sect. 2 we present the linguistic characteristics of Vietnamese; in Sect. 3 the proposed model is described; in Sect. 4 we report our experiments; and finally, we conclude the paper and discuss possibilities for future work.

2 Linguistic Characteristics of Vietnamese

Vietnamese is an isolating language with lexical tones and monosyllabic word structure. These characteristics are evident in all aspects: phonetic, vocabulary, and grammar. In this paper, we focus on adjectives and adverbs.

2.1 Vietnamese Adjectives

Adjectives are words that indicate the features, qualities, and characteristics of things, such as the shape, color, and size, etc. Based on the meaning and the ability to be combined with others, we can divide adjectives into three types: relational adjectives, degree adjectives (relative adjectives), and non-degree adjectives (absolute adjectives).

Relational Adjective. In Vietnamese, relational adjectives: (1) take a role as a common noun: tác phong (rất) công nghiệp industrial style. (2) take a role as a proper noun: thái độ (rất) Chí Phèo Chi Pheo attitude. However, some linguists like Than [6] believes that Vietnamese does not have relational adjectives.

Degree Adjectives. These adjectives are characterized by determining the nature of things. These adjectives can be combined with degree adverbs or they can make a comparable structure.

- Quality adjectives for colors: xanh, đỏ, vàng *(green, red, yellow)*.
- Quality adjectives for tastes: chua, cay, mặn, ngọt *(sour, spicy, salty, sweet)*.
- Quality adjectives for physical appearances: cứng, mềm *(hard, soft)*.
- Quality adjectives for sounds: ồn ào, im lìm, yên lặng *(noisy, quiet, silent)*.
- Intellectual adjectives: thông minh, sáng suốt, ngu đần *(smart, wise, stupid)*.
- Mental adjectives: vui vẻ, buồn, chán *(happy, sad, bored)*.
- Qualitative adjectives: morality: tốt *(good);* physiological characteristics: ốm yếu, khỏe mạnh *(sick, healthy);* manners (actions): nhanh nhẹn *(agile)*.
- Adjectives for temperature: nóng, lạnh, ấm *(hot, cold, warm)*.
- Quality adjectives for sizes: to, nhỏ, ít, nhiều *(big, small, little, many)*.
- Quality adjectives for shapes: thẳng, vuông, gầy, *(straight, square, thin)*.
- Quality adjectives for intensities: mạnh, yếu *(strong, weak)*.
- These adjectives can often create opposite meaning pairs: vui/buồn *(happy/sad)*, mạnh/yếu *(strong/weak)*, or giàu/nghèo *(rich/poor)*.

Non-Degree Adjectives. These adjectives are characterized by themselves and contain significant levels of nature (usually at the absolute level), so they are not normally combining with degree modifiers.

2.2   Vietnamese Adverbs

Adverbs are words that modify or describe verbs, adjectives, clauses, sentences, and other adverbs. For Vietnamese, a selection of the types of adverbs and their ability to combine with adjectives [5] are presented in Table 1.

Table 1. Vietnamese adverbs and their ability to combine with adjectives.

Types	Terms	Examples
Request	đừng, chớ *(don't)*	đừng làm *(don't do (it))*
Degree	quá, lắm *(too)*	đẹp lắm *(very beautiful)*
Finish, completion	rồi *(already)*	tốt rồi *(good enough)*
Circumstance	mãi, luôn *(forever always)*	trẻ mãi *(young forever)*
Consequence	lên, lại, đi *(up, again, rather)*	già đi *(rather old)*
Invert	không, chẳng, chưa, chả *(not)*	không ngoan *(not good)*
Continuity	cứ, vẫn *(just, still)*	vẫn giỏi *(still good)*

3   Proposed Model

In this model, we try to compute the sentiment scores for word phrases that include adjectives and adverbs based on Vietnamese linguistic characteristics. By combining with some adverbs, the adjective phrases will have a smoother sentiment scaling.

3.1   System Architecture

Our system architect is presented in Fig. 1. We used the English sentiment dictionary, SentiWordNet, and the translate tools Vdict[1] and Google Translate[2] to build the core adjective lexicons with sentiment polarities for Vietnamese. The fuzzy rules then computed the sentiment scores for the whole phrase that included the adjectives and associated adverbs.

Building Core Adjectives. We constructed a handcrafted opinion dictionary containing approximately 1,500 adjectives. The number of words was large enough to cater to the problem we sought to solve. These words:

1. appeared in the review corpus that we obtained from [9, 10].
2. are matched with corresponding English words in SentiWordNet, we used Vdict and Google Translate to check this. We assigned opinion words scores that were the same as the scores of words in SentiWordNet.

In Table 2 we describe some of the opinion words that appear in this core dictionary.

Table 2. Fragment of core opinion dictionary.

Name	Positive score	Negative score	POS	Tag
đẹp beautiful	0.75	0	Adjective	Qualitative
xấu urly	0	0.75	Adjective	Qualitative
tốt fine	0.625	0	Adjective	Morality
xấu xa bad	0	0.625	Adjective	Morality

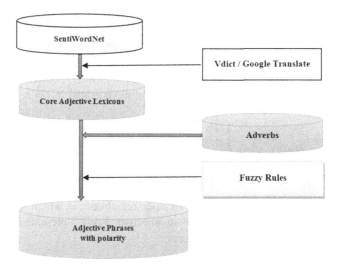

Fig. 1. System architecture.

3.2 Fuzzy Rules

Overall sentiment scores for the adjective phrases were calculated thanks to fuzzy rules that were associated with the combination between the adjective (denotes x) and the adverb (denotes y). We used fuzzy functions to incorporate the effect of the adverbs in the adjective phrases. We considered the sentiment score of an adjective to be its initial fuzzy score $\mu(x)$. Based on Vietnamese characteristics, we realized five sentiment shifting scalings for adverbs, and these were intensifier, booster, diminisher, minimizer, and modifier. Some Vietnamese adverbs are presented by Table 3.

Table 3. Some Vietnamese adverbs with their scalings.

Intensifier	Booster	Diminisher	Minimizer	Modifier
cực kỳ extremely	rất very	khá rather	cũng seemingly	không no
cực strongly	quá too	tương đối relatively	hơi a bit	chẳng no
siêu super	lắm much	tạm rather	rồi already	chả no

In our system, Vietnamese adverbs are also organized in a database. In Table 4 we describe some of the adverbs that appear in our adverb database. In there, "Tag" is the scaling category to which an adverb can belong.

Table 4. Some Vietnamese adverbs with their tags.

Adverbs	Types	Tag
cực kỳ	degree	intensifier
không	invert	modifier
rồi	finish	minimizer

Similar to Zadeh's proposition [13], if the adjective had a preceding adverb, its modified fuzzy score was computed by (1).

$$f(\mu(x)) = 1 - (1 - \mu(x))^\delta \tag{1}$$

We chose δ = 4, 2, 1/2, or 1/4 if the adverb was a(n) intensifier, booster, diminisher, or minimizer, which gives us a modified fuzzy score as indicated in (2).

$$f(\mu(x),\ y) = \begin{cases} 1 - \sqrt[4]{1 - \mu(x)} & y \in Minimizer \\ 1 - \sqrt[2]{1 - \mu(x)} & y \in Diminisher \\ 1 - (1 - \mu(x))^2 & y \in Booster \\ 1 - (1 - \mu(x))^4 & y \in Intensifier \end{cases} \tag{2}$$

With $f(\mu(x),y)$ is the sentiment score of an adjective phrase, in which x: adjective, y: adverb; and $\mu(x)$ is the sentiment score of an adjective. Table 5 presents an example of adjective phrases and their sentiment scores.

Table 5. Sentiment score of adjective phrases.

f(μ(x),y)					μ(x)
Intensifier	Booster	Diminisher	Minimizer	Modifier	Adjective
cực kỳ (đẹp)	rất (đẹp)	khá (đẹp)	cũng (đẹp)	không (đẹp)	đẹp
0.99	0.94	0.5	0.29	0	0.75

Exceptions

(a) **Negation.** Adjectives for morality, manner, and nature: Van Ban [1] believes that the negatives of these adjectives are inversions if they are positive, and the others are neutral, as in (3).

$$f(\mu(x),\ y) \in Modifier) = \begin{cases} -\mu(x) & x \in Morality\ AND\ x\ is\ positive \\ 0 & x \notin Morality\ OR\ x\ is\ negative \end{cases} \tag{3}$$

For example: $f(không\ đẹp\ _{no\ beautiful}) = 0$, but $f(không\ tốt\ _{no\ good}) = -f(tốt\ _{good}) = -0.625$

(b) **Position between adverb and adjective in sentence.** We believe that the position interchange of the two words will usually emphasize the adjective meaning. For example. $f(đẹp\ cực\ kỳ) > f(cực\ kỳ\ đẹp\ _{extremely\ beautiful})$ and $f(hay\ tuyệt) > f(tuyệt\ hay\ _{extremely\ interesting})$ etc.

In this case (the adjective precedes the adverb), $f(\mu(x),y)$ equals the mean of the original sentiment score (the adjective follows the adverb) and the next sentiment level score in the formula (2). If the original sentiment score is the highest value, then we assign the next sentiment score to 1, as in (4). For example: $f(đẹp\ cực\ kỳ) = (f(\mu(đẹp),\ cực\ kỳ) + 1)/2 = (0.99 + 1)/2 = 0.995$.

$$f(\mu(x),\ y) = \begin{cases} \frac{(1-\sqrt{41-\mu(x)})+(1-\sqrt[2]{1-\mu(x)})}{2} & y \in Minimizer \\ \frac{(1-\sqrt[2]{1-\mu(x)})+(1-(1-\mu(x))^2)}{2} & y \in Diminisher \\ \frac{(1-(1-\mu(x))^2)+(1-(1-\mu(x))^4)}{2} & y \in Booster \\ \frac{(1-(1-\mu(x))^4+1)}{2} & y \in Intensifier \end{cases} \tag{4}$$

(c) **Adverbs for finish, completion.** There is not too much sentiment for a finishing adverb (denotes z), for example, "*đẹp rồi*", "*sáng rồi*". The adverb "*rồi*" is only equivalent to the Minimizer level [1], but they can combine with another adverb (denotes y): "*đẹp lắm rồi*", "*tốt quá rồi*", and "*cực đẹp rồi*". In this case, $f(\mu(x), y,z)$ equals the mean of the original sentiment score (phrase without finishing adverb, z) and the preceding sentiment level score in the formula (2). If the original sentiment level is the lowest level, then we select the preceding sentiment score as being equal to 0. If y is a Modifier, we choose a parameter $\beta = 3/4$ to scale down the score of the function $f(\mu(x),y,z)$, as in (5).

For example: $f(cũng\ đẹp\ rồi\ _{pretty\ well\ already}) = (f(cũng\ đẹp\ _{pretty\ well}) +0)/ 2 = (0.29 + 0)/2 = 0.145$.

$$f(\mu(x),\ y,\ z) = \begin{cases} \frac{(1-\sqrt{41}-\mu(x))+0}{2} & y \in Minimizer \\ \frac{(1-\sqrt{41}-\mu(x))+(1-\sqrt[2]{1-\mu(x)})}{2} & y \in Diminisher \\ \frac{(1-\sqrt[2]{1-\mu(x)})+(1-(1-\mu(x))^2)}{2} & y \in Booster \\ \frac{(1-(1-\mu(x))^2)+(1-(1-\mu(x))^4)}{2} & y \in Intensifier \\ -\mu(x) \times \frac{3}{4} & y \in Modifier \end{cases} \tag{5}$$

with z is an adverb with Finish type.

4 Experiments

We randomly collected 100 sentences from Agoda[3] to evaluate the system performance. These sentences included 252 adjective phrases, and the system was capable of handling 245 phrases. The highest score was 0.99 (*cực kỳ đẹp* _{extremely beautiful}), and the lowest

---

[3] https://www.agoda.com/vi-vn.

score was -0.99 *(vô cùng tệ $_{extremely\ worse}$)*. After that, we examined the effect of the valence shifters on classifying the sentences to three classes: positive/neutral/negative. Then we compared the results with the SVM approach that was proposed in [11]. To classify the sentences, we simply counted the mean scores of negative and positive phrases in each sentence. If the final score was more than +0.1 the sentence was considered to show a positive emotion. If the score was less than -0.1 the sentence was considered to show a negative emotion. Otherwise, the sentence was considered to show a neutral emotion.

Contextual Situation. There was an exception if the word "nhưng $_{but}$" appeared in the sentence, as the total sentiment valence did not include the score of the preceding words/phrases.

For example: *Bãi biển khá đẹp nhưng nhân viên không thân thiện (The beach quite nice but the staff is unfriendly).* Total score: f(khá đẹp $_{quite\ nice}$ (nhưng $_{but}$)) + f(không thân thiện $_{unfriendly}$) = 0 + f(không thân thiện $_{unfriendly}$) = -0.75. Therefore this sentence is a negative emotion.

To compare with the SVM approach, we made use of the Accuracy rate (Acc), which is computed as:

$$\text{Acc} = 100\% * (1 - |P_S - A_P|/P_S) \tag{6}$$

where, Acc is the percent accuracy, P_S is the standard value, and A_P is the value we measured.

Cohen's Kappa Coefficient. Two judges participated in categorizing the sentences as "positive," "negative" or "neutral". To compute the "between judges agreement," we used the Cohen's kappa coefficient [3], as follows:

$$k = \frac{\text{Pr}(a) - \text{Pr}(e)}{1 - \text{Pr}(e)} \tag{7}$$

where Pr(a) is the relative observed agreement among the judges and Pr(e) is the hypothetical probability of a chance agreement. The Cohen's kappa coefficient of our corpus was $k = 0.88$ *(Pr(a) = 0.94 and Pr(e) = 0.5).*

We report the classification results from the three systems in Table 6. The results showed that the system with contextual valence shifter achieved the best accuracy rate.

Table 6. Test results for sentiment classification.

System	Accuracy rate
SVM method proposed in [11]	84 %
System without contextual valence shifters	85 %
System with contextual valence shifters	88 %

5 Conclusions

In this paper we presented a mechanism for computing the sentiment scores of adjective phrases by mining the Vietnamese language characteristics and using fuzzy functions. We have shown this approach to be effective. By identifying the opinion phrase polarity automatically, the method can be used to deal with many sentiment analysis problems. In our future research, we will adopt this approach for developing Vietnamese sentiment lexicons with noun phrases and verb phrases as well as increasing the number of adjectives.

Acknowledgment. This paper was supported by the research project C2016-20-32 funded by Vietnam National University Ho Chi Minh City (VNU-HCM).

References

1. Van Ban, D., Van Thung, H.: Ngữ pháp tiếng Việt, Vietnamese Grammar. Vietnam Education Publishing House, Hanoi (1998)
2. Nguyen, H.N., Van Le, T., Le, H.S., Pham, T.V.: Domain specific sentiment dictionary for opinion mining of vietnamese text. In: The 8th Multi-Disciplinary International Workshop on Artificial Intelligence (MIWAI 2014), pp. 136–148 (2014)
3. Carletta, J.: Assessing agreement on classification tasks: the Kappa statistic. Comput. Linguist. **22**, 249–254 (1996)
4. Zadeh, L.A.: The concept of a linguistic variable and its application to approximate reasoning-II. Inf. Sci. **8**(4), 301–357 (1975). Part 3
5. Van Ly, L.: Sơ thảo ngữ pháp Việt Nam, Vietnamese Essentials: Grammar. Vietnam Education Publishing House, Hanoi (1972)
6. Than, N.K.: Nghiên cứu ngữ pháp tiếng Việt, Vietnamese Grammar. Vietnam Education Publishing House, Hanoi (1997)
7. Trinh, S., Nguyen, L., Vo, M., Do, P.: Lexicon-based sentiment analysis of Facebook comments in Vietnamese language. In: Król, D., Madeyski, L., Nguyen, N.T. (eds.) Recent Developments in Intelligent Information and Database Systems, pp 263–276. Studies in Computational Intelligence, vol. 642. Springer, Heidelberg (2016)
8. Baccianella, S., Esuli, A., Sebastiani, F.: Sentiwordnet 3.0: an enhanced lexical resource for sentiment analysis and opinion mining. In: LREC 2010, May 2010
9. Tran, T.K., Phan, T.T.: An upgrading SentiVoice – a system for querying hotel service reviews via phone. In: Proceedings of the 19th International Conference on Asian Language Processing (IALP 2015), Suzhou, China, pp. 115–118 (2015)
10. Tran, T.K., Phan, T.T.: Constructing sentiment ontology for Vietnamese reviews. In: Proceedings of the 17th International Conference on Information Integration and Web-Based Applications and Services (iiWAS 2015), Brussels, Belgium, 11–13 December 2015, pp. 281–285 (2015). ISBN:978-1-4503-3491
11. Tran, T.K., Phan, T.T.: Multi-class opinion classification for Vietnamese hotel reviews. Int. J. Intell. Technol. Appl. Stat. **9**(1), 7–18 (2016)

12. Vu, T.-T., Pham, H.-T., Luu, C.-T., Ha, Q.-T.: A feature-based opinion mining model on product reviews in Vietnamese. In: Katarzyniak, R., Chiu, T.-F., Hong, C.-F., Nguyen, N.T. (eds.) Semantic Methods for Knowledge Discovery and Communication, Polish-Taiwanese Workshop. Studies in Computational Intelligence, vol. 381, pp. 22–23. Springer, Heidelberg (2011)
13. Huynh, V.N., Ho, T.B., Nakamori, Y.: A parametric representation of linguistic hedges in Zadeh's fuzzy logic. Int. J. Approx. Reason. **30**(3), 203–223 (2002)

Sentiment Analysis Using Anaphoric Coreference Resolution and Ontology Inference

Thi Thuy Le$^{(\boxtimes)}$, Thanh Hung Vo, Duc Trung Mai,
Than Tho Quan, and Tuoi Thi Phan

Ho Chi Minh University of Technology, Ho Chi Minh City, Viet Nam
{thuylt, vthung, mdtrung, qttho, tuoi}@cse.hcmut.edu.vn

Abstract. Aspect-based sentiment analysis is an emerging trend in the NLP area nowadays. One of the major tasks in this work is to identify corresponding aspects for rating sentiment. Ontology is considered highly useful to cope with this issue, due to its capability of capturing and representing concepts in a certain domain. However, ontology-based sentiment analysis suffers from the difficulty when handling anaphoric coreference of mentioned entities, which commonly occurs in textual documents. This paper addresses this problem by introducing an approach combining coreference resolution with ontology inference. The initial results are quite promising.

Keywords: Aspect-level sentiment analysis · Aspect-oriented sentiment ontology · Anaphoric coreference resolution

1 Introduction

Sentiment analysis [1] is an emerging trend nowadays in the Natural Language Processing (NLP) field. There are three levels of sentiment analysis: (i) *document-based* level; (ii) *sentence-based* level; and (iii) *aspect-based* level. In this paper, we focus on the problem of *aspect-level* [2]. To be more precise, apart from rating positive/negative sense of a mention, the objects (or *aspects*) targeted by the mention must also be identified. *Ontology* is often used to represent the aspects in a machine-readable form. However, ontology-based approaches usually suffer from the difficulty of handling *anaphoric coreference* problem commonly occurring in natural languages. To illustrate it, let us consider the following example.

(S1) *I consider an iPhone 6S. Unlike Samsung S7, it is unfortunately not really affordable for students. However, the design looks nice and eye-catching.*

When analyzing the above text, Stanford CoreNLP toolkit [3], one of the most popular tools for natural language processing, returns the result shown in Fig. 1. Here one can observe an example of *anaphoric coreference*, when the pronoun *it* in the second sentence implies the aspect *iPhone6* in the first sentence. However, the problems which are still left in this example for further processing are as follows. In the second sentence, the negative term "*is not really affordable*" has already been detected by Stanford CoreNLP Toolkit. Similarity, the positive terms of "*nice and eye-catching*"

© Springer International Publishing AG 2016
C. Sombattheera et al. (Eds.): MIWAI 2016, LNAI 10053, pp. 297–303, 2016.
DOI: 10.1007/978-3-319-49397-8_26

in the third sentence have also been recognized. However, these terms are not connected to the corresponding aspects, which are *iPhone 6* and *design* respectively. Moreover, as *design* is an *attribute* of *iPhone6* in this context, those positive terms should also be connected to *iPhone6* as well.

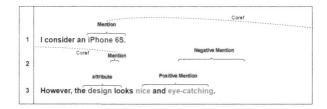

Fig. 1. The processing result of Stanford CoreNLP Toolkit for (S1)

In this paper, we address those problems by the following approach. Firstly, *conceptual graphs* (CGs) are used to represent individual sentences. Then, we perform *anaphoric conference resolution* to link the coreference nodes on the generated CGs together. *Ontology inference* is also used to make connection from the attributes with corresponding aspects. Eventually, we conduct sentiment analysis on the final resultant combined CGs.

2   Related Works

2.1   Ontology-Based Aspect-Level Sentiment Analysis

Ontology is often used more when conducting sentiment analysis at aspect level. Kontopoulos et al. [4] used ontology for analyzing the twitter posts. In [5], a fuzzy product ontology has been proposed. Hierarchical structures of ontology have been proved effective for handling online reviews [6–9]. In NETOWL [10] tool, the ontology-based problem for aspect sentiment and coreference resolution have also been mentioned. However, there is no concrete work reported to combine these two approaches.

2.2   Coreference Resolution

Research on the anaphoric coreference problem primarily focused on the resolution based on *noun*, *pronoun* (anaphora) and *name entities* (NEs). The approaches include using the supervised [11]; semi supervised or unsupervised [12] and machine learning techniques [11]. Work on semantic characteristic such as vocabulary and syntax [13, 14] are also reported. Other remarkable approaches include graph algorithm [15] and rule-based approaches [13]. In particular, *Stanford CoreNLP Toolkit* [3] has emerged recently as the most notable system for anaphoric coreference resolution.

3   Aspect-Oriented Sentiment Ontology

To adopt ontology for sentiment analysis, we firstly introduce formal definition of *Aspect-oriented Sentiment Ontology* as follows.

Definition 1 (Aspect-Oriented Sentiment Ontology). An aspect sentiment ontology SO is a pair of $\{C, R\}$; where $C = (C^A, C^S)$ represents a set of concepts, which consists of 2 elements: C^A is a set of aspect concepts, and C^S is a set of sentiment concepts; $R = (R^T, R^N, R^S)$ represents a set of relationships, which consists of 3 elements: R^N is a set of non-taxonomic relationships, R^T is a set of taxonomic relationships, R^S is a sentiment relationship. Each concept c_i in C represents a set of objects, or instances, of the same kind, denoted as *instance-of* (c_i). Each relationship $r_i(c_p, c_q)$ in R represents a binary association between concepts c_p and c_q, and the instances of such a relationship, denoted as *instance-of* (r_i), are pairs of (c_p, c_q) concept objects. Especially, an instance $r_i^s(a, s)$ in R^S implies a relationship between an aspect $a \in A$ and a sentiment term $s \in S$.

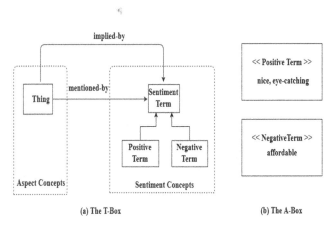

Fig. 2. An example of Generic Ontology

To graphically visualize an ontology, we rely on the idea of *T-Box* and *A-Box* [16]. Basically, a T-Box captures the relations between *concepts* and an A-Box describes *instances* of concepts. Figure 2 presents the T-Box and A-Box of Generic Ontology G_o. Generally speaking, G_o consists of one aspect concept of *Thing*, whose instances can be any real-life concepts. An instance of *Thing* can be *mentioned* or *implied* by a *Sentiment Term,* which can be either *Positive Term* or *Negative Term*. We also introduce two specific sentiment relationships for sentiment ontology, known as *mentioned-by* and *implied-by*. An aspect instance c can be *mentioned-by* a sentiment term s, meaning that c is being positive or negative rated, depending on whether s belongs to *Positive Term* or *Negative Term* classes, respectively. Moreover, *implied-by* is similar to *mentioned-by,* but carrying on more specific meaning. An aspect instance c can be *implied-by* a sentiment term s, that is s is only applicable for c, not for other aspects. Thus, when s occurs in a textual statement ϑ, one can infer that c is also implied in ϑ, without explicit mention.

4 Proposed Method

Our proposed method is carried out as follows. Firstly, we base on the work presented in [17] to construct conceptual graphs (CG) for the sentences in the given context. For example, the CGs of sentences in (S1) are generated as illustrated in Fig. 3.

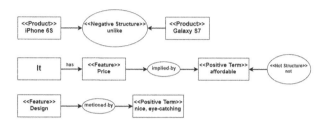

Fig. 3. Conceptual graphs are generated from sentences in (S1)

Then, we perform anaphoric coreference resolution and ontology inference to combine the corresponding nodes on the separate CGs. Ontology inference is performed in a heuristic-based manner. That is, once the system captures the occurrences of two related aspects in the same or two consecutive sentences, it infers that two aspects refer to the same entity. For instance, in the statement (S1), *S7 camera* and *design* are inferred as the same entity object where *design* is an attribute of *S7 camera*.

Finally, based on the discovered *mentioned-by* and *implied-by* as previously discussed, the system can establish relationships between aspects and sentiment terms. Depending on the opinion orientation of the sentiment terms, the rating (*positive/ neutral/negative*) will be evaluated accordingly (Fig. 4).

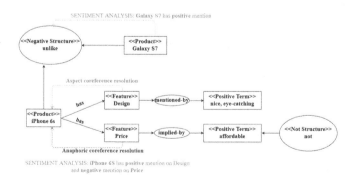

Fig. 4. An example of sentiment analysis on conceptual graph

5   Experimental Result

In order to conduct an experiment from real-life data, we obtain real datasets of user reviews on smartphone products from YouNet Media (YNM), a company dedicated for social listening and market intelligence[1]. The dataset covers 32 smartphone brands, 234 products. It consists of 2809, 3098 and 365 *negative, neutral* and *positive mentions*, respectively. The experts of YNM also helped us to define aspect (attribute) of the Smartphone domain, as depicted in Table 1.

Table 1. Some example of aspect and sentiment terms

Attribute	Aspect term	Sent. term (positive)	Sent. term (negative)
Design	Design, shape	attractive, eye-catching	cloddish, flat
Screen	inch, pixel	sharp, anti-glare	opaque, stained screen
Camera	lens, autofocus	wide, bright	blur, light-interference

We then measured the accuracy of our sentiment analysis approach. We did compare the performance of four sentiment analysis strategies as follows. The first strategy, SEN-FULL applied our full framework. SEN-NO-ONT and SEN-NO-RULES did not use Aspect-oriented Sentiment Ontology and Anaphoric Coreference Resolution respectively. Eventually, we used SVM for sentiment classification, as this technique was employed by various related works.

Figure 5 presents the accuracy percentage when we applied those analysis strategies on the collected datasets. It can be observed that in general our proposed method gains better performance.

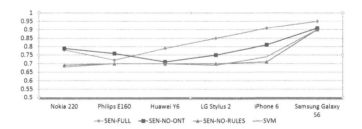

Fig. 5. Accuracy performance of sentiment analysis strategies

6   Conclusion

In this paper, we discuss an approach combining anaphoric coreference resolution with ontology inference for aspect-level sentiment analysis. As a result, the Aspect-oriented Sentiment Ontology is proposed, around which the coreference resolution and

---

[1] http://www.younetmedia.com/.

aspect-based sentiment analysis are centered. Our experiments on real datasets acquired from the actual discussion on social channels have achieved promising performance.

Acknowledgments. This research is funded by Vietnam National University Ho Chi Minh City (VNU-HCM) under grant number C2016-20-36. We are also grateful to YouNet Media for supporting real datasets for our experiment.

References

1. Padmaja, S., Fatima, S.: Opinion mining and sentiment analysis - an assessment of peoples' belief: a survey. Int. J. Ad hoc Sens. Ubiquitous Comput. **4**(1), 21–33 (2013)
2. Hu, M., Liu, B.: Mining and summarizing customer reviews. In: Kim, W., Kohavi, R., Gehrke, J., DuMouchel, W. (eds.) Proceedings of the 10th ACM SIGKDD International Conference on Knowledge Discovery and Data Mining, Seattle, Washington, USA, 22–25 August 2004, pp. 168–177. ACM (2004)
3. Manning, C.D., Surdeanu, M., Bauer, J., Finkel, J.R., Bethard, S., McClosky, D.: The Stanford CoreNLP natural language processing toolkit. In: ACL (System Demonstrations), pp. 55–60, June 2014
4. Kontopoulos, E., Berberidis, C., Dergiades, T., Bassiliades, N.: Ontology-based sentiment analysis of Twitter posts. Expert Syst. Appl. **40**(10), 4065–4074 (2013)
5. Lau, R.Y., Li, C., Liao, S.S.: Social analytics: learning fuzzy product ontologies for aspect-oriented sentiment analysis. Decis. Support Syst. **65**, 80–94 (2014)
6. Thet, T.T., Na, J.C., Khoo, C.S.: Aspect-based sentiment analysis of movie reviews on discussion boards. J. Inf. Sci. (2010). doi:10.1177/0165551510388123
7. Kim, S., Zhang, J., Chen, Z., Oh, A. H., Liu, S.: A hierarchical aspect-sentiment model for online reviews. In: AAAI, July 2013
8. Titov, I., McDonald, R.: A joint model of text and aspect ratings for sentiment summarization. In: McKeown, K., Moore, J.D., Teufel, S., Allan, J., Furui, S. (eds.), Proceedings of the 46th Annual Meeting of the Association for Computational Linguistics, Columbus, Ohio, USA, 15–20 June 2008, ACL 2008, pp. 308–316. The Association for Computer Linguistics (2008)
9. Wei, W., Gulla, J.A.: Sentiment learning on product reviews via sentiment ontology tree. In: Proceedings of the 48th Annual Meeting of the Association for Computational Linguistics, pp. 404–413. Association for Computational Linguistics, July 2010
10. Netowl About. Net Owl. N.p. (2016). https://www.netowl.com/our-story/. Accessed 7 Sept 2016
11. Kobdani, H., Schütze, H., Schiehlen, M., Kamp, H.: Bootstrapping coreference resolution using word associations. In: Proceedings of the 49th Annual Meeting of the Association for Computational Linguistics: Human Language Technologies, vol. 1, pp. 783–792. Association for Computational Linguistics, June 2011
12. Haghighi, A., Klein, D.: Unsupervised coreference resolution in a nonparametric Bayesian model. In: Annual meeting-Association for Computational Linguistics, vol. 45, no. 1, p. 848, June 2007
13. Haghighi, A., Klein, D.: Simple coreference resolution with rich syntactic and semantic features. In: Proceedings of the 2009 Conference on Empirical Methods in Natural Language Processing: Volume 3-Volume 3, pp. 1152–1161. Association for Computational Linguistics, August 2009

14. Bengtson, E., Roth, D.: Understanding the value of features for coreference resolution. In: Proceedings of the Conference on Empirical Methods in Natural Language Processing, pp. 294–303. Association for Computational Linguistics, October 2008
15. Nicolae, C., Nicolae, G.: BESTCUT: a graph algorithm for coreference resolution. In: Proceedings of the 2006 Conference on Empirical Methods in Natural Language Processing, pp. 275–283. Association for Computational Linguistics, July 2006
16. Quan, T.T., Hui, S.C.: Ontology-based natural query retrieval using conceptual graphs. In: Ho, T.-B., Zhou, Z.-H. (eds.) PRICAI 2008. LNCS (LNAI), vol. 5351, pp. 309–320. Springer, Heidelberg (2008). doi:10.1007/978-3-540-89197-0_30
17. Soon, W.M., Ng, H.T., Lim, D.C.Y.: A machine learning approach to coreference resolution of noun phrases. Comput. Linguist. **27**(4), 521–544 (2001)

A GA-SSO Based Intelligent Channel Assignment Approach for MR-MC Wireless Sensors Networks

Zhiping Xu, Zhicong Chen[(✉)], Lijun Wu, Peijie Lin,
and Shuying Cheng[(✉)]

College of Physics and Information Engineering,
Fuzhou University, Fuzhou 350116, China
{zhicong.chen,sycheng}@fzu.edu.cn

Abstract. In this paper, based on the simplified swarm optimization (SSO) algorithm and genetic algorithm (GA), a novel GA-SSO hybrid optimization algorithm is proposed to solve the NP-hard channel assignment problem of multi-radio multi-channel (MR-MC) wireless sensors networks (WSN) which is promising for data intensive application. The aim of channel assignment is to minimize total network interference so as to maximize network throughput. In the GA-SSO based channel assignment, an improved channel merging method is proposed to satisfy the interface constraint condition. Matlab based simulation results show that the proposed GA-SSO features better global search capacity and can reduce the total network interference more effectively, compared to the SSO and DPSO-CA algorithms.

Keywords: GA-SSO · Wireless sensors network · Multi-radio multi-channel · Channel merging

1 Introduction

In recent years, wireless sensors networks (WSNs) are broadly adopted in data-intensive applications, such as structural health monitoring, earthquake monitoring, volcano monitoring and so on [1]. These applications usually need to transfer massive data through the WSN. For these kinds of applications, how to maximize throughput of the WSN is a critical issue that deserves to be solved.

Conventional WSNs are based on single-radio single-channel and work in contention-based mode, which is hard to meet the requirement of massive data transmission in data-intensive applications. This problem can be well solved by the multi-radio multi-channel architecture. However, there only exist few researches about the MR-MC WSNs [2–6]. Gao et al. [2] proposed a data gathering method for WSN, which was composed of multi-radio sink node and double-radio relay nodes, which can effectively improve the WSN performance. Ji et al. [3] proposed a multi-path scheduling algorithm for snapshot data collection in single-radio multi-channel WSN and a continuous data collection method for dual-radio multi-channel WSN, which improved the network capacity. Li et al. [4] studied a jointly cross-level method used in

© Springer International Publishing AG 2016
C. Sombattheera et al. (Eds.): MIWAI 2016, LNAI 10053, pp. 304–311, 2016.
DOI: 10.1007/978-3-319-49397-8_27

dual-radio multi-channel WSN, which is composed of routing, scheduling and channel assignment. Athota et al. [5] proposed two kinds of channel assignment algorithms for MR-MC wireless mesh networks, which can minimize network aggregate interference and assure connectivity.

In this paper, we propose a multi-radio multi-channel assignment method in WSN based on GA-SSO algorithm, which mainly combines the simplified swarm optimization algorithm (SSO) [7] with genetic algorithm (GA). The aim of this research is to minimize the total interference and maximize the throughput of MR-MC WSNs for data intensive applications.

2   Problem Formulation

In this section, firstly the network and interference model is described. And then, the MR-MC channel assignment problem in WSNs is discussed in detail.

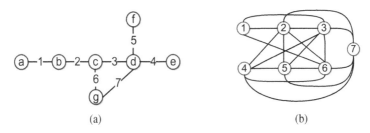

(a)   (b)

Fig. 1. A MR-MC network with 7 nodes and 7 links. (a) The communication graph consisting of 7 nodes and 7 links. (b) The conflict graph corresponding to (a).

2.1   Network Model

The studied WSN model consists of a set of wireless sensor nodes with fixed position in a plane, and every node is configured with a set of radios. We assume that all radios work in half-duplex mode, the antenna is omni-directional, and the transmission distance is L. The WSN topology can be modeled as an undirected graph $G = (V, E)$, where V represents the set of wireless sensor nodes and E the set of communication links. If node i and node j are in the communication range of each other, a link is defined between node i and node j, denoted as (i, j). Figure 1(a) shows an MR-MC network topology, which has seven node a, b, c, d, e, f, g and seven communication links l_{ab}, l_{bc}, l_{cd}, l_{de}, l_{df}, l_{cg}, l_{dg}, which are numbered as 1, 2, 3, 4, 5, 6, 7 respectively.

To simplify the problem, the available channels are assumed to be completely orthogonal. The set of available channels is denoted by $K = \{1, 2, \ldots, k_{max}\}$. R_i represents the number of radios, and K_i represents the set of available channels for corresponding node i.

2.2   Interference Model

In this paper, the protocol model is adopted. In this model, two links interfere with each other if their physical distance is less than the interference distance. When two interfering links transfer data in a same channel at the same time, communication conflict will happen, leading to transmission failure. A conflict graph $G_c = (V_c, E_c)$ can represent the interference model, and a set of vertices V_c correspond to the communication links in the graph G, as shown in Eq. (1).

$$V_c = \{l_{ij} | (i,j) \text{ is communication link, } l_{ij} \in E\} \tag{1}$$

In the conflict graph G_c, V_c is the set of the links in the communication graph; A conflict edge (l_{ab}, l_{cd}) in E_c denotes that links (a, b) and (c, d) interfere with each other if they work simultaneously in the same channel. In this paper, interference distance is configured to be identical with the communication distance L. Figure 1(b) is the corresponding conflict graph of Fig. 1(a).

2.3   Multi-radio Multi-channel Assignment Problem

Assuming that every node in the WSN is configured with a certain number of radios, we need to assign different channel to every link in WSN, and ensure that the number of node's operating channels is less than or equal to the node's radio number. Our goal is to minimize the total interference number so as to improve the whole network capacity, so we define the sum of the interfering links' number as the total interference number.

Provided that the WSN consists of N sensor nodes, the problem is to find the optimal function: $f : V_c \rightarrow K$ which can minimize the overall network interference $I(f)$ and satisfy the interface constraint, which are defined as below:

$$\text{Interface Constraint: } \forall i \in N, |\{k | f(e) = k, e \in E(i) \}| \leq Ri \tag{2}$$

$$\text{Network Interference: } I(f) = \sum_{e \in V_c} \text{interference number of } e \tag{3}$$

2.4   Channel Merging

In the optimization process of the network interference, the interface constraint is very likely to be violated. In order to handle the violation, the following 'channel merging' procedure is applied. The channel merging procedure proposed by [9] will introduce new channel for relevant nodes, which might further violate the constraint. To solve the issue, an improved channel merging operation is proposed. As shown in Fig. 2, Fig. 2 (a) is the network before merging operation, and Fig. 2(b) is after it.

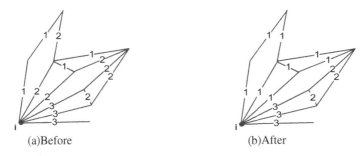

(a)Before   (b)After

Fig. 2. The communication graphs of node i in the network before and after merging operation

The details of channel merging operation are described as follows:

```
Algorithm 1 : Channel Merging
Input    : Channel Assignment Scheme X,Xe,N,Ri
Output   : Channel Assignment Scheme after Merging Xm
1  Change  X  to  Xv,  which  stands  for  the  Channel  Assignment  Scheme
   corresponding to node;
2 Find node i(i∈N),which has max channel number,the number is m.
3  If m<=Ri
4     Xm=X;
5     Return;
6  Endif
7  While (m>Ri)
8     Find the two channel number c1 and c2,which correspond to mim
      number edges incident on node I;
9     Channel merging.Change the edges with channel c1 to c2,the edge
      incident on node i,and do the same operation to the edges connected
      to the changed edges aforementioned ,then Xi;
10    Do step from 1 to 6 for Xi
11 Endwhile
12 Xm=Xi .
```

3   GA-SSO Algorithm

In this paper, based on GA and SSO optimization algorithms, a novel hybrid algorithm GA-SSO is proposed to solve the channel assignment problem for MR-MC WSNs. The GA-SSO's process is detailed described as follows.

3.1   Encoding

Assume that the number of communication links in the WSN is L, and the links are numbered by $\{1, 2, \ldots, L\}$. The particles of the GA-SSO algorithm represent a channel assignment scheme. The particle i'position at current time step t is represented by $X_i^t = (x_{i1}^t, x_{i2}^t, \ldots, x_{iL}^t)$, where x_{ik}^t denotes the channel assignment for node k at time t. The best position of the particles which is achieved so far is represented by $Pbest_i^t = (p_{i1}^t, p_{i2}^t, \ldots, p_{iL}^t)$. The global best position in the population is represented by $Gbest^t = (g_1^t, g_2^t, \ldots, g_L^t)$.

3.2   Population Initialization

In the initialization of the GA-SSO algorithm, channel merging operation is performed after assigning a random position value to each particle in order to satisfy the interface constraint. The particle's position initialization process for the topology in Fig. 1(a) can be shown in Fig. 3, in which each node owns two radios and there are three available orthogonal channels.

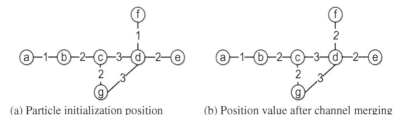

(a) Particle initialization position   (b) Position value after channel merging

Fig. 3. Particle initialization process

3.3   Particle Position Update

The GA-SSO algorithm mainly consists of two parts: one is that Pbest and Gbest have effects on particle position by generic crossover operation, and the other is that particles change position by generic mutation operation. The basic iterative formula of GA-SSO algorithm is presented as follows:

$$X_i^t = \begin{cases} X_i^{t-1}, & \text{if } r \in [0, C_w), \\ \text{Cross}(X_i^{t-1}, \text{Pbest}_i^{t-1}), & \text{if } r \in [C_w, C_p), \\ \text{Cross}(X_i^{t-1}, \text{Gbest}^{t-1}), & \text{if } r \in [C_p, C_g), \\ \text{Mutate}(X_i^{t-1}), & \text{if } r \in [C_g, 1). \end{cases} \quad (4)$$

where $i = 1, 2, \ldots, m$, m is the swarm population. X_i^t is the current time position value of the i - th particle, and X_i^{t+1} is the next time position value; Pbest_i^t is the i - th particle's personal best position until current time; Gbest^t is the global best position in the current whole population; C_w, C_p and C_g are three predetermined positive constants with $0 < C_w < C_p < C_g \leq 1$, C_w is the probability to remain the same, C_p is the crossover factor with Pbest, C_g is the crossover factor with Gbest; r is a random number with $0 < r < 1$. The position update strategy of the GA-SSO can be summarized as follows: in each iteration of position update, the particles choose one of the four operations according to the value of r. The four operations include: (1) remaining the same, (2) crossover with Pbest, (3) crossover with Gbest, and (4) mutation.

3.4   Fitness Function

The main objective of our algorithm is to minimize the total network interference so as to maximize network throughput. Therefore, the total network interference number is used as the GA-SSO algorithm's fitness value. Fitness Function can be represented as $I(f)$, as shown in Eq. 3.

3.5   Algorithm Pseudocode

The algorithm pseudocode is shown as follows. Assume that each node has the same number of radios. The termination condition in algorithm is "iteration times".

```
Algorithm 2 : GA-SSO
Input: G,K={1,2,...,kₘₐₓ},N,Rᵢ,sizepop,maxgen,Cw,Cp,Cg,iteration times
Output: The channel assignment result Kᵢ for each i∈Vc
1   Convert G to Gc=(Vc,Ec);
2   Initialize the GA-SSO's parameters;
3   Initialize population;
4   Calculate fitness function;
5   While(termination)
8      Generate a random number r∈ (0,1)
9       If (r < Cw) Keep position；
10      Elseif   (r <Cp)  Crossover operation with Pbest;
11      Elseif   (r <Cg)  Crossover operation with Gbest;
12      Else  mutate operation ;
13      Endif
14    Update Pbest,Gbest,the current position X;
15  Endwhile.
```

4   Experimental Results

In this section, the performance of the designed GA-SSO algorithms is verified by Matlab based simulation. In order to evaluate the GA-SSO algorithm and channel merging approach proposed in paper, it is compared with the SSO [7] and DPSO-CA [8]. The simulation experiments perform in finite random networks, which have variable number of radios and channels. In a square meters area 1000 m × 1000 m, we consider that the networks have 25 nodes deployed randomly. It is supposed that the range of transmission and interference are all 250 m, and the WSN nodes havethe same number of radios.

The metric Interference Percentage is used to evaluate our algorithms' performance, which is defined by Eq. 5 as follows.

$$\text{Interf}(f) = I(f)/(L \bullet (L-1)) \tag{5}$$

where f is a feasible assignment solution, $I(f)$ is the interference value in network, and L is total interference links in G_c. This metric refers to the thought of normalization method.

Figure 4 shows the simulation results in a network (consisting of 25 nodes), where every node has 12 available channels and radios' number varying from 2 to 12. With the increasing number of radios, the interference percentage reduces. From Fig. 4, it can be observed that when the radio number is less than 5, the interference percentage can be obviously reduced by increasing the number of radios. Overall speaking, the performance of the proposed GA-SSO is better than the other two algorithms.

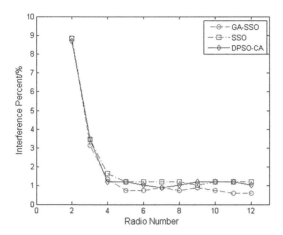

Fig. 4. Interference percentage in WSN with 25 nodes (12 available channels)

Table 1 shows the experimental results in the network (25 nodes) with channel number as 3 and radio number as 5. The running speed of GA-SSO is superior to SSO, but slightly inferior to DPSO-CA. The iteration process of GA-SSO can inherit the last iteration, which can reduce the time from channel merging. But the inheritance property of SSO is worse than GA-SSO's. Moreover, the GA-SSO can effectively solve the premature convergence problem and improve the global convergence performance, which benefits from generic algorithm's property.

Overall, GA-SSO algorithm can obtain better result among the three algorithms.

Table 1. Comparision results (3 channels and 5 radios)

Parameter	SSO	DPSO-CA	GA-SSO
Iteration time per round/s	0.278	0.184	0.192
Interference percentage/%	6.42	5.75	5.55

5   Conclusion

In this paper, a novel hybrid optimization algorithm named GA-SSO is proposed to solve the NP-hard channel assignment problem for data-intensive MR-MC WSN. Moreover, an improved channel merging method is proposed to satisfy interface constraint condition. Compared with SSO and DPSO-CA algorithm through Matlab simulation, the proposed GA-SSO has a better global search and thus can improve the network capacity. In future, the authors will consider how to apply this algorithm to real MR-MC WSNs.

Acknowledgement. The authors would like to acknowledge the supports by the National Natural Science Foundation of China (Grant Nos. 61601127, 51508105, and 61574038), the Fujian Provincial Department of Science and Technology of China (Grant Nos. 2015H0021, 2015J05124 and 2016H6012), the Fujian Provincial Economic and Information Technology Commission of China (Grant No. 830020) and the Fujian Provincial Department of Education of China (Grant No. JA14038).

References

1. Kuroiwa, T., Suzuki, M., Yamashita, Y., Saruwatari, S., Nagayama, T., Morikawa, H.: A multi-channel bulk data collection for structural health monitoring using wireless sensor networks. Communications **96**, 295–299 (2012)
2. Gao, S., Yuan, S., Qiu, L., Ling, B., Ren, Y.: A high-throughput multi-hop WSN for structural health monitoring. J. Vibroengineering **18**(2), 781–800 (2016)
3. Ji, S., Cai, Z., Li, Y., Jia, X.: Continuous data collection capacity of dual-radio multichannel wireless sensor networks. Proc. - IEEE INFOCOM **23**(10), 1062–1070 (2011)
4. Li, J., Guo, X., Guo, L., Ji, S., Han, M., Cai, Z.: Optimal routing with scheduling and channel assignment in multi-power multi-radio wireless sensor networks. Ad Hoc Netw. **31**, 45–62 (2015)
5. Athota, K., Negi, A., Rao, C.R.: Interference-traffic aware channel assignment for MRMC WMNs. In: 2010 IEEE 2nd International Advance Computing Conference (IACC), pp. 273–278. IEEE (2010)
6. Soua, R., Minet, P.: Multichannel assignment protocols in wireless sensor networks: a comprehensive survey. Pervasive Mob. Comput. **16**, 2–21 (2015)
7. Bae, C., Yeh, W.C., Wahid, N., Chung, Y.Y., Liu, Y.: A new Simplified Swarm Optimization (SSO) using exchange local search scheme. Int. J. Innovative Comput. Inf. Control Ijicic **8**(6), 4391–4406 (2012)
8. Cheng, H., Xiong, N., Vasilakos, A.V., Yang, L.T., Chen, G., Zhuang, X.: Nodes organization for channel assignment with topology preservation in multi-radio wireless mesh networks. Ad Hoc Netw. **10**(5), 760–773 (2012)
9. Subramanian, A.P., Gupta, H., Das, S.R.: Minimum interference channel assignment in multi-radio wireless mesh networks. IEEE Trans. Mob. Comput. **7**(12), 1459–1473 (2008)

Author Index

Printed in the United States
By Bookmasters